LLYFRGELL
LIBRARY
ABERYSTWYTH

D1142165

CONTROL MECHANISMS IN ANIMAL CELLS
Specific Growth Factors

Control Mechanisms in Animal Cells

Specific Growth Factors

Editors

Luis Jimenez de Asua, M.D., Ph.D.
Group Leader
Regulatory Events of the Cell Cycle Unit
Cell Biology Group
Friedrich Miescher-Institute
Basel, Switzerland

Rita Levi-Montalcini, M.D.
Professor and Director
Laboratory of Cellular Biology
CNR
Rome, Italy

Robert Shields, D.Phil.
Colworth Laboratory
Unilever Research
Sharnbrook, Bedford, England

Stefano Iacobelli, M.D.
Chief
Laboratory of Molecular Endocrinology
Catholic University
Rome, Italy

Raven Press ■ New York

Raven Press, 1140 Avenue of the Americas, New York, New York 10036

Made in the United States of America

Library of Congress Cataloging in Publication Data

Main entry under title:

Control mechanisms in animal cells.

"Based on papers delivered at the Round Table on the Regulation of the Initiation of DNA Synthesis as well as the EMBO Workshop on Specific Growth Factors held in Rome in October 1979."
Includes bibliographical references and index.
1. Cells-Growth-Regulation-Congresses. 2. Deoxyribonucleic acid synthesis-Congresses. 3. Cellular control mechanisms-Congresses. I. Jimenez de Asua, Luis.
QH604.C65 591.87'61 79-5293
ISBN 0-89004-509-7

Preface

This volume contains chapters based on papers delivered at the Round Table on the Regulation of the Initiation of DNA Synthesis as well as the EMBO Workshop on Specific Growth Factors held in Rome in October 1979.

The term "growth factor" is used to describe a number of extremely active molecules that have cellular receptors of high affinity and act mitogenically. As such, the term embraces conventional hormones as well as other molecules that are not usually thought of as hormones. However, the mechanisms of action of the growth factors appear to be similar, so that results obtained with one factor may be relevant to the study of another.

The purpose of this volume is to highlight certain areas of active research rather than to present a comprehensive overview of the field. Although it is still not possible to give a review as to how growth factors work, some features are becoming clear. The introduction to this volume is an attempt to draw some of these common threads together and to point out areas of knowledge that are firmly established and areas in which future research is needed.

Another purpose of this book is to show that seemingly disparate fields (such as the possible action of growth factors inside the cell and the action of SV40 antigen) may have features in common.

The material in this volume should therefore act as a catalyst for scientists to examine work going on in fields other than their own. For example, endocrinology and tumour biology may have a common ground because many tumours produce growth factors ectopically. This secretion of growth factors by tumours may lead to the discovery of new molecules that play important roles in normal growth and development.

This volume should provide a valuable source of material for molecular endocrinologists, developmental biologists, and those interested in oncogenic transformation.

Luis Jimenez de Asua
Rita Levi-Montalcini
Robert Shields
Stefano Iacobelli

Contents

Contributors

K. Adachi
Department of Biochemistry
Kanazawa Medical University
Kahoku-gun, Ishikawa 920-02, Japan

S. C. Aitken
National Cancer Institute
National Institutes of Health
Bethesda, Maryland 20205

F. S. Ambesi-Impiombato
CNR Endocrinology and Experimental Oncology Center
Institute of General Pathology
Faculty of Medicine
University of Naples
Naples, Italy

S. Barlati
CNR Laboratory of Genetic Biochemistry and Evolution
University of Pavia
27100 Pavia, Italy

D. C. Bennett
The Salk Institute
San Diego, California 92138

S. Biocca
CNR Laboratory of Cell Biology
00196 Rome, Italy

J. Brachet
Department of Molecular Biology
University of Brussels
Brussels, Belgium

R. R. Bürk
Friedrich Miescher-Institute
CH-4002 Basel, Switzerland

P. Calissano
CNR Laboratory of Cell Biology
00196 Rome, Italy

D. Cirillo
Department of Human Anatomy
University of Turin School of Medicine
10126 Turin, Italy

J. Conscience
Friedrich Miescher-Institute
CH-4002 Basel, Switzerland

H. G. Coon
Laboratory of Cell Biology
National Cancer Institute
National Institutes of Health
Bethesda, Maryland 20205

J. E. De Larco
Laboratory of Viral Carcinogenesis
National Cancer Institute
National Institutes of Health
Bethesda, Maryland 20205

T. Ebendal
Department of Zoology
Uppsala University
S-751 22 Uppsala, Sweden

J. E. Errick
Department of Molecular, Cellular, and Developmental Biology
University of Colorado
Boulder, Colorado 80309

M. Gabbay
Weizmann Institute of Science
Rehovot, Israel

D. Gospodarowicz
Cancer Research Institute
University of California Medical Center
San Francisco, California 94143

A. Graessmann
Institute for Molecular Biology and Biochemistry
Free University of Berlin
1 Berlin 33, West Germany

M. Graessmann
Institute for Molecular Biology and Biochemistry
Free University of Berlin
1 Berlin 33, West Germany

F. Grummt
Max-Planck-Institute for Biochemistry
D-8033 Munich, West Germany

R. T. Hamilton
Cell Biology Laboratory
The Salk Institute
San Diego, California 92138

K. O. Hedlund
Department of Zoology
Uppsala University
S-751 22 Uppsala, Sweden

R. Hinnen
Friedrich Miescher-Institute
CH-4002 Basel, Switzerland

R. W. Holley
Molecular Biology Laboratory
The Salk Institute
San Diego, California 92138

S. Iacobelli
Laboratory of Molecular Endocrinology
Catholic University
00168 Rome, Italy

C. O. Jacobson
Department of Zoology
Uppsala University
S-751 22 Uppsala, Sweden

P. James
Specialized Center for Thrombosis Research
Temple University School of Medicine
Philadelphia, Pennsylvania 19140

L. Jimenez de Asua
Friedrich Miescher-Institute
CH-4002 Basel, Switzerland

L. K. Johnson
Cancer Research Institute
University of California Medical Center
San Francisco, California 94143

T. Kano-Sueoka
Department of Molecular, Cellular, and Developmental Biology
University of Colorado
Boulder, Colorado 80309

W. R. Kidwell
Laboratory of Pathophysiology
National Cancer Institute
National Institutes of Health
Bethesda, Maryland 20205

P. Leuthard
Friedrich Miescher-Institute
CH-4002 Basel, Switzerland

A. Levi
Weizmann Institute of Science
Rehovot, Israel

A. Levi
CNR Laboratory of Cell Biology
00196 Rome, Italy

L. A. Liotta
Laboratory of Pathophysiology
National Cancer Institute
National Institutes of Health
Bethesda, Maryland 20205

M. E. Lippman
National Cancer Institute
National Institutes of Health
Bethesda, Maryland 20205

P. Longo
Laboratory of Molecular Endocrinology
Catholic University
00168 Rome, Italy

P. C. Marchisio
Department of Human Anatomy
University of Turin School of Medicine
10126 Turin, Italy

B. R. McAuslan
CSIRO Molecular and Cellular Biology Unit
North Ryde, N.S.W. 2113, Australia

W. Meier
Friedrich Miescher-Institute
CH-4002 Basel, Switzerland

P. Mignatti
Institute of Genetics
University of Pavia
27100 Pavia, Italy

D. Monard
Friedrich Miescher-Institute
CH-4002 Basel, Switzerland

C. Mueller
Institute for Molecular Biology and Biochemistry
Free University of Berlin
1 Berlin 33, West Germany

L. Naldini
Department of Human Anatomy
University of Turin School of Medicine
10126 Turin, Italy

C. Natoli
Laboratory of Molecular Endocrinology
Catholic University
00168 Rome, Italy

R. A. Newman
Imperial Cancer Research Fund
London WC2A 3PX, England

S. Niewiarowski
Specialized Center for Thrombosis Research
Temple University School of Medicine
Philadelphia, Pennsylvania 19140

M. Nilsen-Hamilton
Cell Biology Laboratory
The Salk Institute
San Diego, California 92138

K. Nishikawa
Department of Biochemistry
Kanazawa Medical University
Kahoku-gun, Ishikawa 920 02, Japan

G. Norrgren
Department of Zoology
Uppsala University
S-751 22 Uppsala, Sweden

C. Okitsu
Department of Biochemistry
Kanazawa Medical University
Kahoku-gun, Ishikawa 920 02, Japan

A. M. Otto
Friedrich Miescher-Institute
CH-4002 Basel, Switzerland

D. Paul
Department of Pharmacology and Toxicology
Hamburg University Medical School
Hamburg 20, West Germany

W. J. Pledger
Department of Pharmacology
University of North Carolina School of Medicine
Chapel Hill, North Carolina 27514

F. O. Ranelletti
Laboratory of Molecular Endocrinology
Catholic University
00168 Rome, Italy

K. M. V. Richmond
Friedrich Miescher-Institute
CH-4002 Basel, Switzerland

M. A. Ritter
Imperial Cancer Research Fund
London WC2A 3PX, England

B. Rucinski
Specialized Center for Thrombosis Research
Temple University School of Medicine
Philadelphia, Pennsylvania 19140

S. Rucker
Fels Research Institute
Temple University School of Medicine
Philadelphia, Pennsylvania 19140

P. S. Rudland
Ludwig Institute for Cancer Research
Haddow Laboratories
Sutton, Surrey SM2 5BX, England

D. Salomon
Laboratory of Pathophysiology
National Cancer Institute
National Institutes of Health
Bethesda, Maryland 20205

G. H. Sato
Department of Biology
University of California, San Diego
La Jolla, California 92093

N. Savion
Cancer Research Institute
University of California Medical Center
San Francisco, California 94143

J. Schlessinger
Weizmann Institute of Science
Rehovot, Israel

Y. Shechter
Weizmann Institute of Science
Rehovot, Israel

R. Shields
Unilever Research
Bedfordshire MK44 1LQ, England

D. A. Sirbasku
Department of Biochemistry and Molecular Biology
University of Texas Medical School at Houston
Houston, Texas 77030

G. Steck
Friedrich Miescher-Institute
CH-4002 Basel, Switzerland

G. J. Todaro
Laboratory of Viral Carcinogenesis
National Cancer Institute
National Institutes of Health
Bethesda, Maryland 20205

D. Tramontano
CNR Endocrinology and Experimental Oncology Center
Institute of General Pathology
Faculty of Medicine
University of Naples
Naples, Italy

M. Ulrich
Friedrich Miescher-Institute
CH-4002 Basel, Switzerland

K. G. Varma
Fels Research Institute
Temple University School of Medicine
Philadelphia, Pennsylvania 19140

I. Vlodavsky
Cancer Research Institute
University of California Medical Center
San Francisco, California 94143

M. J. Warburton
Ludwig Institute for Cancer Research
Haddow Laboratories
Sutton, Surrey SM2 5PX, England

W. Wharton
Department of Pharmacology
University of North Carolina School of Medicine
Chapel Hill, North Carolina 27514

M. S. Wicha
Laboratory of Pathophysiology
National Cancer Institute
National Institutes of Health
Bethesda, Maryland 20205

Y. Yoshitake
Department of Biochemistry
Kanazawa Medical University
Kahoku-gun, Ishikawa 920 02, Japan

I. Yuli
Weizmann Institute of Science
Rehovot, Israel

Introduction

Robert Shields

What are growth factors? A facetious answer is that they are molecules that make cells grow and divide, but this definition covers many nutrients as well as substances considered to be true growth factors. Probably the best answer comes from Gordon Sato: Growth factors are hormones; they share with hormones their high specific activities, which are due in turn to their high-affinity receptors in the cell.

Where do growth factors come from? Conventional hormones are synthesized by glands and carried by the blood to their target cells, which may be far removed from their site of origin. Very few growth factors have been shown to have a unique site of synthesis. While the insulin-like growth factor IGF (formerly known as NSILA) is probably made in the liver (not generally regarded as a gland), NGF appears to be made systemically. Many other growth factors appear in certain organs in high concentrations (FGF in the pituitary and EGF in the submaxillary gland), but it is unclear where they are synthesized (they too may be systemic). And what about the platelet-derived growth factors discussed by Pledger and Wharton and Paul et al.? Where are they made?

The absence of any well-defined locale for their synthesis means that the classic technique of endocrinology—removal of the gland—is impossible. Their function must be studied in tissue culture. This is well illustrated in Sato's chapter. He shows that growth factors are interesting not merely because they help cells grow and sometimes differentiate, but also in a practical sense. By their judicious use it may be possible to culture any differentiated tissue in a functional state. This is illustrated by many examples in Sato's chapter and also in that of Ambesi-Impiombato et al., who show how to grow normal functioning thyroid cells *in vitro.*

While most of the growth factors discovered to date make cells grow, inhibitory growth factors do exist. Probably the best known negative factor is interferon, and the chapter by Holley describes another molecule that may turn out to be a chalone. Also, it should be remembered that growth factors have many actions besides promoting (or inhibiting) growth. Some (NGF is a good example) promote differentiation. Neither do growth factors work in isolation. Many act in conjunction with other hormones, and the target cell shape and the proximity of other cells also plays a role. This having been said, what is known about how growth factors work?

There seems to be general agreement that the initial interaction of peptide growth factors with their target cells occurs at specific cell surface receptors which are evenly distributed over the cell surface. After interaction with the hormone, the hormone–receptor complexes form small aggregates which then coalesce into larger "patches" on the cell surface and are subsequently internalized. The disappearance of the receptor from the surface means the cell becomes temporarily refractile to further stimulus by the same hormone (down regulation). Despite intensive research in a number of laboratories, it is still not known whether peptide hormones act entirely at the cell surface or whether some of their effects require internalization. What is known is that the mere possession of receptors and the internalization of the hormone–receptor complex is not sufficient for the cell to be responsive (Gospodarowicz et al.).

At least three possible mechanisms (not necessarily exclusive) for peptide hormone action can be entertained: (a) The hormone acts at the cell surface to produce its effects either directly (e.g., by interacting with membrane transport sites) or indirectly via second messengers that act within the cell. (b) The hormone is internalized and acts inside the cell to produce its effects (either directly or via one or several subsidiary messenger molecules). According to this model the receptor is merely a vehicle for transporting the hormone. (c) The receptor itself (or fragments or derivatives of the receptor) are the active species. The function of the hormone is then to internalize the receptor.

Several chapters in this volume discuss these models. That by Schlessinger et al. shows that the rapid effects of EGF and insulin on membrane transport do not involve large scale patching or internalization of their receptors. Indeed, many of the so-called "early events" following mitogenic stimulation occur too rapidly for hormone internalization to be involved. Whether all the effects of growth factors may be explained by their action at the cell surface is unclear. While there is fairly convincing evidence that the protease thrombin does not need to be internalized to be mitogenic, the enzyme appears to act by cleaving its own receptor, and these fragments could act inside the cell (1). Recent studies using inhibitors of receptor internalization have shown that the mitogenic effects of EGF are undiminished if internalization is prevented (3). Against this, it should be pointed out that the degree of EGF receptor internalization and the mitogenic effects of the peptide are closely paralleled (2). So while it is clear that at least some of the effects of protein hormones are a result of cell surface events, internalization of the hormone may be required for others.

Several cellular events are known to be initiated after the binding of macromolecules to their receptors and their internalization. Among these are the control of cholesterol transport following internalization of low-density lipoproteins and their receptors and the toxic effects of ricin and cholera toxin. Against this background the idea that some of the actions of peptide hormones requires their internalization seems quite reasonable. The internalization and intracellular distribution of NGF and EGF are discussed in chapters in this volume (Biocca et al., Marchisio et al., and Gospodarowicz et al.). It is shown that the internalization of NGF and EGF leads to down regulation and the internalized hormones become localized in and around the cell nucleus, where they could possibly exert some of their effects. This nuclear localization of growth factors may explain reports that it is possible to extract growth factors from the nuclei of tumour cells (Nishikawa et al.).

There is some precedent for the action of peptide growth factors in the cell nucleus. The T antigen of SV40, when microinjected, can act as a growth factor for quiescent cells. It too has a largely nuclear location (see Graessman et al., *this volume*), although it must be admitted that it is unknown whether the molecules detected by immunofluorescence are the same ones that are acting mitogenically. The direct microinjection technique used in these studies will no doubt prove to be very valuable in the study of growth factor–cell interactions. It has already proved possible to demonstrate that EGF microinjected directly into cells is nonmitogenic (A. Graessman, *personal communication*), which shows that either EGF works at the cell surface or that it must be suitably modified or bound to its receptors before it will work in the cell. So while it is clear that while some macromolecules bind to receptors, are internalized, and then act inside the cell, it is not yet clear whether this applies for growth factors. It may even be the case that the mitogenic signal is delivered from the cell surface, but changes such as differentiation (which require altered gene expression) require factor internalization.

One possibility that has been rather overlooked is that the presence of the growth factor inside the cell may not be as important as the presence of its receptor. The function of peptide growth factors may be to promote internalization of the receptor which acts within the cell. Steroid hormones seem to work in this way; most of the responses to steroids depend on the translocation of the receptor to the nucleus (see Iacobelli et al. and Aitken and Lippman, *this volume*). Could it be that the steroid receptor acts on the nucleus in the same way as the T antigen of SV40? Once growth factor receptors have been purified it may be possible to answer this question.

Many different mitogens act on a variety of cell types to produce a rather similar range of metabolic responses (dubbed "pleiotypic" by Gordon Tomkins). These events include rapid changes in nutrient and ion transport, changes in protein synthesis and degradation, and ultimately cell division itself. A central question raised by Robert Holley is which of the metabolic events induced by growth factors are necessary for cell division to occur. Do peptide growth factors act on the cell membrane to produce a single pleiotypic mediator which orchestrates intracellular events? Do the changes in membrane transport allow in nutrients rate limiting for DNA synthesis? Are there several "second messengers" that interact on some central cellular process to form an initiator of DNA synthesis? Or do growth factors merely "jazz up" the cell machinery which makes DNA synthesis more probable?

While the idea of a single pleiotypic mediator is attractive, no candidate for such a molecule has yet stood the test of time; however, Grummt provides a new possible mediator. A fruitful way to search for such a substance is to examine an intracellular event which occurs rapidly, before the growth factor is internalized. The study of the phosphorylation of the ribosomal protein S6, which correlates well with the growth factor-induced stimulation of protein synthesis, may provide such a system (Nilsen-Hamilton and Hamilton). Even if no single mediator is discovered, at least such experiments may reveal part of the mechanism of the increase in protein synthesis, which is one of the very few cellular events that appears to be uniquely correlated with DNA synthesis.

An alternative approach to the problem of how mitogens work is to focus on the kinetics of cell division. The cell cycle seems to be adequately described as being controlled by two random events separated by a lag (Shields). Viewed in this way the control of cell division reduces to how growth factors control the probability of undergoing these transitions and what the events between these transitions are and what influence growth factors have on them. These problems are discussed in a number of chapters in this volume. What appears clear is that these two transitions are differentially affected by different growth factors, some influencing the first transition (which makes the cell competent), some affecting commitment of competent cells, and some affecting both transitions and events between (Pledger et al., Jimenez de Asua, Richmond, and Otto et al.). If it is accepted that growth factors influence the *probability* of a cell's dividing, then a cell may be regarded as similar to a car. If conditions are suboptimal (the oil thick, the spark plugs dirty, the battery old, and the brakes seized), then the probability that the car will proceed will be low (but finite). The improvement of any one of these parameters will increase the probability of motion. Viewed in this way there is no unique path leading to cell division and growth factors may be additive, synergistic, or inhibitory. The function of growth factors may be to "jazz up" many features of cell metabolism that interact to make the probability of division high; the search for a unique pathway to growth may therefore prove fruitless. This may be the reason why attempts to uniquely correlate events at the cell membrane with cell division have been unsuccessful.

Although the normal sites of synthesis of many growth factors are unknown it has become clear that a number of tumour cell lines produce growth factors in culture. Why is this? An attractive explanation is that cells that produce their own growth factors have an advantage over other cells. The production of growth factors may then be one of the events that leads to the successful establishment of a tumour (Todaro and De Larco and Bürk). It seems unlikely that tumour cells are producing entirely novel growth factors, but rather that these factors are another example of ectopic hormone production by tumours. A study of the proteins released from cultured tumour cells may reveal many as yet undiscovered factors that are normal developmental and growth hormones. These ectopically produced growth factors may be responsible for many of the characteristics of transformed cells, such as their morphology and their ability to grow in agar (Todaro and De Larco and Barlati and Mignatti). If these factors act on the cells that produce them and are internalized and transported to the perinuclear area (like other growth factors), it may be possible to extract growth factors from the nuclei of tumour cells (Nishikawa et al.).

Tumour cells might also gain a selective advantage not by manufacturing their own growth factors but by increasing the supply of growth factors by promoting tumour vascularization. This may be done by secreting a tumour angiogenesis factor (TAF) which is a growth factor for endothelial cells. Alternatively, tumours could release a chemotactic agent that encourages migration of endothelial cells, with division occurring subsequent to migration. There is no reason to suppose that such a factor need be a protein; an ion could be active (McAuslan).

Many of the growth factors discussed in this volume have been discovered by their actions on cells cultured *in vitro.* While it is quite clear that a number of tumours are hormone dependent *in vivo,* hormone dependency cannot always be demonstrated *in vitro*. This could be due to shortcomings of the *in vitro* technique, or, more interestingly, the hormone *in vivo* may elicit the production of a second substance that acts as the ultimate growth factor. Two contributions in this book give examples of such a situation (Sirbasku and Kano-Sueoka and Errick): In one case a high-molecular weight factor (estromedin) mediates the effects of estrogens; in another case (MGF) is shown to be phosphoethanolamine. It should be remembered that the idea of secondary factors mediating the effects of hormones is not new. For instance, glucocorticoids mediate some of the physiological effects of ACTH, and growth hormone (an *in vivo* growth factor par excellence) exerts many of its trophic effects via somatomedin.

The term "growth factor" implies that such substances are concerned only with promoting cell division. This is an understatement. Many of these factors can cause or modify differentiation in suitable target tissues. It might be thought that differentiation was too complicated to study in pure cell culture, as cell–cell interaction and cell factor interaction are often involved. However, a number of chapters in this volume lead me to suspect that the study of differentiation *in vitro* might not be as intractable as had first been thought. The approach (which proved so successful with the mitogenic action of growth factors) has been to produce clonal lines of cells which differentiate in pure culture. Initially, only tumour cells could be cloned. This had the doubtful advantage that their differentiation potential was limited and the advantage that one part of differentiation (e.g., neurite extension in neuroblastomas or globin induction in Friend cells) could be studied in isolation. The disadvantage was that it is unclear whether these events had any relevance to differentiation of non-tumourous tissue *in vivo*. Conscience and Meier show that it is possible to "map" the state of differentiation of Friend cells so that one can be reasonably sure that the induction of globin synthesis seen in these cells represents events occurring during the normal development of erythroid cells. Having a clonal cell line that differentiates *in vitro* does not mean that the problem of what causes differentiation will soon be solved. Hinnen and Monard describe attempts to find out the glial factor (GF) acting on a single cell type (neuroblastoma) elicits a simple response (neurite extension). The possibility that NGF causes membrane depolarization is interesting, as it immediately suggests how growth factors can elicit pleiotropic responses through changes in ion transport and intracellular pH. NGF does not appear to be the only factor that causes neurite growth; the situation *in vivo* may be rather more complicated (Jacobson et al.).

Cell–cell interactions have long been claimed to be involved in differentiation. Kidwell et al. show that the function of one cell in this putative interaction may be to produce a suitable support on which differentiation of the other cell type occurs. Thus it may be possible to study the differentiation of a single cell type if it is grown on the correct substrate. Rudland et al. describe just such a system where a single stem cell type produces both myoepithelial and alveolar cells. This promising system may provide the basic tool to study the action of growth factors on both the growth and development of the mammary gland. Finally, anyone who believes that a suitable *in vitro* system is the limiting factor in understanding differentiation should read the chapter by Brachet. Although much of the phenomenology of oögenesis and the events following fertilization in amphibia is well documented, after many years of study we still have little understanding of the mechanisms involved.

REFERENCES

1. Cunningham, D. D., Carney, D. H., and Glenn, K. C. (1979): In: *Cold Spring Harbor Conference on Cell Proliferation, Vol. 6,* pp. 199–215. Cold Spring Harbor Laboratories, Cold Spring Harbor, New York.
2. Fox, C. F., Vale, R., Peterson, S. W., and Das, M. (1979): in *Cold Spring Harbor Conference on Cell Proliferation, Vol. 6,* pp. 143–157. Cold Spring Harbor Laboratories, Cold Spring Harbor, New York.
3. Maxfield, F. R., Davies, P. J. A., Klempner, L., Willingham, M. C., and Pastan, I. (1979): *Proc. Natl. Acad. Sci., USA*, 76:5731–5735.

Control Mechanisms in Animal Cells,
edited by L. Jimenez de Asua et al.
Raven Press, New York © 1980.

Cell Culture, Hormones, and Growth Factors

Gordon H. Sato

Department of Biology, University of California, San Diego, La Jolla, California 92093, U.S.A.

The period around the 1930's was an exciting time in the history of biochemistry. A great number of the vitamins were either discovered, isolated or identified during this period (7). Although most of the amino acids had been discovered much earlier threonine was not discovered until 1934 (8). This also happened to coincide with a period of active research in tissue culture nutrition.

The Lewises, in 1911, had already clearly forseen the usefulness of defined media and had shown that the survival of cells in simple culture media was prolonged by the addition of glucose and unspecified amino acids (10).

Albert Fischer, in 1941, showed the requirement for glutamine for cells growing in dialysed serum containing media (5). It was not until the 1950's that the logical extension of Fischer's work was carried out by H. Eagle and associates (4) culminating in the well-known media, MEM.

My reason for this rambling introduction is that I believe a historical parallel can be drawn between the 1930 period of nutritional research and the present day work on growth factors. I believe that many of the vitamins and some of the amino acids could have been discovered by tissue culture techniques had the development of technology been only slightly accelerated and had the early culturists a broader perspective of their times.

Hindsight is easy. How can we be sure that our own vision of the future is not blurred by narrow specialization. I believe that bridging the concepts and techniques of endocrine physiology and cell culture can give us a sufficiently broad perspective and technical flexibility to avoid some of the mistakes and lost opportunities of the past. Growth factors can be readily discovered in culture because of the availability today of a large variety of cell types in culture and because our knowledge of the cellular requirements for growth is much more sophisticated than it was a few short years ago. Growth factors discovered today should become hormones tomorrow - important in normal physiology, implicated in the etiology of disease and useful in their treatment.

The elimination of serum from culture medium is a key step in speeding up the process of discovery of new growth factors. We have been working over the past several years in developing methodology for replacing serum with complexes of hormones, growth factors, attachment factors and transport proteins (Table 1) (6).

TABLE 1. Media formulations for established cell lines (2)

GH_3 (rat pituitary): insulin, 5 μg/ml; transferrin, 5 μg/ml; FGF, 1 ng/ml; TRH, 1 ng/ml; PTH, .5 ng/ml; somatomedin C, 1 ng/ml.

HeLa (human cervical carcinoma): insulin, 5 μg/ml; transferrin, 5 μg/ml; EGF, 10 ng/ml; FGF, 50 ng/ml; hydrocortisone, 50 nM.

MDCK (dog kidney): insulin, 5 μg/ml; transferrin, 5 μg/ml; hydrocortisone, 50 nM; T3, .005 nM; prostaglandin E_1, 25 ng/ml. Supports primary cultures of normal kidneys.

B104 (rat neuroblastoma): insulin, 5 μg/ml; transferrin, 100 μg/ml; CIg, 10 μg/ml; progesterone, 20 nM; putrescine, 100 μM. Supports primary cultures of nervous tissue and neuroblastomas from many species and rat pheochromocytoma.

M2R (Cloudman mouse melanoma): insulin, 5 μg/ml; transferrin, 5 μg/ml; testosterone, 10 nM; FSH, .4 μg/ml; LHRH, 10 ng/ml; NGF, 3 ng/ml. Supports other melanomas and melanomas in primary culture.

C_6 (rat glioma): insulin, 2 μg/ml; transferrin, 5 μg/ml; FGF, 20 ng/ml; gimmel factor, 5 μg/ml.

MCF-7 (human mammary carcinoma): insulin, 0.1 μg/ml; transferrin, 25 μg/ml; EGF, 100 ng/ml; CIg, 7.5 μg/ml; prostaglandin $F_2\alpha$, 100 ng/ml, α-1 spreading protein, 1 μg/ml. Medium also supports growth of BT20 human mammary cells.

RF-1 (normal rat ovarian cells): insulin, 2 μg/ml; transferrin, 5 μg/ml; hydrocortisone, 10 nM; T3, 0.3 nM; CIg, 8 μg/ml. Medium supports normal granulosa cells in primary culture.

F9 (mouse teratocarcinoma): insulin, 1 μg/ml; transferrin, 5 μg/ml; CIg, 5 μg/ml.

TM4 (normal mouse testicular cells): insulin, 5 μg/ml; transferrin, 5 μg/ml; EGF, 3 ng/ml; T3, 0.5 nM; FSH, .5 μg/ml; growth hormone, 100 ng/ml; somatomedin C, 1 ng/ml; retinoic acid, 50 ng/ml.

T84 (human colonic adenocarcinoma): insulin, 2 μg/ml; transferrin, 2 μg/ml; EGF, 1 ng/ml; hydrocortisone, 50 nM; glucagon, 0.2 μg/ml; ascorbic acid, 10 μg/ml.

SV40-3T3: insulin, 0.25 μg/ml; transferrin, 0.5 μg/ml; CIg, 7.5 μg/ml; fatty-acid-free BSA, 1 mg/ml; linoleic acid, 5 μg/ml.

The presence of serum in the medium would, of course, obscure most of the requirements listed in Table 1.

Let me draw some generalizations from the data of Table 1:

1. Each cell type requires a different complex of factors for growth in serum-free medium.

2. The complex of factors seems to be cell type-specific because, for instance, a medium which was designed for a rat neuroblastoma cell line supports the growth of neuroblastomas from mice and humans as well as a rat pheochromcytoma (2).

3. Defined media designed for an established cell line, whether normal or cancerous, can support the growth of the normal tissue of origin in primary culture. In theory, such media should be superior to serum-based media for primary culture because the specific factors required by the cell type under study can be provided at the optimal concentrations. The conventional procedure of diluting serum in synthetic media cannot be expected to produce this optimal environment. In fact, serum probably provides materials deleterious to the growth and function of cells as well as factors stimulatory to contaminating fibroblasts.

In practice, the use of hormones in place of serum has already allowed the establishment of a cell type (normal rat thyroid follicular cells) in culture which could never be cultured before by conventional means (Ambesi and Coon, this symposium).

An examination of a single medium (for GH_3 pituitary cells) can raise a number of interesting points. These cells require insulin (pancreas), transferrin (liver), T3 (thyroid), parathyroid hormone (parathyroid), TRH (hypothalamus), FGF (pituitary), and somatomedin C (liver). It is noteworthy that so few factors are required to completely replace serum in view of the complexity of serum which contains at least 500 different proteins. On the other hand, it is surprising that so many different hormones elicit a positive growth response from these cells. Surely this is outside the expectations of classical endocrinology, and we must be prepared to revise our ideas about the multiple hormonal regulation of metabolism of every type of animal cell. These data also reveal the immense power of the serum-free approach for revealing hormonal responses and dependencies. To find these responses by classical means, one would have to perform pancreatectomy, hepatectomy, thyroidectomy, parathyroidectomy, hypothalectomy and hypophysectomy. Such an experiment would not be practical. The point being made here is that the removal of serum from cells in culture is the most radical endocrine ablation possible and when this is done, wholly unexpected hormonal dependencies are revealed. Among these are the requirements for growth factors like FGF and somatomedin C. Are these factors hormones? A facetious rejoinder could be that by the operational definitions employed, insulin, T3, PTH, TRH and parathyroid hormone could be called growth factors. Seriously however, evidence is mounting that the somatomedins actually mediate the effects of growth hormone _in vivo_ (6). Of interest to us here is the mode of their discovery. They were discovered by _in vitro_ culture methods (11).

The classical modes of discovery (endocrine ablation or injection of tissue extracts) were not available because no storage organ for the somatomedins exist and the organ of synthesis cannot be ablated.

There is no longer any doubt about the physiological relevance of NGF because antibodies against NGF can prevent the development of the sympathetic nervous system (9). The discovery of NGF by Rita Levi-Montalcini was an important milestone and the brilliant contributions of Stanley Cohen and Ralph Bradshaw of this school of investigators continues to be a major part of growth factor research today.

My own involvement with growth factors began with OGF and FGF and the early criticism of these developments reflects the controversies that are likely to be with us for the next several years. OGF and FGF were discovered in my laboratory as contaminants of a luteinizing hormone preparation, active on ovarian and 3T3 cultures (1, 3). I was informed by very established figures in endocrinology that these must be _in vitro_ artifacts since it was unthinkable that a contaminant of the gonadotrophins could be active on ovarian cells. I was also told that it was foolish to study the growth promoting effects of a hormone mediated by cyclic AMP because cyclic AMP was known to be growth inhibitory. It is ironic in this modern era of science that dogmatic classical truths and ready acceptance of fashionable beliefs can be stumbling blocks to progress. I have two theses. The first is that growth factors are hormones and the second is that hormonal responses in culture not only reflect _in vivo_ physiology but also represent a powerful new method for analysing integrated physiology.

I do not expect physiologists to readily accept these ideas. We have made sufficient progress along this road, however, that we can command their respectful attention while the facts are sorted out.

REFERENCES

1. Armelin, H. A. (1973) _Proc. Natl. Acad. Sci. USA_, 70: 2702.
2. Barnes, D. and Sato, G. (1979) _Anal. Biochem._ (in press).
3. Clark, J. L., Jones, K. L., Gospodarowicz, D. and Sato, G. (1972) _Nature New Biol._, 236: 180.
4. Eagle, H. (1955) _Science_, 122: 43-46.
5. Fischer, A. (1941) _Acta Physiol. Scand._, 2: 143.
6. Fryklund, L., Skottner, A., Forsman, A., Castensson, S. and Hall, K. (1979) In: _Hormones and Cell Culture_, edited by G. Sato and R. Ross, pp. 49-59, Cold Spring Harbor Laboratory Cold Spring Harbor, New York.
7. Hawk, P. B., Oser, B. L. and Summerson, W. H. (1947) _Practical Physiological Chemistry_, pp. 1022-1192, the Blakiston Co. Philadelphia, Toronto.
8. Hawk, P. B., Oser, B. L. and Summerson, W. H. (1947) _Practical Physiological Chemistry_, pp. 121, The Blakiston Co.,

Philadelphia, Toronto.
9. Levi-Montalcini, R. and Booker, B. (1960) Proc. Natl. Acad. Sci. USA, 46: 384-391.
10. Lewis, M. R. and Lewis, W. H. (1911) Anat. Rec., 5: 277-293.
11. Salmon, Jr., W. D. and Daughaday, W. H. (1957) J. Lab. Clin. Med., 49: 825.

ACKNOWLEDGEMENTS

This work was supported in part by grants from National Science Foundation, NSF PCM76-80785 and NIH, USPHS GM 17019.

Control Mechanisms in Animal Cells,
edited by L. Jimenez de Asua et al.
Raven Press, New York © 1980.

Hormones, Cell Growth, and Differentiation *In Vitro:* The Thyroid System

F. S. Ambesi-Impiombato, D. Tramontano, and *H. G. Coon

*Center for Experimental Endocrinology and Oncology of the C.N.R., Institute of General Pathology, Faculty of Medicine, University of Naples, Naples, Italy; and *Laboratory of Cell Biology, National Cancer Institute, National Institutes of Health, Bethesda, Maryland 20205, U.S.A.*

Among the many cell types adapted to grow in vitro as established cell lines, the vast majority consists of only partially differentiated cells, mainly of mesenchimal origin (5). Moreover, the expression of specialized functions in vitro has for long time thought to be possible only in tumor or transformed cell lines and not in long term, untransformed cultures. Recently, an increasing number of epithelial, differentiated cells have been grown in vitro, but in most cases only in short term cultures.

We propose that this deficiency may be due mainly to the composition of the classical serum-supplemented media rather than to intrinsic characteristics of the differentiated cells. This view is supported by the most recent developments of serum-free, hormones-supplemented media (5,10) which provided the starting point for our research.

THYROID CELLS IN VIVO

Thyroid cells can be considered a very good system for the study of hormone-dependent cell growth and differentiation, and for the study of hormones-cells interactions in vivo. In the animal, the thyroid follicular cell has been extersively investigated both morphologically and biochemically (9). It is mainly devoted to the synthesis of a single protein, the thyroglobulin (TG), a large (19 S, 660.000 M.W.) iodinated glycoprotein, whose primary structure and subunit composition has not yet been completely elucidated. TG is secreted from the apical side of the polarized follicular cell into a confined space, the follicular lumen, which is surrounded by a monolayer of the cuboidal follicular cells. The lumen, together with the follicular cells, forms the functional unit of the thyroid gland. Thyroid cells in vivo respond to the trophic action of the specific thyroid stimulating hormone (TSH).

Additionally, iodine metabolism is very characteristic in

thyroid cells which are able to concentrate many fold iodide, a relatively rare ion, from the circulating blood. Following organification, iodine is restricted where it is present in the form of the thyroid hormones T_3 and T_4 (three- and tetra-iodothyronines) and their precursors MIT and DIT (mono- and di- iodothyrosines). When thyroid hormones are required in the organism, TG is readsorbed from the lumen and degraded to release the free hormones which are subsequently secreted into the blood stream.

THYROID CELLS IN VITRO

The availability of thyroid cells in vitro would be of importance because of its being differentiated, hormone-dependent highly specialized epithelial cells. For these reasons frequent attempts have been made to maintain and grow thyroid tissue in vitro, and more recently as cell cultures.

The early work (3) dealt with organ culture followed by freshly isolated cells in short term suspension cultures (14) and most recently with trypsinized cells in monolayer culture. With this latter system interesting work has been done with chick embryo cells (6,13) and with short term primary or early passage cultures from several animal species (6,7,8,12). Until our report, however, continuous cell growth of epithelial cell strains from mammals that showed either thyroglobulin synthesis or iodide uptake have not been described (1,2).

TUMOR CELL LINES

Several experimental rat thyroid tumors have been available, chiefly those produced by Wollman (17) and Volpert (15). From these we have obtained cell lines named 1-5G (as the original Wollman tumor) and FRA (Fischer rat autonomous, by Volpert). These were cultured and cloned in standard culture conditions (modified F-12 medium supplemented with 5% calf or fetal calf serum), but - apart from a protein that is apparently immunologically related to TG, found in FRA cells - no differentiated functions persisted in the cell culture strains from tumors.

NORMAL THYROID CELL LINES

Cells grown in conventional culture conditions

Following the isolation of the tumor cell lines normal glands from syngeneic animals (Fischer 344 rats) were chosen as starting material. It soon became evident that normal epithelial cells could not be grown in standard serum-supplemented media alone. The first line of continuously growing, thyroid epithelial cells (FRT: Fischer rat thyroid cell) was obtained from primaries kept in medium conditioned by rat thyroid fibroblasts to which dibutyryl 3':5'-cyclic AMP (dbc-AMP) or Thyrotropin (TSH) was first added to the conditioned medium producers subsequently, FRT cells were adapted to grow in our standard medium without prior conditioning, dbc-AMP or TSH.

Characterization of the thyroid specific functions expressed in vitro by FRT proved negative as far as TG production and iodine metabolism were concerned.

It was then assumed that some component has to be eliminated or drastically reduced from the culture medium to grow differentiated cells. At this point, calf serum was the first choice. When normal rat thyroid primaries were again attempted in low serum (0.5%) it was found that the addition of growth promoting substances was necessary to obtain cell attachment and sustain cell growth. It was found that the combination of six hormones or factors (6H), among the many combinations tested, was particularly successful (TABLE 1).

TABLE 1. Hormonal supplement for FRTL cells

Hormones or Factors	Source	Concentration
Thyrotropin	Bovine	10 mU/ml
Insulin	Bovine	10 μg/ml
Hydrocortisone	Synthetic	10^{-8} M
Transferrin	Human	5 μg/ml
Somatostatin	Synthetic	10 ng/ml
Glycyl-L-histidyl-L-lysine acetate	Synthetic	10 ng/ml

Under these low serum conditions epithelial cells were easily grown free of fibroblast contamination due to the low serum concentration. After several passages and colonial isolations, a new homogeneous cell strain was obtained, which was named FRTL (Fischer rat thyroid cells in low serum), (2). This line has now been cultured continuously in our laboratories for nearly three years and characterized periodically throughout this time. It is a very slow growing cell strain, with a population doubling time of 5-7 days, but it still maintains the normal diploid chromosome complement and differentiated thyroid functions like the biosynthesis of a 19 S iodoprotein fully immunoprecipitable with anti-rat TG purified antibodies, and the ability to trap iodide, which was taken up by the cells with a C/M value of up to 100 (C/M: ratio of intra / extracellular concentration). Individual cells also tend to form intracellular lumen-like areas (FIG. 1 A) while bigger colonies form follicle-like areas (FIG. 1 B) which are PAS positive.

Hormonal requirements of FRTL cells

From early primary cultures, FRTL cells were grown in almost serum free, hormone supplemented medium without a noticeable lag phase, or adaptation period that is usually seen in primary cell cultures. Cell growth of FRTL-like cells without fibroblasts has occurred in all primary cultures rather than being a rare event.

Early after FRTL cells were first cultured their dependence on all six hormones were assayed. Results at that time seemed to indicate that cell growth would stop whenever any one of the six hormones was missing. Cell growth would resume if the deleted hormone was added back within a couple of weeks. From more recent and extensive experiments, however, the following conclusions could be drawn:

1) rat thyroid epithelial cells do not grow in our media without both some serum and hormone supplement.

2) the small amount of calf serum is still necessary in these cell cultures that are continuously passaged by trypsinization; it may be more important for cell attachment, rather than for growth. FRTL cultures may, in fact, continue to divide in medium plus hormones without serum for weeks or even months.

3) hormonal requirements also depend upon cell densities: at high density FRTL cells may survive and even grow very slowly with only three hormones (insulin, TSH, and transferrin).

4) at low densities colonies may survive but individually attached cells may lyse following the withdrawal of any one of the six hormones.

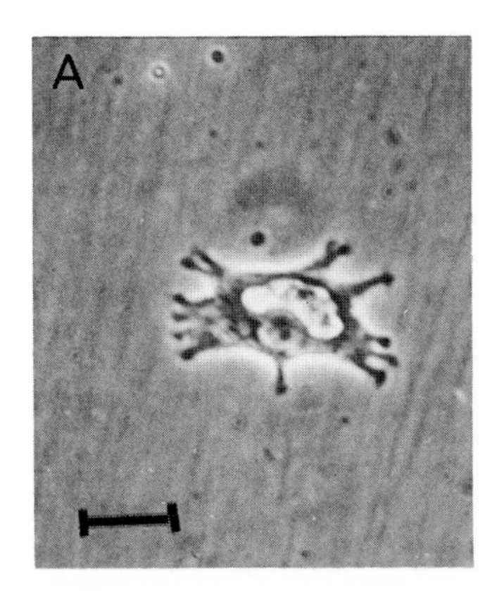

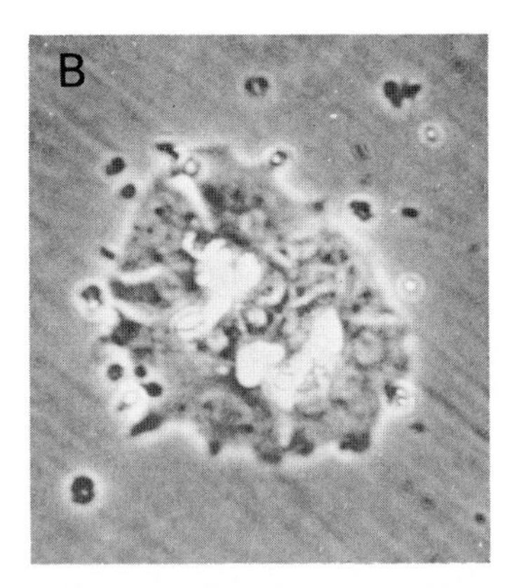

FIG. 1. Phase contrast microphotograph of a single cell (A) and of a small colony (4-6 cells) (B) of FRTL. Although of epithelial nature, cells appear stretched in various directions due to incomplete attachment to the plastic substrate, because of the low concentration of serum present (see text). Big vacuoli, probably equivalent to intracellular lumens can be seen in the cytoplasm (A). Such structures tend to become confluent among neighbour cells inside colonies (B). Bar represents 50 μM.

5) some requirements (especially insulin and TSH) show up very quickly, others are very slowly manifested after withdrawal (Glycyl-L-Histidyl-L-lysine acetate (GHL) and hydrocortisone).

6) sparse cultures and sparse regions of more dense cultures grow best when all six hormones are present; FRTL cells are routinely grown in 0.5% calf serum plus six hormones (modified F-12 + 6H) (2).

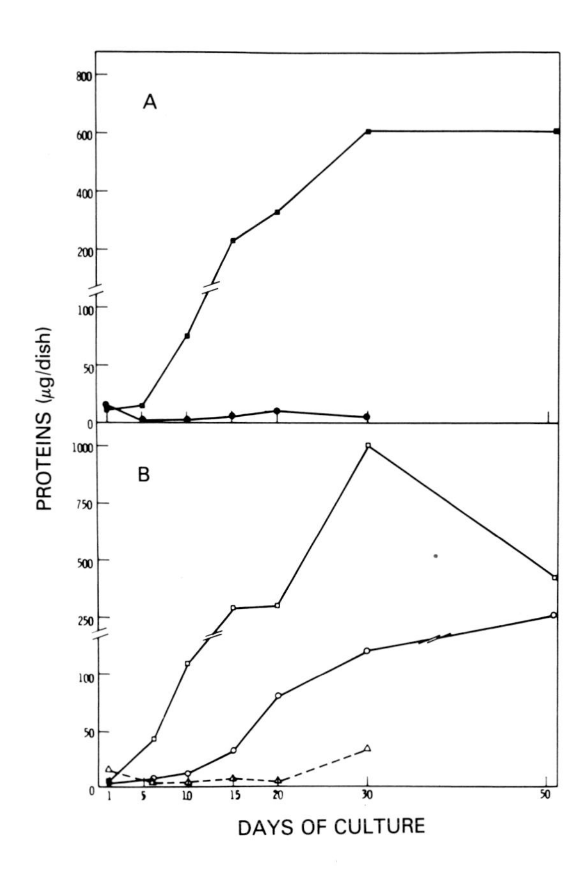

FIG. 2. Growth curve of rat thyroid primary cells: without (A), and with (B) the 6 hormones (see text). Rapid growth is acheived in 10 % fetal calf serum without (A, ■——■) and with hormones (B, □——□); however, in (A) growth is mainly accounted for by the fibroblast component. In the presence of 0.5 % serum, slow but steady preferential epithelial cell growth is obtained in the presence (B, ○——○) but not in the absence (A, ●——●) of the hormones. Without serum, only a limited growth was acheived with the hormones present (A, △---△), while not even primary cell attachment was obtained in the absence of hormones.

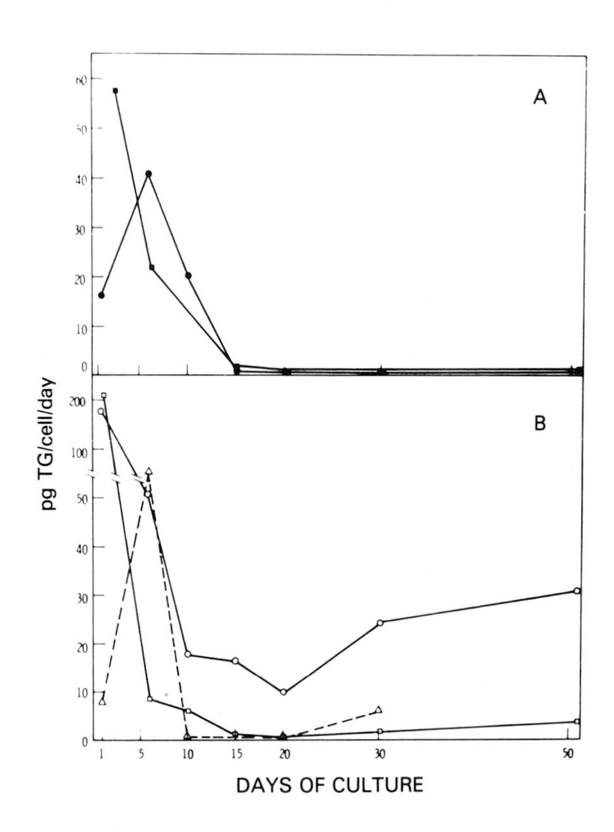

FIG. 3. Thyroglobulin production in primary cultures. TG was measured quantitatively by immunological method (4) and all results were normalized for the number of cells present in the culture. Cultures were kept in the same conditions as in FIG. 2., in the absence (A) and in the presence (B) of hormones. TG was found to be released at the onset of all cultures, probably from dying cells (see text). Later, TG has not been found in the absence of hormones in high serum (A, ■——■) or in low serum (A, ●——●). However, when hormones were present, TG was produced steadily throughout the culture and in maximal amounts from cells in 0.5 % serum (B, ○——○), and only later and in significantly lower amounts from cells in high serum (B, □——□). Probably, cells in the absence of serum, due to the initial very low plating efficiency, never grew enough to produce significant amounts of TG (B, △---△).

Primary cultures

Cell growth may not be the only parameter to be taken into consideration in evaluating the requirements of differentiated cells like FRTL. Thyroid functions expressed in vitro may even be affected indipendently of cell growth.To follow differentiation in relation to medium composition, primary cells were preferred to even the FRTL line which was already adapted to the modified F-12 + 6H medium. While using the same basic medium formulation (mF-12) the following variables were tested: A) serum concentrations (0; 0.5; 10%). B) animal source (fetal calf or calf). C) presence or absence of the 6H, on: 1) Cell growth; and 2) Differentiation.

1) Cell growth (FIG. 2.)

a) Cell growth was more rapid in high serum concentrations, but fibroblasts contributed massively from day 15-20 of culture.

b) Low serum concentrations only allowed epithelial cell growth. Also cell densities influenced fibroblast survival and growth; in especially dense primaries (more than 3×10^5 cells/60mm dish) fibroblasts could persist for weeks and were even found in secondary cultures, however, they never overgrew the epithelial cells.

2) Differentiation (FIG. 3.)

a) An initial burst of TG released in the medium was noted in all conditions tested and was interpreted as intracellular TG released into the medium from dying cells. This is probably a common event in the first days of culture, and also occurs after the subsequent cell passages, due to the low plating (attachment) efficiency of these cells (less than 30% in primaries and less than 50% in later passages). It was, therefore, considered that TG was produced by cells in vitro only if it persisted after day 10 of culture.

b) TG was never detected in epithelial cultures where the six hormones were not present. Our previous work (MS in preparation) indicates, however, that TG can be produced by mixed populations of epithelial colonies surviving among thryoid fibroblasts grown with serum supplement only.

c) High serum levels (more than 5%) affected TG production negatively, suppressing it completely during the first weeks of culture, and only permitting it - at a reduced rate - after a long adaptation period. At that time many fibroblats were present and again complex phenomena of parabiosis cannot be excluded.

CONCLUSIONS

It is possible that the loss of differentiation in vitro (11) was the consequence of the composition of the conventional media which were designed to meet the requirements of mesenchimal or tumor cells which are not dipendent on stimulation by hormones not found in adequate concentration in sera. At the onset of tissue culture serum (plasma clot) was the 'medium' in which organs were first maintained. To date, animal sera are still added in generous proportions (5-20%) to otherwise chemically defined media. This may be a quite unphysiological environment for most animal cells, with the exception of endothelial cells and the fibroblast-like cells active in wound healing. The serum could, therefore, act as a strong selective pressure favoring the mesenchymal cell types. If serum can be deleted or reduced in primary cultures, it is likely that new cell lines will be grown which will more resemble the original cell type in vivo.

The approach described here which has led to cultures of epithelial cells from rat thyroid with maintenance of characteristic differentiation, may be applied with appropriate modifications of media or hormonal supplement to other epithelial cell types form different animal species and organs.

Part of this work has been supported by the NIH grant N. 1 R01 AM21689-01.

REFERENCES

1. Ambesi-Impiombato, F. S., and Coon, H. G. (1979): Intern. Rev. of Cytology, in press.
2. Ambesi-Impiombato, F. S., Parks, L. A. M., and Coon, H. G.: submitted for publication.
3. Carrell, A., and Burrows, M. T. (1910): J. Am. Med. Ass., 55: 1379-81.
4. Fontana, S., and Rossi, G.: Submitted for publication.
5. Hatt, H. D., and Gantt, M. J., Editors (1979): The American Type Culture Collection Catalogue of strains II, 2. ed., ATCC Rockville, Maryland (USA).
6. Hayashi, I., and Sato, G. H. (1976): Nature, 259:132-34.
7. Hilfer, R. S. (1962): Developmental Biol., 4:1-21.
8. Kalderon, A. F., and Wittner, M. (1967): Endocrinology, 80: 797-800.
9. Rapoport, B. (1976): Endocrinology, 98:1189-97.
10. Salvatore, G., and Edelhoch, H. (1973): In: Hormonal proteins and peptides, Vol. 1, edited by C. H. Li, pp. 201-241. Acad. Press, New York.
11. Sato, G.H. (1975): In: Biochemical actions of hormones, edited by G. Litwack, pp. 391-409. Academic Press, New York.
12. Sato, G. H., Zaroff, L., and Mills, S. E. (1960): Proc. Nat. Acad. Sci. (USA), 46:963-72.
13. Siegel, E. (1971): J. Cell Sci., 9:49-60.
14. Spooner, S. B. (1970): J. Cell Physiol., 75:33-48.
15. Tong, W. (1974): In: Methods in Enzymology 32 (part B), edited by S. P. Colowick and N. O. Kaplan, pp. 745-758. Academic Press, New York.
16. Volpert, E. M., and Presyma, A. P. (1977): Acta Endocrinol., 85:93-101.
17. Winand, R. J., and Kohn, L. D. (1975): J. Biol. Chem., 250: 6534-40.
18. Wollman, S. H., (1963): Rec. Progr. in Horm. Res., 19:579-618.

Control Mechanisms in Animal Cells,
edited by L. Jimenez de Asua et al.
Raven Press, New York © 1980.

Control of Growth of Kidney Epithelial Cells

Robert W. Holley

Molecular Biology Laboratory, Salk Institute for Biological Studies, San Diego, California 92138, U.S.A.

Those interested in the control of growth of mammalian cells are concerned with the conditions outside the cell, and the events inside the cell, that lead to the initiation or the arrest of growth. Present knowledge indicates that matters are complicated, both outside and inside the cell. In this paper I plan to discuss both situations, illustrating certain points with our recent studies of the control of growth of kidney epithelial cells.

GROWTH-CONTROLLING FACTORS OUTSIDE THE CELL

First, consider the situation outside the cell. Work in many laboratories has shown that the growth control mechanisms respond to numerous different types of factors. Synergisms between different growth-controlling factors are often observed. Many types of materials can initiate growth, when added to the culture medium of quiescent cells. The list includes such varied materials as: polypeptide growth factors (9), corticosteroids (32), prostaglandins (17), cholera toxin (23), cyclic nucleotides (21), lithium ions (16), vasopressin (26), common nutrients (19,31), proteolytic enzymes (5,10), lectins (33), and tetradecanoyl phorbol acetate (7).

The conclusion seems inescapable that the mechanisms that control growth are complex.

The factors that are probably of greatest interest *in vivo* are the polypeptide growth factors, including epidermal growth factor (EGF) and fibroblast growth factor (FGF). These factors are small proteins, similar to the polypeptide hormones. Under proper conditions they stimulate the growth of cells at ng/ml concentrations. For full activity, they require the presence of other essential materials, typically insulin and nutrients, in the culture medium.

Present information suggests that many different polypeptide growth factors exist and that many of them remain unpurified and unidentified. Serum appears to contain numerous different growth factors, and medium conditioned by growth of cell cultures is also a source of growth factors (9,24).

Although our understanding has progressed to the point that several cell lines can now be grown in completely defined medium, without serum (2,22), the situation in vivo remains poorly defined. Partly this is because the purified growth factors have broad, though differing, specificities. For example, EGF and FGF stimulate growth of large, overlapping, but different, sets of cells (9). The overlapping specificities, as well as the synergisms between different factors, make it very difficult to determine which growth factors act in natural situations. This is not of concern to those of us who wish to understand the broad mechanisms of growth control, but it remains a serious practical problem for those who want to know which growth factors are important in different situations in vivo.

The growth factors act by binding to specific cell surface receptor sites (3). Whether a growth factor acts on a certain type of cell depends on the number of receptor sites on the cell and on the affinity of the receptor sites for the growth factor. Also, the responsiveness of cells to bound growth factor can vary in ways that we do not understand. The interaction of the growth factors with cells leads to a series of events inside the cell that culminates in the initiation of DNA synthesis.

Before turning to the growth-controlling events inside the cell, I want to discuss two other poorly understood influences on events outside the cell. One is the role of endogenous growth inhibitors in controlling growth, and the other is the importance of cell surface area and/or shape. Growth inhibitors and the surface area of cells are both important in some situations. For example, both appear to be involved, along with growth factors, in the control of growth of the BSC-1 cell, an African Green monkey kidney epithelial cell line that we have been studying. This work will be summarized briefly.

DENSITY-DEPENDENT REGULATION OF GROWTH

One of the very interesting phenomena of growth control observed in cell culture is the slowing of growth of "normal" cells when the culture becomes crowded. We have studied this phenomenon, density-dependent regulation of growth, first in 3T3 mouse embryo fibroblasts (14,15) and more recently in BSC-1 African Green monkey kidney epithelial cells (11-13). We have concluded that density-dependent regulation of growth is a complex phenomenon that results from a combination of causes, that it has different causes in different cell types, and that it can have different causes in different situations involving the same cell type.

TABLE 1. Density dependent regulation of growth of 3T3 and BSC-1 cells

Growth controlling factor	Contribution to density dependent regulation of growth	
	3T3 (fibroblast)	BSC-1 cells (epithelial)
Serum factors,		
Inactivation	> 90%	< 10%
Decrease in receptor sites	Not observed	∿ 50%
Concentration of low molecular weight nutrients	Mechanisms present, but not important at normal concentrations	10% or less
Inhibitors produced by the cells	< 10%	∿ 50%

Table 1 contrasts the causes of density-dependent regulation of growth in 3T3 cells and BSC-1 cells. With both cell types one can show responses to the same general types of growth controls - serum growth factors, nutrient concentrations, and growth inhibitors produced in the culture medium. However, growth restriction has very different causes in the two different cell cultures. With 3T3 cells, under normal culture conditions, the primary growth restriction is due to depletion of serum growth factors from the medium (14). This is shown most clearly by experiments in which the volume of medium is increased relative to the area of cells (see Fig. 1). When a very large volume of medium is present, relative to the area of the cells, 3T3 cells grow to ten times the cell density attained under "normal" culture conditions (14). Using the large volume of medium, which gives culture conditions approaching equilibrium conditions, the cell density at which restriction of growth takes place still depends on the serum concentration, but a given cell density is attained at a much lower serum concentration. Growth restriction presumably now results, at least in part, from a diffusion boundary layer (30) of serum factors caused by cellular destruction of the serum factors next to the cell layer.

With BSC-1 cells, the situation is very different. Although serum is needed for the growth of BSC-1 cells, it requires a 20-fold increase in the serum concentration to double the cell density (Fig. 2) (13). Approximately half the serum can be replaced by EGF, to attain the same cell density (13). With BSC-1 cells, in contrast to 3T3 cells, it can be shown that the number of receptor sites for EGF decreases dramatically as the cells become crowded (Fig. 3) (13). Also, maximum binding of EGF to sparse BSC-1 cells is attained much sooner after the addition

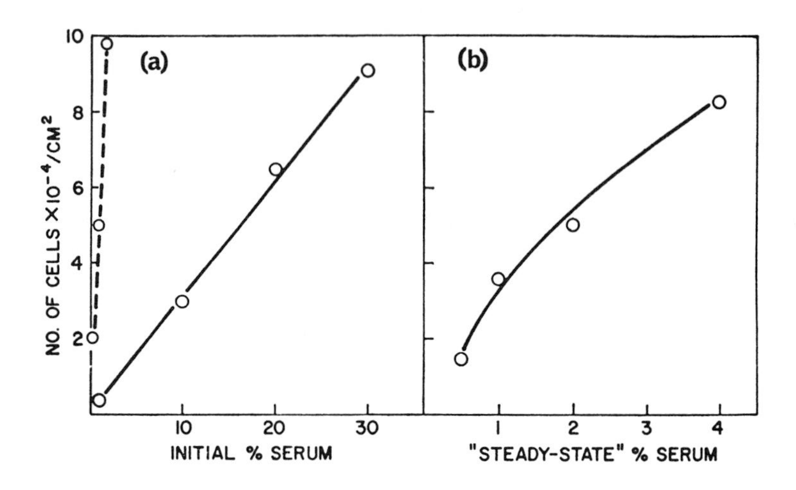

FIG. 1. *Dependence of 3T3 cell density on serun concentration.* In Fig. 1*a* (*left*), the cell density is that attained after growth had largely ceased (4 days) in media with the serum concentration shown. ———, 3T3 cells; - - -, SV3T3 cells. In Fig. 1*b* (*right*), the cell density is that attained by 3T3 cells on a 12-mm coverslip after growth had ceased (5 days) in 10 ml medium, fluid being changed daily, with the serum concentration shown. (Reprinted with permission from ref.14.)

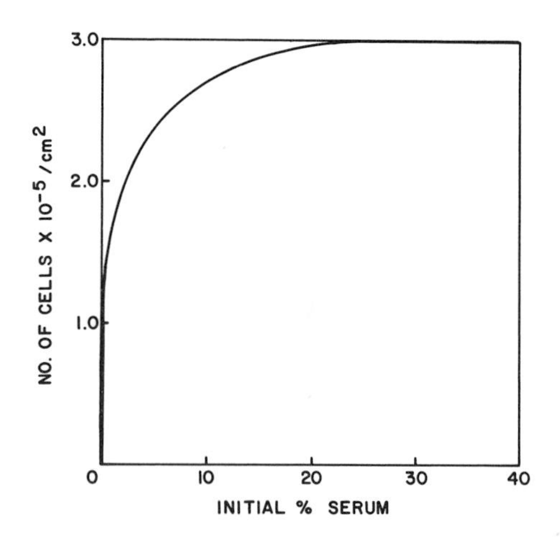

FIG. 2. *Dependence of BSC-1 cell density on serum concentrations.*

of EGF than it is with crowded cells (Fig. 4). Sparse cells have a larger fraction of the bound EGF present on the surface of the cells (13). It seems likely that all of these differences favor growth of sparse BSC-1 cells relative to crowded cells.

Also, in contrast to the situation with 3T3 cells, BSC-1 cell density is sensitive to the nutrient concentrations at the concentrations of nutrients that are present in common media used for cell culture. Increasing the glucose concentration in Dulbecco-modified Eagle's medium 4-fold increases the cell density approximately 60% (11). It is not that the lower concentration provides inadequate glucose; excess glucose is always present. Instead, it is the concentration of glucose that has a growth-regulatory effect. The "normal" cell density of BSC-1 cultures can also be decreased by lowering the concentrations of many nutrients (11). These results indicate that the mechanism of density-dependent regulation of growth is sensitive to the concentrations of many different nutrients.

Another major difference between 3T3 cells and BSC-1 cells is the production of growth inhibitors by BSC-1 cells (12). With 3T3 cells, the only growth inhibitors we have detected in the culture medium are ammonium ion and lactic acid, and these are found at concentrations low enough that they have little effect on growth. In contrast, BSC-1 cells produce at least one high molecular weight inhibitor and at least one low molecular weight inhibitor, in addition to ammonium ion and lactic acid. The high molecular weight growth inhibitor, which seems to be of most interest, is concentrated by ultrafiltration, is purified initially by gel filtration, and finally is purified by high performance liquid chromatography. A dose-response curve of the purified inhibitor obtained from high performance liquid chromatography is shown in Fig. 5. The purified material is growth inhibitory at ng/ml concentrations. Preliminary results indicate that this growth inhibitor is relatively specific for kidney epithelial cells. Flow microfluorometric studies indicate that it arrests the growth of BSC-1 cells in the G_1 phase of the cell cycle. We are interested in the action of this growth inhibitor *in vivo* and we hope it can be used, in a practical way, in the control of cell growth *in vivo*.

Many endogenous growth inhibitors have been reported in the literature and a number of inhibitors have been postulated (for a review see ref. 20). Because of technical difficulties, the study of endogenous growth inhibitors is less advanced than the study of the polypeptide growth factors. Most of the inhibitor preparations described in the literature have very low specific activities and this has raised doubts about their significance. Nevertheless, my expectation is that growth inhibitors will turn out to be important components of the growth control system *in vivo*. The best studied polypeptide materials that are active as growth inhibitors are ACTH (35) and interferon (1,34), which inhibit the growth of cells in culture at ng/ml concentrations.

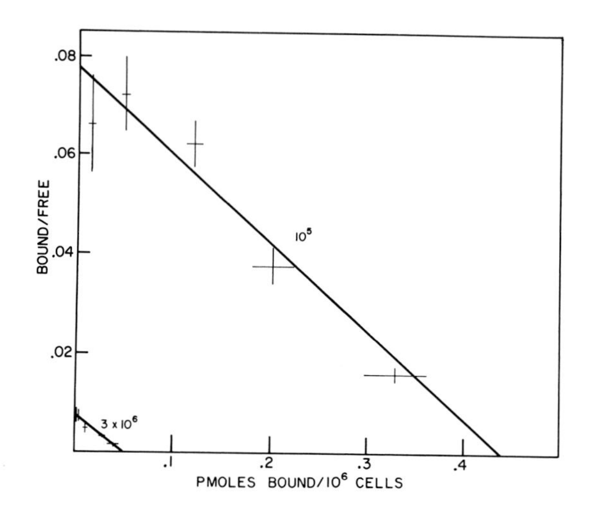

FIG. 3. *Scatchard plots of ^{125}I-EGF binding to BSC-1 cells for 1 hour at 0°-4°.* The cell densities (10^5 and 3×10^6) are for 5-cm plates. The concentrations of EGF at the various data points correspond to approximately 0.9, 2.8, 9, 28, and 90 ng/ml, from left to right. The results are the mean (± SEM) of six different experiments for each point. (Reprinted from ref. 13.)

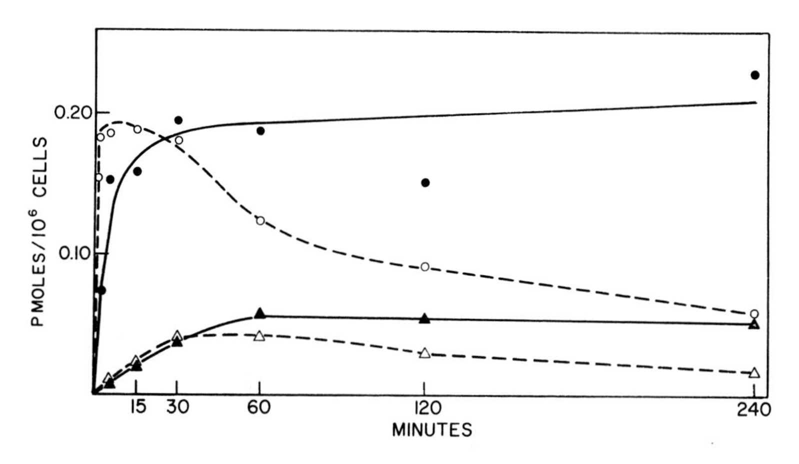

FIG. 4. *Time course of binding of ^{125}I-EGF to BSC-1 cells.* O, 10^5 cells per 5-cm plate at 37°; Δ, 3×10^6 cells at 37°; ●, 10^5 cells at 0°-4°; ▲, 3×10^6 cells at 0°-4°. EGF was added at 70 ng/ml.
(Reprinted from ref. 13.)

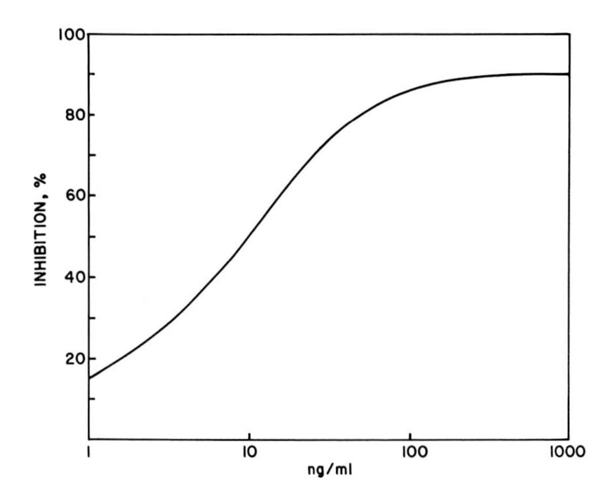

FIG. 5. *Dose-response curve of the inhibition of DNA synthesis in BSC-1 cells by partially purified kidney epithelial cell growth inhibitor.*

Another fascinating question that remains about growth control in cell culture is the role of cell surface area. There are a number of situations in which the rate of growth of cells in culture correlates well with the apparent surface area of the cells (4,8,37). This is observed with BSC-1 cells also. Although it is only one of many possibilities, the hypothesis that appeals to me is that decreasing the surface area of a cell has a combination of effects: it could decrease the number of receptor sites for some growth factors and it could decrease the activities of numerous "pumping" mechanisms and nutrient uptake mechanisms. All of these could interact to decrease the rate of growth.

One hypothesis for density-dependent regulation of growth that has been popular is the possibility that contacts between cells provide a direct signal for growth suppression. It has been proposed that the growth-inhibitory membrane proteins that can be extracted from 3T3 cell membranes represent a component of such a growth regulatory system (36). Except for their effect on cell surface area and/or shape, I am not aware of any convincing evidence that contacts between cells are important in growth control. Many cells, particularly epithelial cells, grow in patches in culture with contacts maintained between cells even when the cells are growing rapidly. In the specific instance of 3T3 cells, under normal culture conditions, arrest of growth takes place in medium with a low concentration of serum when the cells are still sparse, but the cells grow to high density in the same medium when the volume of medium is large (14). This argues against contacts being a growth-controlling factor.

GROWTH-CONTROLLING EVENTS INSIDE THE CELL

If we now shift our attention to events inside the cell, what can be said about the events that lead to the initiation, or the arrest, of DNA synthesis? When growth factors stimulate the initiation of DNA synthesis, what do they do? How do the other factors influence these processes?

Considering growth factors first, a major question is whether the growth factors act on the cell surface or whether they are internalized before they act. The earliest changes that are known to follow the addition of growth factors to quiescent cells are changes at the cell surface membranes: changes in ion fluxes (25,28) and uptake processes (6) and membrane changes such as phospholipase A_2 activation (27). These changes take place within minutes, long before internalization and degradation of the growth factors achieve maximum rates. Work with insulin suggests that early transport changes do not require internalization of the growth factor (18). Unfortunately, this does not prove that the growth factors act at the cell surface since there is no proof that any of these early changes at the cell membrane are required for the initiation of DNA synthesis. To provide evidence that a growth factor acts inside the cell, one would like to initiate DNA synthesis by injecting the factor into a cell. As far as I am aware no one has succeeded with this experiment, and I doubt that it would be considered conclusive unless the factor is unusually active when injected. It is difficult to see how to exclude the possibility that internalization of the growth factor is essential. Delivery inside the cell of even one molecule of the growth factor, or of something related to the specific receptor for the growth factor, could be all that is required, and there seems no possibility of proving that this does not take place.

One can ask whether the earliest events observed at the cell surface membrane are such that they might be sufficient to lead to the initiation of DNA synthesis. The changes that are observed, in ion fluxes and uptake processes, and in activation of membrane phospholipase A_2, are such that they could lead to a cascade of effects, altering a variety of intracellular processes. Though we do not know whether these early changes are crucial, and whether they lead to the initiation of DNA synthesis, it seems to me to be reasonable working hypothesis at present to assume that they are.

We are still left with the question of how combinations of different growth factors can interact and how the polypeptide growth factors can interact with the diverse types of growth-controlling substances listed earlier, to control the initiation of DNA synthesis. The hypothesis one chooses here is influenced by one's view of a different fundamental question: What is the process, or series of processes, that directly controls the initiation of DNA synthesis? The answer is, of course, unknown, but two distinctly different simple models can be postulated (Fig. 6). In one model (Fig. 6a), the growth-controlling factors act directly to influence a series of processes that lead to the initiation of DNA synthesis. In the second model (Fig. 6b), the growth-controlling factors act on general cellular processes and these, indirectly, lead to the initiation of DNA synthesis. A more complicated possibility is a combination of (a) and (b), in which some growth-controlling factors act directly and some act indirectly.

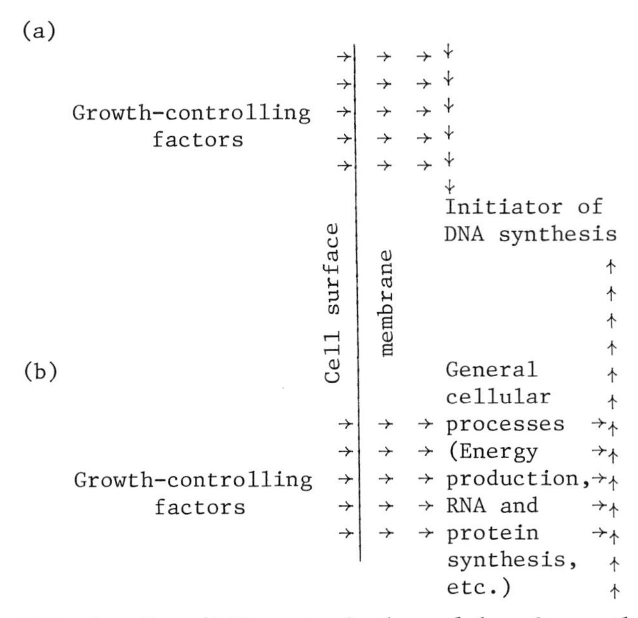

FIG. 6. *Two different simple models of growth factor action.*

I am not aware of any evidence that clearly favors one possibility over the other. The synergisms that are observed between different growth-controlling factors suggest that different factors act in different ways, but this is possible in either Fig. 6(a) or 6(b). The nature of certain of the growth-controlling factors, e.g. ions, amino acids and glucose, suggest that general cellular processes play a role, but this role could be direct or indirect. My guess at present is that the control system will turn out to be very complex and will combine features of both models in Fig. 6(a) and 6(b). Considering the complexity of DNA replication in bacteriophage (29), it seems quite possible that all of our models are over-simplications.

ACKNOWLEDGMENTS

This work was supported in part by Grant Number CA11176, awarded by the National Cancer Institute, DHEW, and Grant Number VC-304-F from the American Cancer Society. R. W. Holley is an American Cancer Society Professor of Molecular Biology.

REFERENCES

1. Balkwill, F., and Taylor-Papadimitriou, J. (1978): Nature, 274:798.
2. Bottenstein, J. A., and Sato, G. H. (1979): Proc. Natl. Acad. Sci. USA, 76:514.
3. Carpenter, G., Lembach, K. J., Morrison, M. M., and Cohen,S. (1975): J. Biol. Chem., 250:4297.
4. Castor, L. N. (1968): J. Cell. Physiol., 72:161.
5. Chen, L. B., and Buchanan, J. M. (1975): Proc. Natl. Acad. Sci. USA, 72:131.
6. Cunningham, D. D., and Pardee, A. B. (1969): Proc. Natl. Acad. Sci. USA, 64:1049.
7. Dicker, P., and Rozengurt, E. (1978): Nature, 276:723.
8. Folkman, J., and Moscona, A. (1978): Nature, 273:345.
9. Gospodarowicz, D., and Moran, J. S. (1976): Ann. Rev. Biochem., 45:531.
10. Grumm, F. G., and Armstrong, P. B. (1979): Exp. Cell Res., 119:317.
11. Holley, R. W., Armour, R., and Baldwin, J. H. (1978): Proc. Natl. Acad. Sci. USA, 75:339.
12. Holley, R. W., Armour, R., and Baldwin, J. H. (1978). Proc. Natl. Acad. Sci. USA, 75:1864.
13. Holley, R. W., Armour, R., Baldwin, J. H., Brown, K. D., and Yeh, Y.-C. (1977) Proc. Natl. Acad. Sci. USA, 74:5046.
14. Holley, R. W., and Kiernan, J. A. (1971): In: Ciba Foundation Symposium on Growth Control in Cell Culture, edited by G. E. W. Wolstenholme, and J. Knight, pp. 3-10. Churchill Livingstone, London.
15. Holley, R. W., and Kiernan, J. A. (1974): Proc. Natl. Acad. Sci. USA, 71:2908.
16. Hori, C., and Oka, T. (1979) Proc. Natl. Acad. Sci. USA, 76:2823.
17. Jimenez de Asua, L., O'Farrell, M. K., Clingan, D., and Rudland, P. S. (1977): Proc. Natl. Acad. Sci. USA, 74:3845.
18. Le Cam, A., Maxfield, F., Willingham, M., and Pastan, I. (1979): Biochem. Biophys. Res. Commun., 88:873.
19. Ley, K. D., and Tobey, R. A. (1970): J. Cell. Biol., 47:453.
20. Lozzio, B. B., Lozzio, C. B., Bamberger, E. G., and Lair, S. V. (1975): Internatl. Rev. Cytology, 42:1.
21. Marcelo, C. L. (1979): Exp. Cell Res., 120:201.
22. Mather, J. P., and Sato, G. H. (1979): Exp. Cell Res., 120:191.
23. Pruss, R. M., and Herschman, H. R. (1979): J. Cell. Physiol., 98:469.
24. Rheinwald, J. G., and Green, H. (1975): Cell, 6:331.

25. Rozengurt, E., and Heppel, L. A. (1975): Proc. Natl. Acad. Sci. USA, 72:4492.
26. Rozengurt, E., Legg, A., and Pettican, P. (1979): Proc. Natl. Acad. Sci. USA, 76:1284.
27. Shier, W. T. (1979): Proc. Natl. Acad. Sci. USA, in press.
28. Smith, J. B., and Rozengurt, E. (1978): Proc. Natl. Acad. Sci. USA, 75:5560.
29. Stetler, G. L., King, G. J., and Huang, W. M. (1979): Proc. Natl. Acad. Sci. USA, 76:3737.
30. Stoker, M. G. B. (1973): Nature, 246:200.
31. Taylor-Papadimitriou, J., and Rozengurt, E. (1979): Exp. Cell Res., 119:393.
32. Thrash, C. R., and Cunningham, D. D. (1973): Nature, 242:399.
33. Transplantation Review, edited by G. Möller, p. 11. Munksgaard, Copenhagen, 1972.
34. Watanabe, Y., and Sokawa, Y. (1978): J. Gen. Viol., 41:411.
35. Weidman, E. R., and Gill, G. N. (1976): J. Cell. Physiol., 90:91.
36. Whittenberger, B., Raben, D., Lieberman, M. A., and Glaser, L. (1978): Proc. Natl. Acad. Sci. USA, 75:5457.
37. Zetterberg, A., and Auer, G. (1971): Exp. Cell Res., 67:260.

Control Mechanisms in Animal Cells,
edited by L. Jimenez de Asua et al.
Raven Press, New York © 1980.

Common and Distinct Aspects in the Regulation of Specific Growth Factors

J. Schlessinger, A. Levi, I. Yuli, M. Gabbay, and Y. Shechter

Weizmann Institute of Science, Rehovot, Israel

Introduction

Serum is required for the growth and survival of nearly all animal cells in culture (9, 10). Many cellular processes begin with the binding of a serum protein to a specific cell surface receptor. Among these processes is the control of cholesterol transport by low density lipoprotein (1), the delivery of enzymes to the lysosomes (18), the transport of the iron carrier transferrin(17) and the action of polypeptide hormones (7). All of these processes, with the exception of polypeptide hormones, require the internalization of the protein in order to achieve the biological response. Although more and more information is becoming available concerning the sequential steps that lead to hormone internalization, the relevance of this phenomenon to the mechanism of action of the hormone is not yet understood.

Our approach to investigate the mechanism of action of polypeptide hormones is to study the binding of fluorescent hormone conjugates to their membrane receptors and to investigate their mobility and distribution under various conditions. We tamper with different stages in the processing of the hormone and correlate the observed changes with hormone responsiveness.

A method for studying the mobility and distribution of fluorescent molecules on cells

We have prepared biologically active rhodamine conjugates of insulin (21), epidermal growth factor (EGF) (21) and nerve growth factor (NGF) (14).

The lateral mobility of fluorescent hormones is studied with the method of fluorescence photobleaching recovery (FPR) (6, 11, 13). The localization of the fluorescent hormones is accomplished with a sensitive image intensified camera (19). Both systems are combined on a single microscope which allows the determination of receptor mobility and distribution on a single viable cell.

Our apparatus is described in Fig. 1. The beam from an argon laser is focused through a vertical illuminator and a set of lenses (L and OB) onto the specimen membrane. In a typical experiment the fluorescently labeled cells adhering to the cover

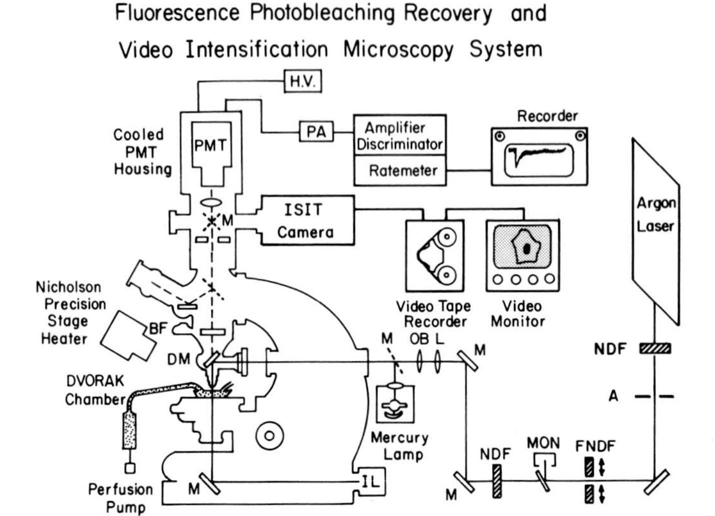

FIG. 1. Schematic diagram which describes our fluorescence photobleaching recovery (FPR) and video intensification system. Dotted lines indicate removable components (see text for notations).

slips in which they were cultured are viewed as viable cells in a Dvorak chamber containing complete growth medium or buffered saline solution. The fluorescence radiation is collected by the same objective lens, passes through the dichroic mirror (DM) of the vertical illuminator, through a field diaphragm onto sensitive photomultiplier tube (PMT) that is connected to photon-counting electronics for measurement of the time dependent fluorescence intensity (13). The illuminated area on the fluorescently labeled membrane is usually a spot of $\sim 1 \mu m$ radius. The appropriate laser line to excite the fluorophore is selected by a prism in the laser cavity. The beam is passed through neutral density filters (NDF), a scatter aperture (A) and a beam splitter that supplies a small portion of the radiation to a photodiode intensity monitor (MON). In order to visualise the cells the mirror (M) and the field diagram can be moved to present an image to viewing eyepiece or an Intensified Silicon Intensified Target (ISIT) camera (RCA TC 1040/H). This camera can detect very low levels of light and is employed to localize the fluorescent hormones in the cell. The phase and fluorescent images are video taped (Panasonic VTR NV-8030) and the micrographs were taken with a Polaroid camera from the television screen (9" National VW-5300) that projects the intensified images. This camera is also used to align and focus the laser beam on the cell surface. The entire apparatus is mounted on a table with pneumatic supports that eliminate essentially all building vibrations. To supply the photobleaching light pulse for FPR experiments, a solenoid operated neutral density filter FF (OD 4 or 3) is momentarily removed from the laser beam and simultaneously the high voltage bias potential is removed from the photomultiplier dynode chain to avoid damaging it. The fluorescence light intensity is measured by sensitive cold operated photomultiplier to reduce dark current. The electronic system includes: photon counting electronics, multichannel digital recording, and recorders.

A common pathway for the internalization of polypeptide hormones

Employing biologically active fluorescent conjugates of insulin, EGF and NGF we have demonstrated a common pathway for the internalization of peptide hormones by their target cells. The pathway involves the following steps (15, 19, 20):

1. A hormone binds to diffusely distributed mobile membrane receptors.
2. The occupied receptors cluster, in a temperature sensitive process, and form visible immobile patches.
3. The patches are endocytosed by an energy dependent process.
4. The internalised endocytic vesicles interact with other intracellular organelles.

For insulin and EGF it was shown that coated regions on the plasma membrane of fibroblasts are involved in the process of hormone internalization (15, 6).

Insulin, EGF and NGF induce rapid and delayed responses on their target cells. The temporal relationship between the processing of the hormone and the various biological responses which are mediated by them are far from being understood.

The significance of receptor aggregation

Studies with autoantibodies against insulin receptor isolated from sera of patients with insulin-resistant diabetes demonstrated that clustering of insulin-receptors plays an important role in the activation of various responses to the hormone (12). Similar conclusions were reached by Shechter *et al.* which showed that a cyanogen bromide cleaved EGF binds to the cell surface but fails to induce receptor clustering and thymidine incorporation into 3T3 fibroblasts (22). The bioactivity of this derivative was completely restored by the addition of divalent anti EGF antibodies while monovalent Fab' fragments did not have the same effect (22).

Based on these and other results we proposed a model for the action of insulin and EGF (20). We have divided the hormone induced aggregation into two processes. A *local* aggregation process involving few receptor molecules is postulated to be required for the rapid membrane responses to insulin and EGF and for the stimulation of thymidine incorporation by EGF. Global aggregation involving the formation of visible immobile patches is related to receptor down regulation and precedes the internalization and degradation of the hormones.

It has been shown that ammonia inhibits the internalization of receptor bound human choriogonadotropin (3) and that ammonia and primary amines inhibit the clustering (16) and the subsequent internalization (4, 16) of EGF on fibroblasts. We have used methyl amine as a diagnostic tool which can specifically block patching of receptors, and the subsequent internalization of hormone receptor complexes. We studied hormone induced uptake of α-amino isobutyrate (AIB) by 3T3 cells in the presence or absence of methyl amine. For EGF both the amine treated and the non-treated cells show essentially the same dependence on time and concentration of the hormone (Fig. 2).

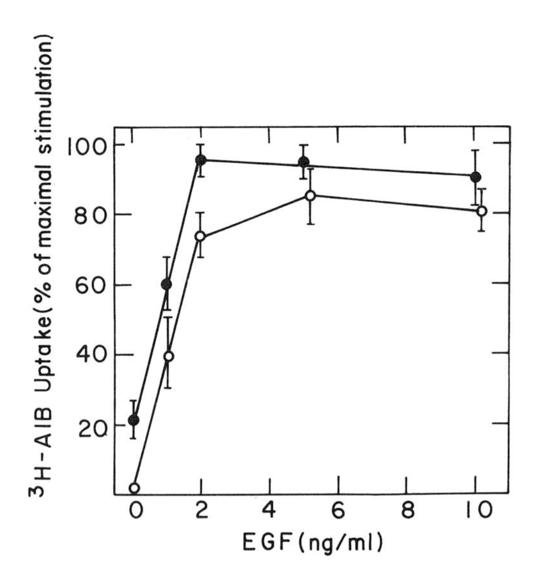

FIG. 2. EGF stimulation of α-aminoisobutyrate (AIB) uptake in the presence or absence of methylamine. Confluent monolayers of 3T3 cells were preincubated in the presence (●) or absence (o) of 5 mM methylamine in Hank's balanced salt solution with 0.1% BSA for 45 min at 37°C. Then the cells were incubated with the indicated concentrations of EGF and for 50 min with 8 μM AIB containing the labeled tracer. AIB uptake was linear within the first hour of incubation. Maximal uptake was 170% of control values in the absence of added mitogen.

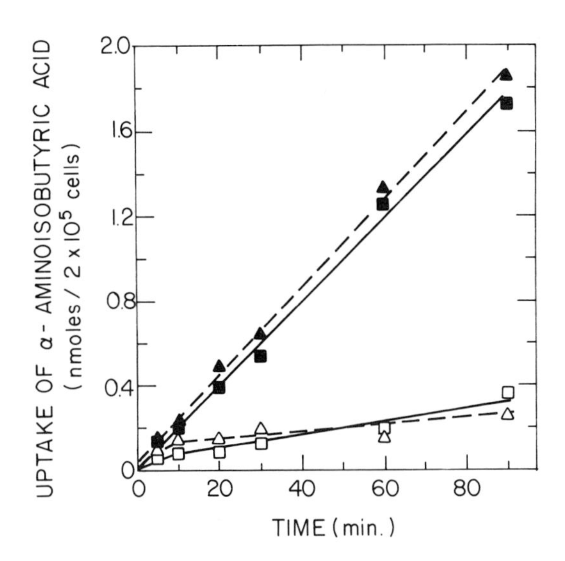

FIG. 3. Insulin stimulation of α-aminoisobutyrate (AIB) uptake in the presence or absence of methylamine. Confluent monolayers of 3T3 cells were preincubated with 10 mM of methylamine for 30 min at 37°C. Insulin (100 ng/ml) was present for 1 hr at 37°C and then 10 μM of (^{3}H)AIB was added for various periods of time at 37°C. The uptake was stopped by washing the cells three times with ice cold PBS. The cells were dissolved in 1 N NaOH and their radioactive content was determined. Basal uptake in the presence (Δ) or absence (■) of methyl amine and insulin-stimulated uptake in the presence (▲) or absence (■) of methylamine.

Similar results were obtained for insulin mediated uptake of AIB by 3T3 cells in the presence or absence of methyl amine (Fig. 3). This result, together with the fact that methyl amine can inhibit patch formation and internalization during the period necessary to obtain AIB uptake mediated by these two hormones indicates the irrelevance of these processes in the activation of AIB uptake.

We have recently shown that EGF or NGF receptor which patch and become immobile at 37°C remain dispersed and mobile in the presence of methyl amine on the target cells (14, 23). The diffusion of a lipid probe is only marginally effected by the presence of methyl amine, indicating that the building of the lipid matrix does not play a role in the process of receptor patching and immobilization. Other effects of methyl amine include (Fig. 4): inhibition of receptor loss (down regulation), dissociation and degradation (23).

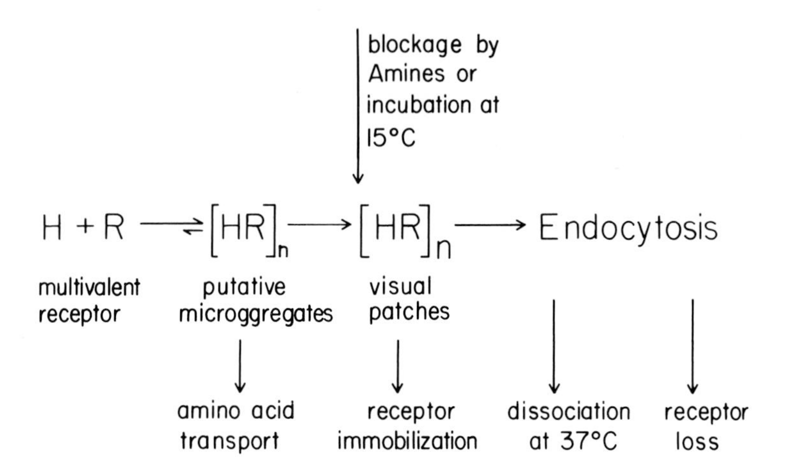

FIG. 4. Schematic diagram which summarizes the effect of methyl amine on various aspects of the regulation of EGF. Methyl amine blocks the formation of visible patches, therefore the occupied receptor remain dispersed and mobile thus preventing the subsequent processes of receptor loss and release ("dissociation") of degraded hormone. EGF stimulated amino acid transport is postulated to be activated at the stage in which microaggregates of hormone-receptor complexes are formed (20).

Recent reports indicate that insulin and EGF can induce covalent modifications of cellular components. The binding of EGF to isolated membranes results in a marked stimulation of phosphorylation of intrinsic proteins in the presence of (^{32}P) ATP (5). It was proposed that the receptor may be a protein kinase or a regulator of phosphoprotein phosphatase.

Insulin or antibodies against insulin receptor stimulate the phosphorylation of intracellular protein in 3T3-L1 preadipocytes This protein was tentatively identified as ribosomal protein S-6 (24). EGF is also able to stimulate this phosphorylation but at higher dose of hormone. The relationship between these hormone-induced chemical modifications and their mode of action remains to be elucidated.

More and more reports indicate that insulin (8) and NGF (2, 25) possess specific receptor on the nucleus of their target cells and that these receptors are responsible for the delayed responses of the two hormones, i.e. stimulation of DNA synthesis for insulin and the long term maintenance of cell viability for NGF. To date, there are no reports which indicate the binding of EGF to the nucleus of its target cells.

Conclusions

Insulin, EGF and NGF share many common and few distinct aspects in their action on their target cells. All three polypeptide hormones bind to diffuse and mobile (D = $(4\text{-}8) \times 10^{-10}$ cm^2/sec) receptor on the membrane of their target cells. Hormone binding induces formation of visible immobile patches of hormone-receptor complexes which in an energy dependent process become endocytosed and degraded. Microaggregation of few insulin or EGF receptors plays an important role in the chain of events which lead to bioactivity. It is not clear whether microaggregation of NGF receptors plays any role in its action.

Methyl amine blocks the formation of visible patches and the subsequent internalization of insulin, EGF and NGF. It does not interfere, however, with insulin or EGF mediated AIB uptake. This indicates that the formation of visible patches of insulin or EGF receptors and their subsequent internalization do not play a role in stimulation of amino acid uptake.

Insulin and NGF have specific receptors on the nucleus of their target cells. EGF and insulin stimulate the phosphorylation of specific cellular components.

References

1. Anderson, R.G.W., Goldstein, J.L., and Brown, N.S. (1977): Nature, 279:695-699.
2. Andres, R.Y., Jeng, I., and Bradshaw, R.A. (1977): Proc. Natl. Acad. Sci. USA, 74:2785-2789.
3. Ascoli, M. (1978): J. Biol. Chem., 253:7839-7843.
4. Carpenter, G., and Cohen, S. (1976): J. Cell. Biol., 71:159-171.
5. Carpenter, G., King, L., Jr., and Cohen, S. (1978): Nature, 276:409-410.
6. Edidin, M., Zagyansky, Y., and Lardner, T.J. (1976): Science, 191:466-468.
7. Fain, J.N. (1974): In: Biochemistry Series One, MTP International Review of Science, edited by H.V. Rickenberg, 8:1-23.
8. Goldfine, I.D., Smith, G.J., Wong, K.Y., and Jones, A.L. (1977): Proc. Natl. Acad. Sci. USA, 74:1368-1372.
9. Gospadarowicz, D., and Moran, J.S. (1976): Ann. Rev. Biochem., 45:531-558.
10. Holley, R.W. (1975): Nature, 258:487-490.
11. Jacobson, K., Wu, E., and Poste, G. (1976): Biochim. Biophys. Acta, 433:215-222.
12. Kahn, C.R., Baird, K.L., Jarrett, D.B., and Flier, J.S. (1978): Proc. Natl. Acad. Sci. USA, 75:4209-4213.
13. Koppel, D.E., Axelrod, D., Schlessinger, J., Elson, E.L., and Webb, W.W. (1976): Biophys. J., 16:1315-1329.

14. Levi, A., Shechter, Y., Neufeld, E.J., and Schlessinger, J. (1979): Proc. Natl. Acad. Sci. USA (submitted).
15. Maxfield, F.R., Schlessinger, J., Shechter, Y., Pastan, I., and Willingham, M.C. (1978): Cell, 14:805-810.
16. Maxfield, F.R., Willingham, M.C., Davies, P.J.A., and Pastan, I. (1979): Nature, 277:661-663.
17. Morgan, E.H. (1974): In: Iron in Biochemistry and Medicine, edited by Jacobs and Worwood), pp. 29-71, Academic Press, New York.
18. Neufeld, E.F., Sando, G.N., Garvin, A.J., and Rome, L.H. (1977): J. Supramolecular Structure, 6:95-101.
19. Schlessinger, J., Shechter, Y., Willingham, M.C., and Pastan, I. (1978): Proc. Natl. Acad. Sci. USA, 75:2659-2663.
20. Schlessinger, J. (1979): In: Physical Chemical Aspects of Cell Surface Events in Cellular Regulation, edited by C. Delisi and R. Blumenthal, pp. 89-111, Elsevier Press, New York.
21. Shechter, Y., Schlessinger, J., Jacobs, S., Chang, K.J., and Cuatrecasas, P. (1978): Proc. Natl. Acad. Sci. USA, 75:2135-2139.
22. Shechter, Y., Hernaez, L., Schlessinger, J., and Cuatrecasas, P. (1979): Nature, 278:835-838.
23. Shechter, Y., Gabbay, M., and Schlessinger, J. (1979): J. Biol. Chem. (submitted).
24. Smith, C.J., Wejksnora, P.J., Warner, J.R., Rubin, S.C., and Rosen, O.M. (1979): Proc. Natl. Acad. Sci. USA, 76:2725-2729.
25. Yanker, B.A., and Shooter, E.M. (1979): Proc. Natl. Acad. Sci. USA, 76:1268-1273.

This work was supported by grants from NIH (CA-25820) and from the United States-Israel Binational Science Foundation.

Control Mechanisms in Animal Cells,
edited by L. Jimenez de Asua et al.
Raven Press, New York © 1980.

Platelet Basic Protein: A Mitogenic Peptide for 3T3 Cells Secreted by Human Platelets

*,**,† Dieter Paul, **Stefan Niewiarowski,
*,**Kodungallore G. Varma, **Boguslaw Rucinski, *Steve Rucker,
and **Pranee James

**Fels Research Institute and Department of Pathology and **Specialized Center for Thrombosis Research, Temple University School of Medicine, Philadelphia, Pennsylvania 19140, U.S.A.*

The proliferation of mammalian cells in culture is usually dependent on the presence of serum in the culture medium. Most of the mitogenic activity in serum appears to be derived from platelets (1,2). Several preparations of platelet derived growth factors have been described and recently two laboratories reported the purification of PDGFs derived from outdated human platelets (3,4). We have focused our attention on a mitogenic cationic peptide ("Platelet Basic Protein" [PBP]) which is secreted by platelets in conjunction with several other proteins in response to specific stimulation, e.g. by thrombin. PBP is one of several structurally and immunologically related secretory peptides which bind to heparin with either high affinity, e.g. platelet factor 4 [PF_4] (5), or with low affinity, e.g. low-affinity platelet factor 4 [LA-PF_4] (6), β-thromboglobulin [βTG] (6), and PBP (7). Here we discuss recent work on the characterization of PBP. Available evidence shows that PBP can be considered as a polypeptide hormone with mitogenic activity which is stored in platelets and can be released in response to specific stimulation.

I. PURIFICATION AND CHARACTERIZATION OF PLATELET BASIC PROTEIN

Material released by washed fresh human platelets in response to thrombin (2.5 μg/ml) was heated (100°C, 5 min) and subjected to fractionation by isoelectric focusing (IEF) (pH 3.5-10) in a sucrose gradient (LKB-column), resulting in the separation of materials with isoelectric point (pI) of 7.0, 8.0 and 10.8 (±0.3), which were identified by radioimmunoassay (RIA) as βTG, LA-PF_4 and PBP, respectively. The platelet supernatant contained LA-PF_4 (72%), βTG (14%) and PBP (13%). After adsorption of dialyzed IEF fractions to heparin-sepharose pure peptides were eluted with 0.5 M NaCl at pH 7.5. According to RIA, PBP has been purified about 83,000-fold with respect to platelet rich plasma (PRP) with a recovery of about 5%. The peptides have been shown to be

---

† Present address: Department of Pharmacology and Toxicology, Hamburg University, Medical School, Hamburg Eppendorf, Martinistrasse 52, Hamburg 20, Germany

$\geq$95% pure as shown by N-terminal analyses and by a combination of IEF and RIA (7). Some properties of the three peptides are summarized in Table I.

TABLE I. PROPERTIES OF LA-PF_4, βTG AND PBP

	LA-PF_4	βTG	PBP
Molecules Weight (daltons)			
reduced:	9,278	8,851	9,000-10,000
non-reduced:	13,500	12,500	11,000-13,000
Isoelectric point (pH)	8.0	7.0	10.8
N-terminal amino acid	Asn	Gly	Gly
Concentration of NaCl required for elution from heparin-sepharose	0.5 M	0.5 M	0.5 M
Multiplication stimulating activity in 3T3 cells	±	-	+++ (1-10 ng/ml)
Inhibitory effect of heparin on mitogenic activity	N.D.	-	+++ (30-1000 ng/ml)
Immunological crossreactivity with anti-PBP-antibody	+	+	+

The molecular weights (MW) of the three platelet peptides were estimated by SDS-PAGE. LA-PF_4 and βTG appeared as single bands on SDS-PAGE with apparent MWs of 13,500 and 12,500 daltons respectively, when compared to conventional MW marker proteins. After reduction with βmercaptoethanol they migrated as expected from their MWs determined by amino acid sequencing (9278 and 8851 daltons respectively). Material in non-reduced PBP preparations ran as a single immunoreactive band with a somewhat lower mobility than contaminating ampholytes, corresponding to an apparent MW of 11-13,000 daltons as estimated by comparison with histones as MW markers which were used since highly basic proteins migrate abnormally in SDS-PAGE (8). Reduced PBP preparations had an apparent molecular weight of 9-10,000 daltons.

As shown by SDS-PAGE the separation of peptides from residual ampholytes has not always been successful. In some cases ampholytes can be separated from peptides by conventional methods. For example, most ampholytes with a pI of 7.0 or 8.0 (± 0.5) are readily dialyzable and bind poorly to heparin-sepharose. Therefore, the extent of ampholyte contamination in LA-PF_4 or βTG preparations is low. However, ampholytes with a pI 9-11 (LKB or Pharmacia) contained 1-2% components which are not readily dialyzable (membranes with a MW cut-off of >6,000 daltons) which were essential for the stabilization of a pH gradient. Also, these ampholyte species adsorbed to heparin-sepharose and were eluted together with PBP (0.5 M NaCl). For these reasons, variable amounts of ampholytes were usually present in most PBP preparations. They have therefore not always been useful for the characterization of the mitogenic activity of PBP but they have been valuable for the biochemical and immunological characterization of the peptide. PBP preparations with potent mitogenic activity have been obtained by exhaustive dialysis of the fractions eluting from heparin-sepharose using membranes to which ampholytes apparently were adsorbed with some selectivity. However, the yield of such highly mitogenic PBP preparations was very much reduced (0.5%). We have not been successful as yet in eluting mitogenic material from SDS-polyacrylamide gels.

The three peptides have been extensively characterized immunologically. They were shown to be immunologically identical by the following criteria: (a) using antisera raised in rabbits against individual peptides it was shown that each one of the pure peptides competed with ^{125}I-LA-PF_4 for antibody binding sites resulting in identical, superimposable competition curves (antibody dilution: $1:10^{-4}$); (b) these antisera precipitated ^{125}I-LA-PF_4 and ^{125}I-PBP equally well in RIA's (c) anti-LA-PF_4 antibody purified by affinity chromatography on LA-PF_4 sepharose precipitated all three peptides equally well yielding competition curves in RIA which were identical and superimposable (K.G. Varma et al., manuscript in preparation).

II. MITOGENIC ACTIVITY

The mitogenic activity of PBP was assayed using sparse quiescent Swiss 3T3 cells in 1.5% serum. As shown in Fig. 1, PBP at levels $\gtrsim$1 ng/ml stimulated [^{3}H]-thymidine incorporation into DNA. The mitogenic activity of PBP was shown to be dose-dependent, was maximal at $\gtrsim$10 ng/ml and was similar to that observed in response to EGF (kindly provided by Dr. R. Savage, Temple University, Medical School). However, in contrast to $\gtrsim$4% serum, which caused 90% of the cells to enter S (autoradiography), PBP or EGF stimulated G1 $\rightarrow$ S transition in only 60-70% of the cells at optimal peptide levels. Pure heparin with either high or low affinity for antithrombin III (kindly provided by Dr. U. Lindahl, Uppsala, Sweden) (9) was shown to block mitogenic activity of PBP but not of EGF at levels of about 1 μg/ml (10) presumably by complexing with the anti-heparin factor PBP thereby rendering it unavailable to cells, rather than interacting with cells rendering them unresponsive to mitogenic stimulation. The growth-inhibitory activity of heparin and other polyanions is well documented (11,12). The mitogenic activity of PBP was comparable to that of 3% serum. Although the evaluation of the apparent specific mitogenic activity of PBP is difficult, it appeared to be about 1-2 x 10^5 fold higher than that of serum. PBP did not stimulate the growth of SV3T3 cells cultured in 0.15% serum.

In separate experiments PBP was shown to promote cell multiplication in 3T3 cells cultured in 5% human plasma. In these experiments cell number per 3 cm dish increased in control cultures from 3 x 10^4 (day 0) to 5.1 ± 0.6 x 10^4 cells (day 5) and in the presence of 100 ng/ml PBP to 14 ± 1.4 x 10^4 cells/dish. Its mitogenic activity was destroyed by treatment with pepsin.

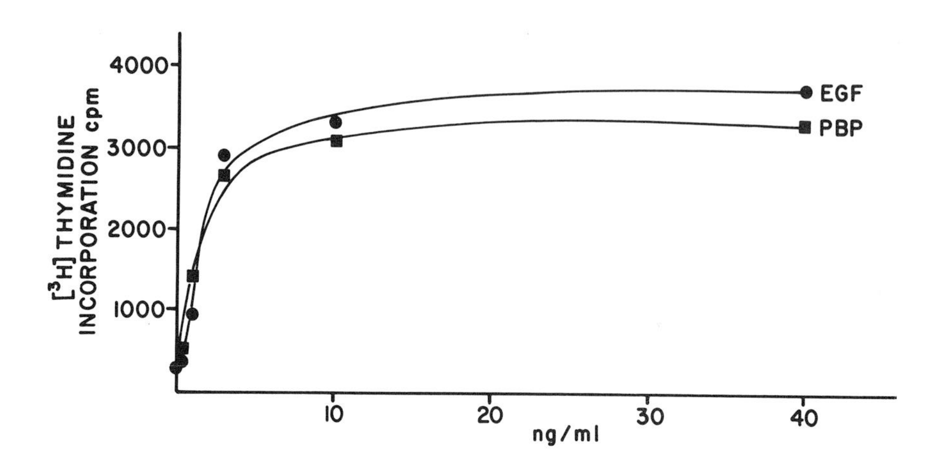

FIG. 1. Mitogenic activities of PBP and EGF in resting 3T3 cell cultures. Cells were seeded in duplicate dishes and incubated in 10% serum at 37°C. The medium was aspirated (day 1) and replaced with fresh medium containing 1.5% serum. Peptides were added at the indicated levels at day 4 and cultures pulsed with [^{3}H]-thymidine (6 μM, 10 μCi/dish) 16-24 hrs later. TCA insoluble radioactivity collected on glass fiber filters was determined. Results are expressed in cpm/culture, the error does not exceed ± 13% (10% serum: 4800 cpm/culture). The labeling index (autoradiography) in cultures which received either 3% or 4% serum or no addition at all (controls) was 62 or 90 or 2% respectively. PBP and EGF stimulated $G_1 \rightarrow S$ transition in 68% and 64% of cells at 10 ng/ml, 60% and 61% of cells at 40 ng/ml, respectively.

The mitogenic activity of PBP varied considerably from one preparation to the other. This could be due to partial proteolysis of PBP during the release reaction of platelets in the presence of thrombin causing a partial inactivation of mitogenic peptides. Also, it could be due to the presence of variable amounts of ampholytes in the preparation which are known to interfere with the mitogenic response of cells. $LAPF_4$ clearly had mitogenic activity in 3T3 cells as assayed under conditions similar to those described in Fig. 1, although at least 20-fold more $LA\text{-}PF_4$ than PBP was required for maximal stimulation. The fraction of labeled nuclei in maximally stimulated cultures was lower

(40%) than observed in response to PBP (Fig. 1). It appears that connective tissue activation peptide III (CTAP III), reported to promote synovial cell proliferation in culture (13) is identical with LA-PF_4 (14).

Thus, it has been shown that several platelet secretory peptides are immunologically closely related with each other. Furthermore, PF_4 shows about 50% sequence homology with βTG (5,6). βTG shows a striking partial amino acid sequence homology with C-terminal domains of hEGF (15) or mEGF (16). For example, eleven of thirty C-terminal amino acid residues in mEGF overlap with sequences in βTG or PF_4, including three cysteine residues.

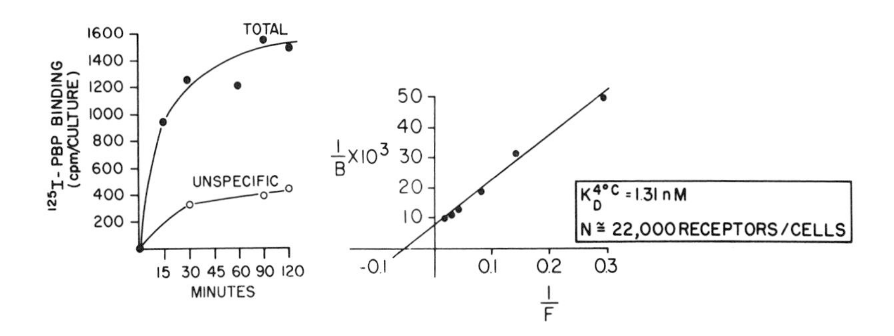

FIG. 2. Binding of immunoreactive ^{125}I-PBP to confluent cultures of 3T3 cells. Cells were cultured in 10% serum and used at day 8-10. The medium was aspirated, the cultures washed 3 X with Tris-buffer/saline/0.2% albumin (pH 7.4) (binding buffer) and cultures incubated at 4°C in 0.7 ml binding buffer in the presence of various amounts of ^{125}I-PBP. (A) Cultures were incubated with 36 ng/ml ^{125}I-PBP for the indicated times. Non-specific binding was determined in the presence of 1 μg/ml unlabeled PBP. (B) Cells were incubated for 90 min with 3,6,12,24,36 or 48 ng/ml ^{125}I-PBP at 4°C with or without 1 μgr/ml unlabeled PBP. After removal of binding buffer cultures were washed 5 X with binding buffer, cells dissolved in 1% SDS and the radioactivity determined in a gamma counter. Cells were counted with a Coulter Counter. Background binding ("0-time-binding") was subtracted from experimental values. Specific binding was determined and data plotted as described (25). The experimental error does not exceed ± 17%.

Available sequence data suggest that βTG differs from (partially sequenced) LA-PF_4 by the lack of 4 N-terminal amino acids and appears to be derived by partial (extracellular) proteolysis from

$LA-PF_4$ (14). Thus, we tentatively conclude that $LA-PF_4$ might show a similar structural relationship to EGF as does βTG. In view of the immunological identity of βTG, $LA-PF_4$ and PBP, which can be taken as an indication for a relatively small number of sequence differences between the peptides (17) PBP might be expected to share some structural similarities with EGF.

III. BINDING OF ^{125}I-PBP TO 3T3 CELLS

PBP was labeled with ^{125}I as dexcribed (18). Pure ^{125}I-PBP was obtained after absorbing the material to heparin-sepharose and eluting with 0.5 M NaCl. Purity was confirmed by SDS-PAGE/autoradiography which showed a single band at the position predicted from previous analyses of native PBP (11-13,000 daltons). When ^{125}I-PBP was precipitated with anti-$LA-PF_4$ gamma globulin, the pellet dissolved in SDS and then analyzed by SDS-PAGE/autoradiography, a single band was present at the predicted position (11-13,000 daltons). This pure immunoreactive ^{125}I-PBP was used for studying the binding to confluent, resting 3T3 cells. The results shown in Fig. 2 indicate that binding of ^{125}I-PBP was reversible and saturable with a K_D of 1.3 nM at 4°C assuming that the MW of PBP is 12,000 daltons. The number of binding sites on the 3T3 cell surface as shown in this particular experiment was about 22,000/cell. However, in similar experiments it was found that the number of binding sites per cell can be as high as 45,000/cell. SV3T3 cells which did not appear to respond to the mitogenic activity of PBP displayed about 50-75,000 ^{125}I-PBP binding sites with a K_D of $1.1-1.5 \times 10^{-9}$ M.

We conclude that PBP has potent mitogenic activity in 3T3 cells at levels of $10^{-10}-10^{-9}$ M, which is mediated by specific high affinity binding sites at the cell surface. PBP is secreted by platelets in response to specific stimuli (e.g. thrombin) and presumably interacts with target cells at some distance from its site of release. Thus, PBP has characteristics of a polypeptide hormone.

In contrast to the preparations of PDGF which have been recently purified in two laboratories from outdated platelet extracts (3,4), PBP is a peptide which is secreted by functionally intact platelets in response to physiologically relevant stimulation, e.g. by thrombin. Although similar to some extent, there appear to be considerable differences between the three preparations of the mitogenic cationic platelet peptides. There appears to be a clear difference between the estimates of the MW of the two PDGF's (13-16,000 daltons [3] vs. ∿30,000 daltons [4]). Also, different methods used to determine MW (SDS-PAGE, gel filtration) lead to different estimates. The apparent molecular weight of the major constituent of PBP preparations is 11-13,000 daltons as compared with histones used as MW markers and is similar to that estimated for PDGF from SDS-PAGE data (3). Thus, the relationship between mitogenic platelet derived peptides if any, remains to be elucidated.

The data presented here indicate that platelets derived from 1 ml PRP (3 x 10^8 platelets) are able to secrete about 840 ng of PBP as determined by RIA (Table I). Thus, the levels of PBP in medium containing 4% serum (which results in maximal mitogenic stimulation of 3T3 cells) are about 30-40 ng/ml. These levels are in reasonably good agreement with the data presented in Fig. 1 showing that maximal mitogenic stimulation by pure PBP was observed at $\sim$10 ng/ml. Thus, the data presented here suggest that human platelets contain a great excess of potentially available cell multiplication stimulating activity which can be released in response to specific stimuli, e.g. at sites of blood vessel damage. In more recent experiments (19) it was shown by RIA that the levels of immunoreactive material (i.e. $LAPF_4$, βTG and PBP) in plasma of patients with chronic renal failure (CRF) are considerably elevated (291 ± 26 ng/ml) as compared to normal plasma (32 ± 3 ng/ml). A recent report confirmed this observation (20). The bulk of this immunoreactive material present in uremic urine was found to focus at pH 10.0-10.8. This cationic urine fraction (28 ng/ml), PBP (10 ng/ml) and pure EGF (3-10 ng/ml) showed similar mitogenic activities as determined by [^{3}H]-thymidine incorporation into 3T3 cells (19). It is possible that the immunoreactive cationic mitogenic material might be identical with PBP and that it is cleared from the circulation in patients with CRF less efficiently than βTG and $LA\text{-}PF_4$, thus leading to its accumulation to abnormally high concentrations in plasma.

It has been suggested that platelet growth factors are involved in the pathogenesis of atherosclerosis (21). The relationship between elevated levels of cationic mitogenic material in plasma of CRF patients (19) and the vastly increased incidence of vascular disorders including premature atherosclerosis in these patients (22,23) remains to be determined.

ACKNOWLEDGMENT

This work was supported in part by research grants from the National Health Service, the American Cancer Society, the American Heart Association, the National Science Foundation and the W.W. Smith Charitable Trust, Ardmore, PA.

REFERENCES

1. Kohler, N. and Lipton, A. (1974):Expt.Cell Res., 87:297-301.
2. Ross, R., Glomset, J., Kariya, B. and Harker, L. (1974): Proc.Nat.Acad.Sci.USA, 71:1207-1210.
3. Antoniades, H.N., Scher, C.D. and Stiles, C.D. (1979):Proc. Nat.Acad.Sci.USA, 76:1809-1813.
4. Heldin, C.H., Westermark, B. and Wasteson, A. (1979):Proc. Nat.Acad.Sci.USA, 76:3722-3726.
5. Levine, S.P. and Wohl, H. (1976):J.Biol.Chem., 251:324-328.
6. Begg, G.S., Pepper, D.S., Chesterman, C.N. and Morgan, F.J. (1978):Biochemistry, 17:1739-1744.

7. Paul,D., Niewiarowski, S., Varma, K.G., Rucker, S. and James, P. submitted for publication.
8. Panyin, S. and Chalkly, R.J. (1971):J.Biol.Chem.,246:7557
9. Höök, M., Björk, I., Hopwood, J. and Lindahl, U. (1976):FEBS Letters, 66:90-93.
10. Paul, D., Niewiarowski, S., Varma, K.G. and Rucker, S. submitted for publication.
11. Clarke, G.D. and Stoker, M.G.P. (1971):In:CIBA Symp.Growth Control in Cell Cultures, edited by G.E.W. Wolstenholme and J. Knight, pp. 17-32. Churchill Livingstone, London.
12. Temin, H. (1966):J.Nat.Cancer Inst., 37:167-175.
13. Castor, C.W., Ritchie, J.C., Scott, M.E. and Whitney, S.L. (1977):Arthr.Rheum., 20:859-868.
14. Rucinski, B., Niewiarowski, S., James, P., Walz, D.A. and Budzinski, A.Z. (1979):Blood, 53:47-62.
15. Gregory, H. (1975):Nature, 257:325-327.
16. Savage, R.C., Ingami, T. and Cohen, S. (1972):J.Biol.Chem. 247:7612-7621.
17. Prager, E.M. and Wilson, A.C. (1971):J.Biol.Chem.,246:7010.
18. Bolton, A.E. and Hunter, W.M. (1973):Biochem.J.,133:529-539.
19. Guzzo, J., Niewiarowski, S., Musial, J., Rao, A.K., Berman, I. and Paul, D., submitted for publication.
20. Deppermann, D., Andrassy, K., Seeling, H., Ritz, E. and Post, D. (1979):In:VII.Congr.Thromb.Haemost.London,Abstract 994. F.K. Schottauer Verlag, Stuttgart, New York.
21. Ross, R. and Glomset, J.A. (1973):Science,180:1332-1339.
22. Lindner, A., Charrer, B., Sherrard, D.J. and Scribner, B.H. (1974):New Engl.J.Med.,290:697-701.
23. Ibels, L.S., Stewart, J.H., Mahoney, J.J., Neale, F.C. and Sheil, A.G.R. (1977):Q.J.Med., 46:197-214.
24. Lowry, O.H., Rosebrough, N.J., Farr, A.L. and Randall, R.J. (1951):J.Biol.Chem., 193:265-275.
25. Steck, T.L. and Wallach, D.F.H. (1965):Biochim.Biophys.Acta 97:510-516.

Control Mechanisms in Animal Cells,
edited by L. Jimenez de Asua et al.
Raven Press, New York © 1980.

Cell Density-Dependent Down Regulation of NGF-Receptor Complexes in PC12 Cells

S. Biocca, A. Levi, and P. Calissano

Laboratory of Cell Biology, C.N.R., 00196 Rome, Italy

INTRODUCTION

It is now widely established that several hormones, as well as epidermal growth factor are internalized into the cell after binding to their receptors (4,6,8). Although the biological significance of this process is still unclear, it is reasonable to assume that it is related to the mechanism of action of the ligand. This is particularly true whenever multiple, long term effects of the hormone or growth factor cannot be reasonably accounted for by a simple, transient interaction with a receptor located on the surface of the cell.

The protein nerve growth factor (NGF) (9,10) induces differentiation of a pheochromocytoma clone (PC12) by promoting arrest of cell division, sprouting of electrically excitable neurites and several other molecular changes (5,7,9,14).

Such multiple effects, which occur in different part of the cell, are presumably achieved through a spatially or temporally diversified action. In order to assess this possibility we have undertaken in our laboratory a series of investigations intended at following the fate of NGF after binding to its receptor. These studies have shown that a large fraction of ^{125}I-NGF after interaction with specific binding sites rapidly becomes irreversibly bound and resistant to trypsin digestion (3). Immunofluorescence studies and autoradiography have shown that a progressively larger portion of bound NGF is detectable in the cytoplasm of PC12 cells and concentrates in the nuclear contour in the form of discrete granules (13). These findings, together with other studies performed in other laboratories (17) have led to the conclusion that NGF is internalized into the cell after binding to receptors located on the plasma membranes.

A common phenomenon accompanijng ligand internalization is the process known as down regulation, i.e. a progressive decrease of total ligand bound to the target cells as function of time of incubation. Such decrease has been interpretated as due to internalization and degradation of the ligand.

We report here experiments showing down regulation of NGF-receptor complexes in PC12 cells and its dependence upon the cell density. This phenomenon has been followed by measuring decrease of NGF bound to the plasma membrane with the use of ^{125}I-labeled antibodies directed against this protein and by measuring total TCA-precipitable ^{125}I-NGF bound to the cells at different times of incubation. The use of these two different techniques has allowed us to distinguish between surface-bound and total NGF bound to the cells.

METHODS

PC12 cells were maintained in collagen coated tissue culture dishes in RPMI-1640 (KC biological Inc. Texas) containing 5% fetal calf serum and 10% heat inactivated horse serum. The day before the experiment cells were removed by pipetting, counted and seeded in 24 wells (16 mm/well diameter Costar Tissue Culture, Cambridge; Mass.) previously coated with collagen.

NGF was purified as described by Bocchini and Angeletti (1969) (1), filtered through a millipore and stored at 2-4°C in sterile vials. Antibodies against NGF were purified with the method of Stöckel et al. (1976) (16).

Labeling of NGF and NGF-antibodies

Iodination of NGF and of purified NGF-antibodies was performed using the procedure of Young et al. (1978) (18), obtaining a specific radioactivity of 70-100.000 cpm/ng protein. ^{125}I-NGF and ^{125}I-Ab were stored at 2-4°C in the presence of 0.01% NaN_3 for a maximum period of 3-4 weeks.

^{125}I-Ab Binding Assay

Cells were washed and incubated with a medium containing RPMI-1640 buffered with 20 mM Hepes pH 7.4, 5% fetal calf serum and 10% heat inactivated horse serum (sol.A). At time intervals 50 ng/ml of unlabeled NGF were added to each well. At the end of the incubation at 37°C in a 7% CO_2 atmosphere the medium was removed by aspiration and the plates were washed with serum free RPMI-1640. Cells were fixed by adding 0.5 ml of 10% paraformaldehyde made up in 120 mM NaCl, 10 mM potassium phospate pH 7.4. After 10 min incubation at room temperature the excess of paraformaldehyde was removed by washing several times with PBS. Fixed cells were then incubated for 1 hour at 37°C in the presence of the binding solution made of Sol.A plus ^{125}I-Ab (200.000 cpm/well). The extent of ^{125}I-antibodies bound to the cells was measured after 3 washes in PBS to remove all the free radioactivity and by dissolving the cells in 1 ml NaOH 1.0 N. Non specific binding of labeled antibodies to the cells was calculated in parallel experiments performed in the absence of NGF and was subtracted from total binding.

^{125}I-NGF Binding Assay

The binding solution made of Sol.A plus ^{125}I-NGF and 50 ng/ml NGF was added to each well. At time intervals the medium was removed and the plates were washed three times with cold PBS. Cells were dissolved in 1 ml 0.1 N NaOH and the proteins were precipitated by addition of 0.2 ml 50% TCA. All data reported are expressed as ^{125}I-NGF TCA-precipitable bound to the cells. Specific binding is calculated from the difference in cell bound radioactivity in the presence and absence of 10 μg/ml unlabeled NGF/ml.

RESULTS

As mentioned in the introduction, PC12 cells have specific binding sites for NGF located on the surface of the plasma membrane. Total binding capacity amounts to 3-4 ng NGF/10^6 cells in PC12 cells and increases to 20-22 ng/10^6 cells after 5-7 days of exposure to NGF (3). When interaction of NGF with these cells is measured as a function of time of incubation, binding reaches a maximum within 1-2 hours at 37°C and then starts decreasing to a minimum after 6-8 hours. Within 24 hours total binding capacity is restored and continue to increase thereafter for several days.

In a preliminary series of experiments, which will be reported in detail elsewhere (Biocca, Levi, Calissano, in preparation) we found that cells can be fixed with paraformaldehyde after incubation with NGF without drastically altering the subsequent capacity of the growth factor of being recognized by its specific iodine-labeled antibodies.

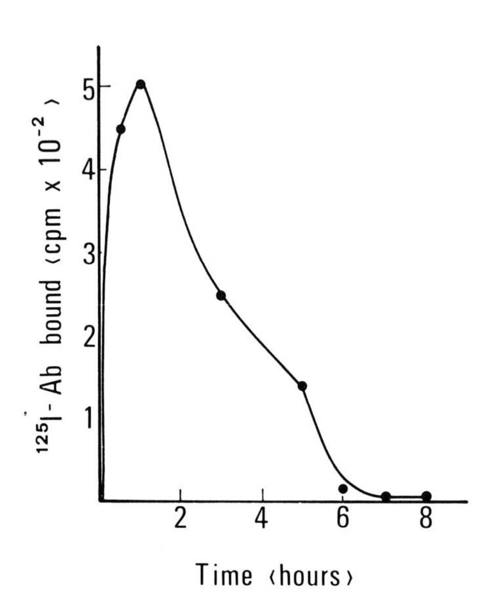

FIG. 1 - ^{125}I-NGF antibody binding to PC12 cells preincubated for different times with NGF. PC12 cells (4 x 10^5 cells/well) were incubated with 50 ng/ml of unlabeled NGF. ^{125}I-Ab binding assay was carried out as described under Methods. Each point represents an average of 3 determinations.

The data reported in Fig. 1 represent a typical experiment, in which it can be seen that the maximum amount of NGF bound to the surface of the cells is detectable after 1 hour of incubation and then NGF starts declining and is almost no more detectable after 6-8 hours. Since experiments carried out with ^{125}I-NGF show the same time course of loss of NGF bound to the cells, all subsequent studies have been performed with the use of labeled-NGF, due to the complex and time-consuming procedure required by the use of labeled-antibodies.

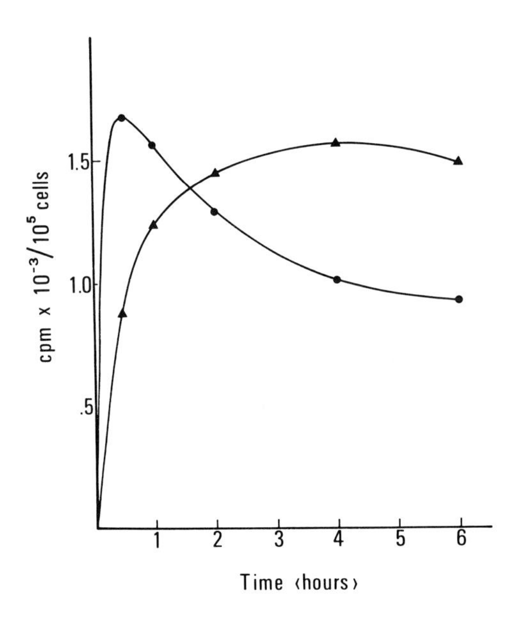

FIG. 2 - Time course of ^{125}I-NGF binding at two different cell densities. PC12 cells were incubated in the presence of 50 ng/ml NGF plus trace amount of ^{125}I-NGF (200.000 cpm/well). At time intervals the radioactivity TCA-precipitable was measured as described under Methods. (●) PC12,10^5 cells/well; (▲) PC12,10^6 cells/well. Each point represent an average of two determinations.

Previous studies (3) had shown that after 2 hours incubation with PC12 cells a portion of bound NGF becomes resistant to trypsin digestion and not exchangeable with free NGF present

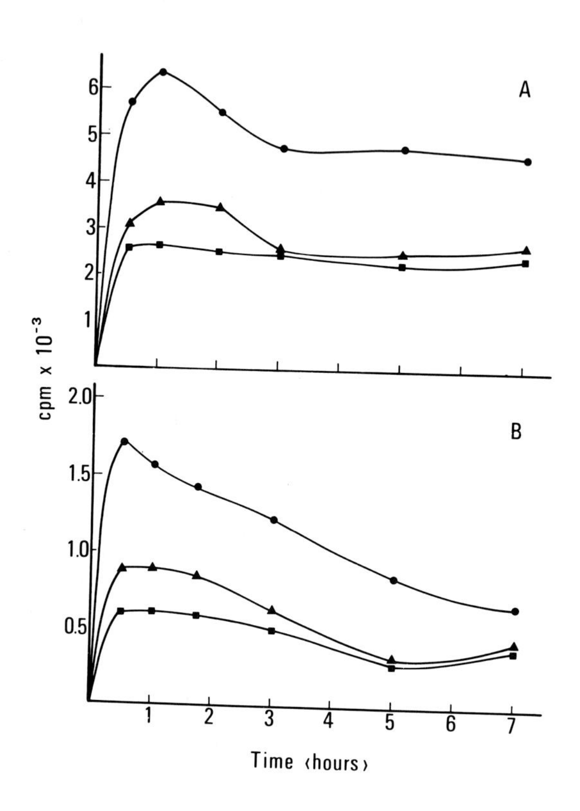

FIG. 3 - ^{125}I-NGF binding at two cell densities in the presence of different concentrations of NGF.
A. PC12, 4×10^5 cells/well. B. PC12, 10^5 cells/well. ^{125}I-NGF alone (●), or plus 50 ng/ml (▲) or 100 ng/ml (■) of unlabeled NGF.

in solution. Moreover, it was found that the extent of this "tight binding" increased with the cell density, i.e. the higher the number of cells per dish, the larger the fraction of bound NGF that, after two hours, remained sequestered either somewhere within the context of the membrane or into the interior of the cell. In order to relate the phenomenon of "tight binding" with internalization and subsequent processing of NGF, we investigated whether the cell density also affects the down regulation of NGF-receptor complexes.

Fig. 2 shows that while at cell density of 0.5×10^6 cells/cm^2, when cells are confluent, NGF binding reaches a plateau and does not decrease even after 6 hours of incubation, at 0.5×10^5 cells/cm^2 PC12 cells exhibit the typical down regulation. It is worth mentioning that the experiments reported in Fig. 1, where maximum decrease in binding is observed, were performed at very low cell density. At variance with experiments reported in Fig. 1, however, down regulation is never more than 50-60% of maximum binding. It should be noted that while antibodies detect only the amount of the growth factor which, at any given time, is present on the surface of the cells, measures with ^{125}I-NGF detect not only this fraction but also that being internalized and on its way of being degraded. Therefore, the extent of down regulation not only reflects a measure of internalization but also of NGF turnover.

Fig. 3 shows that the extent of down regulation is also dependent upon the concentration of NGF. While at a concentration of 2 ng/ml of NGF the down regulation is 60% of maximum binding and is also noticeable (although to a lower extent) at high cell density, at higher NGF concentrations (50 and 100 ng/ml) it becomes almost undetectable at the high cell density.

Finally, in PC12 cells previously exposed to the growth factor for 3 days, down regulation is no more detectable even at low cell density (Fig. 4). This finding suggests that the first input of NGF molecules into the cells triggers a mechanism which abolishes the subsequent massive, synchronous entrance of the ligand.

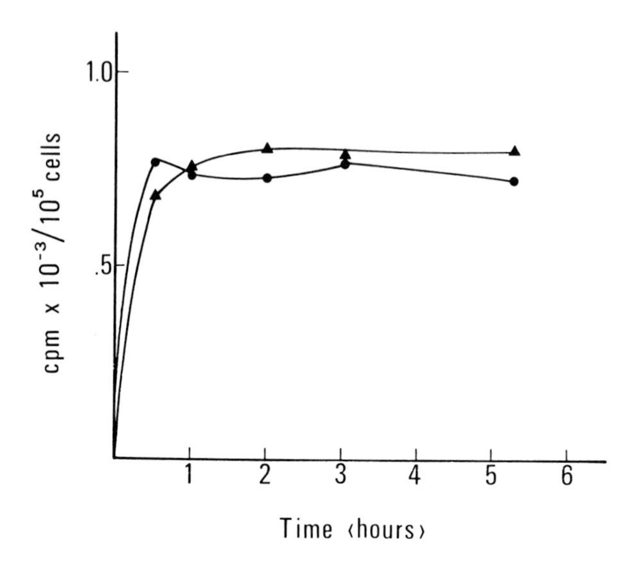

FIG. 4 - Time course of ^{125}I-NGF binding to cells preincubated with NGF. PC12 were incubated in the presence of 50 ng/ml NGF for three days. The medium was removed, the cells were washed and an ^{125}I-NGF binding assay was performed as described under Methods. (●) PC12,10^5 cells/well; (▲) PC12,10^6 cells/well.

DISCUSSION

The data presented show that interaction of PC12 cells with NGF leads to a progressive disappearance of the growth factor from the surface of the cells (fig.1). This event is demonstrated with two distinct experimental approaches, binding of ^{125}I-NGF antibodies to cells previously exposed to unlabeled NGF and direct binding of this ligand to the target cells. Together with previous findings (11) demonstrating a progressive, time-dependent, accumulation of NGF in a perinuclear disposition the overall evidence indicates that this growth factor is internalized into the cell.

Moreover, experiments reported in this paper prospect a modulation of the process of down regulation, taken as an indirect measure of ligand internalization by the density of PC12 cells.

When cells are at a density close to confluency, total NGF bound does not decrease after reaching a plateau at 1-2 hours incubation. On the contrary at lower cell density NGF binding, after reaching a maximum, diminishes by 50-60%. Although it is premature to postulate a precise molecular mechanism to explain such cell-density-dependent down regulation, we may hypothesize that it is a property of some membrane constituent, either the NGF receptor itself or some other membrane component functionally connected with the NGF receptor.

The previous demonstration of the existence of a portion of NGF referred to as tightly bound increasing with cell density suggests two alternative fates of this growth factor after interaction with PC12 cells. At high cell densities part of bound NGF is confined to the plasma membrane (tightly bound) where it plays some role probably connected with the increased cell adhesiveness to each other, to the substratum, and perhaps also in conferring directionality to the growing neurites. At low cell density NGF is internalized and/or degraded at a much faster rate. It is interesting to note that PC12 cells already partially differentiated after 3 days of exposure to NGF behave as undifferentiated cells at high cell density i.e. they do not show down regulation of NGF receptors. This behaviour may be explained by assuming that the first, massive entrance of NGF molecules triggers, by some feedback mechanism, a regulation of subsequent NGF uptake and degradation. Alternatively, we may postulate that priming PC12 cell with NGF stimulate the synthesis of some membrane component(s) endowed with the property of "anchoring" NGF receptors. This substance(s) would also be operative at high but not at low cell density.

The experiments reported suggest that if internalization of the ligand is instrumental to some NGF effects but not to others (e.g. arrest of cell division vs. neurite outgrowth, enzyme activation vs. inhibition, etc.) it should be possible to dissect them out by measuring each of them at different cell density. The possible modulation of NGF action as a function of cell density may also occur to other ligand-target cell systems and prospect a new parameter regulating the action of hormones or growth factors.

REFERENCES

1. Bocchini, V., and Angeletti, P.U. (1969): Proc. Natl. Acad. Sci., 64: 787-794.

2. Brown, K.D., Yun-Chi Yeh, and Holley, R.W. (1979): J. Cell Physiol., 100: 227-238.
3. Calissano, P., and Shelanski, M.L. (1980): Neuroscience, in press.
4. Carpenter, G., and Cohen, S. (1976): J. Cell Biol., 71: 159-171.
5. Dichter, M.A., Tischler, A.S., and Greene, L.A. (1977): Nature, 268: 501-504.
6. Gavin, J.R.III, Roth, J., Neville, D.M. jr., De Meyts, P., and Buell, D.N. (1974): Proc. Natl. Acad. Sci., 71: 84-88.
7. Greene, L.A., and Tischler, A.S. (1976): Proc. Natl. Acad. Sci., 73: 2424-2428.
8. Hinkle, P.H., and Tashian, A.H. jr. (1975): Biochemistry, 14: 3845-3851.
9. Lee, V., Shelanski, M., and Greene, L.A. (1977): Proc. Natl. Acad. Sci., 74: 5021-5025.
10. Levi-Montalcini, R., and Angeletii, P.U. (1968): Physiol. Rev. 48: 534-569.
11. Levi-Montalcini, R., and Booker, B. (1960): Proc. Natl. Acad. Sci., 46: 373-384.
12. Lucas, C.A., Edgar, D., and Thoenen, H. (1979): Exp. Cell Res., 121: 79-86.
13. Marchisio, P.C., Naldini, L., Calissano, P. (1980): Proc. Natl. Acad. Sci., in press.
14. Schubert, D., and Klier, F.G. (1977): Proc. Natl. Acad. Sci. 74: 5184-5188.
15. Schubert, D., and Whitlock, L. (1977): Proc. Natl. Acad. Sci., 74: 4055-4058.
16. Stöekel, K., Gagnon, C., Guroff, G., and Thoenen, H. (1976): J. Neurochem. 26: 1207-1211.
17. Yanker, B.A., and Shooter, E.M. (1979): Proc. Natl. Acad. Sci., 76: 1269-1273.
18. Young, M.Y., Blanchard, M.H., and Saide, J.D. (1978): In: Methods of Hormone Radioimmunoassay, edited by B.H. Jaffe, and H.R. Behrman, pp. 941-957. Academic Press, New York.

Control Mechanisms in Animal Cells,
edited by L. Jimenez de Asua et al.
Raven Press, New York © 1980.

The Intracellular Localization of Nerve Growth Factor in Target Cells

P. C. Marchisio, D. Cirillo, L. Naldini, and *P. Calissano

*Department of Human Anatomy, University of Turin School of Medicine, 10126 Turin, Italy; and *Laboratory of Cell Biology, C.N.R., 00196 Rome, Italy*

Nerve growth factor (NGF) provides an essential trophic and differentiating role on sympathetic and sensory neurons (6). NGF is active also on a cell line derived from a rat adrenal medullary tumor and known as PC12 (4). However, in the latter cells NGF is not absolutely required for survival in vitro but induces cells to change into a population which later shows morphological and functional properties similar to sympathetic cells. For this reason, PC12 cells have been adopted as a suitable model for studying the biological properties of NGF.

In previous studies it has been reported that a large fraction of ^{125}I-NGF, after binding to specific receptors, remains tightly bound to PC12 cells and becomes resistant to trypsin digestion (2). Further studies also suggest that the tight binding of NGF is the first step leading to its internalization into the inner cell compartment.

Recently we have reported immunofluorescence and autoradiographic studies showing that NGF is indeed internalized into PC12 cells and is subsequently localized in the peri- and intranuclear compartments (8). Parallel studies have shown that the internalization of NGF occurs also in sympathetic neurons explanted from the chick embryo and cultured in vitro (9).

THE LOCALIZATION OF NGF IN PC12 CELLS

PC12 cells possess specific binding sites for NGF which account for about 90% of the total NGF binding while the remaining aliquot results from non-specific adsorption of the polypeptide (2). Such specific binding may be visualized by indirect immunofluorescence microscopy using an affinity-purified NGF antibody on cells exposed to 50 ng/ml at +4°C; such binding appears in the form of minute patches all around the cellular outline and is somewhat more intense at cell-to-cell contacts. When the temperature of the incubation medium is raised to 37 C the pattern of the indirect immunofluorescence reaction (performed after making the cell membrane permeable to antibodies) changes considerably and shows tiny fluorescent granules appearing first in the outer rim of the cell and then progressively accumulating between 4 and 12 h around the nucleus (Fig. 1 a). At later exposure times to NGF (12-48 h) a material which is intensely stained by NGF antibodies appears in the form of one or two discrete dots also within the nuclear compartment and very often in close contiguity to the nucleolus (Fig. 1 b). After 24 h ex-

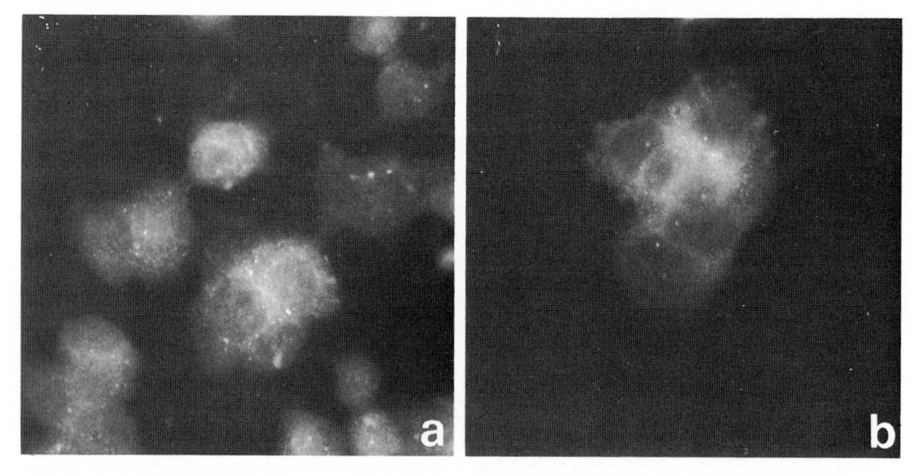

Fig. 1 PC12 cells exposed for 12 h (a) and 24 h (b) to 50 ng/ml NGF and processed for indirect immuno fluorescence microscopy employing NGF antibodies. In (a) many fluorescent granules are found around the nuclear contour while in (b) intranuclear fluorescent granules may be observed in most cells.
Magnification x 600.

posure to NGF most PC12 cells show this brightly fluorescent granule in the nucleus as well as a row of fluorescent granules all around the nuclear contour.

Since the location of NGF immunoreactive material within the nucleus was quite unexpected, we wanted to ascertain whether the brightly fluorescent dots were indeed intranuclear and not just adjacent to the nuclear outer surface. In order to get information about this point we incubated monolayer PC12 cells with a buffer containing 0.5% of the non-ionic detergent NP40. The nuclear monolayer which resulted after solubilization of most cytoplasmic structures still showed some perinuclear granules and left the intranuclear granule unaffected. This results gave reasonable certainty that the strongly fluorescent paranucleolar dot is indeed intranuclear.

Cytoplasmic and intranuclear fluorescence was never found i) in cells which had never been exposed to NGF, ii) in cells exposed to NGF and stained with preimmune rabbit IgGs and iii) in cells which, after exposure to NGF for 2d, were further incubated in its absence for two more days. This latter control suggests that this portion of NGF immunoreactive material is in a state of continuous turnover.

The data obtained by indirect immunofluorescence microscopy were substantially confirmed by autoradiography of PC12 cells which had been exposed to ^{125}I-NGF (8). After progressively longer times of incubation with the labelled factor, cells were chased with cold NGF for a period long enough to deplete most surface-bound NGF. Under these conditions silver grains were located mostly over the cytoplasm after 2-4 h exposure to the tracer. At longer exposure times, silver grains were found over the nucleus in a position which was consistent with that observed by indirect immunofluorescence microscopy.

THE LOCALIZATION OF NGF IN SYMPATHETIC NEURONS IN VITRO

At variance with PC12 cells, sympathetic neurons obtained from 9 d chick embryos and cultured in vitro show an absolute requirement for NGF. In other words, they do not survive when the factor is omitted from the culture medium (6).

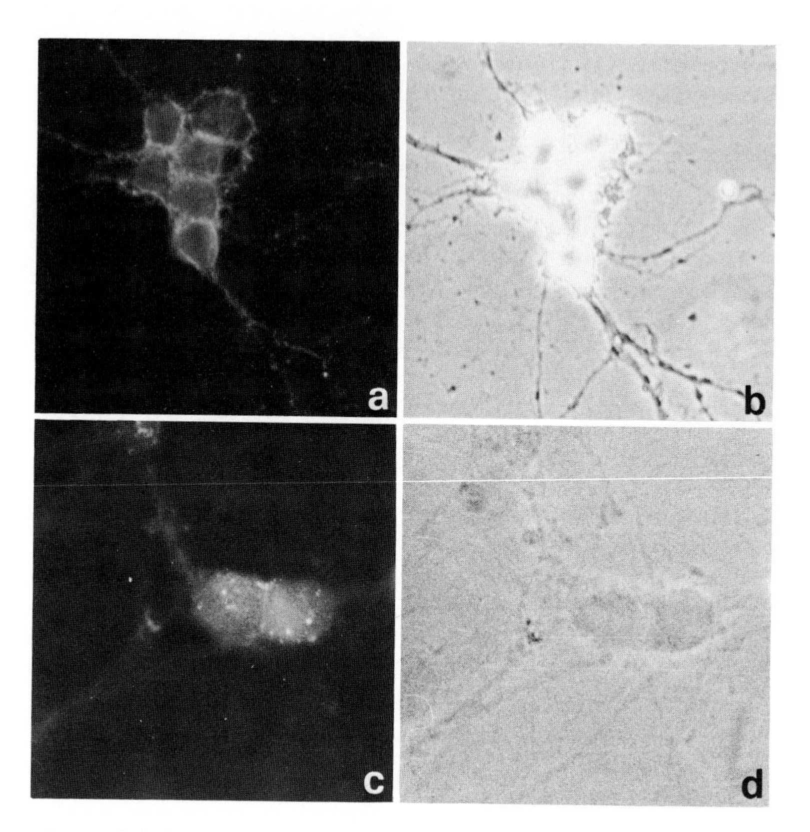

Fig. 2 Chick embryo sympathetic neurons cultured in vitro in the presence of NGF and stained with NGF antibodies to show surfac-bound NGF (a) and internalized NGF-like material (c). Fig. 1 b and d show the corresponding phase-contrast pictures. Intense fluorescence at intercellular contacts is seen in (a) while in (c) perinuclear fluorescent granules are visible.

One day after explantation, mild trypsin dissociation, and culture in a medium containing 50 ng/ml NGF, sympathetic neurons show abundant fluorecent granules over perikarya and along processes when stained with indirect immunofluorescence microscopy employing purified NGF antibodies (Fig. 2 a and b) (9). Surface fluorescence seems to be particularly intense at the sites of cell-to-cell contact. When the same cells are fixed and made permeable to IgGs, immunofluorescence shows

NGF antibody reactive material also in the cytoplasm. Such a material appears within the long processes as well as in perikarya and notably around the nuclear profile (Fig. 2 c and d). We have not been able to clearly show intranuclear granules in these cells as we did in PC12 cells. However, even if we cannot exclude that an occasional migration of NGF-like material into the nucleus occurs also in sympathetic neurons, the phenomenon is certainly not as constant and extensive as in PC12 cells. At longer times of continuous exposure to NGF (7-10 d), sympathetic neurons form complicated networks of neurites and tend to collect in small ganglion-like clusters. With immunofluorescence microscopy we have observed that NGF-like material tends to decrease from processes and to increase in perikarya while maintaining a preferential location around the nucleus.

In parallel experiments we have studied the autoradiographic location of NGF-like material after exposure of sympathetic cells to ^{125}I-NGF. We could confirm the data of immunofluorescence microscopy also with this technique: a marked accumulation of silver grains was indeed found over 1 um sections of sympathetic axons and perikarya. Occasionally silver grains were seen also over the nuclear area.

DISCUSSION AND CONCLUSIONS

From this series of studies (8,9) we may suggest that either NGF itself or a fragment of the NGF molecule, after binding to the surface of target cells, is internalized by an energy-dependent mechanism which is similar or identical to the receptor-mediated endocytosis reported for other growth factors and some hormones (for review see 3). Once the factor has been internalized it migrates toward the immediate perinuclear area and forms rows of granules around the nuclear contour. A working hypothesis suggests that the perinuclear clusters of NGF-like material are in contact with a pool of tubulin molecules which previous in vitro studies have shown to interact with NGF in a rather specific way by inducing the formation of large bundles of microtubules (5). Some support to this hypothesis comes from the recent demonstration that the early events preceding neurite outgrowth in neuroblastoma cells involve the appearance of multiple microtubule organizing centers located around the nucleus and in the cytoplasm (7, 11). Thus, it is not unconceivable that NGF may come in contact with similar centers which initiate neurite outgrowth in NGF target cells.

Somewhat more puzzling is the intranuclear location of NGF in PC12 cells. The mechanism whereby the factor gets into the nucleus is unknown and so is the biological function which is carried out by NGF within the nucleoplasm. However, one possibility is that the intranuclear migration of NGF is related to the "priming" mechanism of NGF target cells which has been shown to depend on de novo synthesis of RNA (1). This tentative interpretation may be indirectly supported by the results on sympathetic neurons which, having been already "primed" in vivo by the NGF-rich embryonic environment, do not show any lag period in neurite sprouting in vitro and, at least initially, do not require RNA synthesis for extending neurites (10). The fact that primary sympathetic neurons fail to show a constant intranuclear location of NGF, as unprimed PC12 cells do, may be relevant to associate the phenomenon of "priming" to a direct action of NGF on a mechanism based in the nucleus.

Our laboratories are now jointly investigating the possible biological significance of the intracellular pool of NGF in target cells.

ACKNOWLEDGMENTS

The work summarized in this paper was partially supported by C.N.R. contract 78.02172.04 to P.C.M. We are indebted to Mr. C. Cozzari for the affinity purification of NGF antibodies.

REFERENCES

1. Burstein, D.E. and Greene, L.A. (1978): Evidence for RNA synthesis-dependent and independent pathways in stimulation of neurite outgrowth by nerve growth factor. Proc. Natl. Acad. Sci. U.S.A. 75: 6059-6063.
2. Calissano, P. and Shelanski, M.L. (1979): Interac-

tion of nerve growth factor with pheochromocytoma cells. Evidence for tight binding and sequestration. Neuroscience, in the press.

3. Goldstein,J.L., Anderson, R.G.W., and Brown, M.S. (1979): Coated pits, coated vesicles,and receptor-mediated endocytosis. Nature(London) 279:679-685.
4. Greene, L.A. and Tischler, A.S. (1976): Establishment of a noradrenergic clonal line of rat adrenal pheochromocytoma cells which respond to nerve growth factor. Proc. Natl. Acad. Sci. U.S.A. 73: 2424-2428.
5. Levi, A., Cimino, M., Mercanti, D., Chen, J.S., and Calissano, P. (1975): Interaction of nerve growth factor with tubulin. Studies on binding and induced polymerization. Biochim. Biophys. Acta 399:50-60.
6. Levi-Montalcini, R. (1966): The nerve growth factor: its mode of action on sensory and sympathetic nerve cells. Harvey Lect. 60:217-259.
7. Marchisio, P.C., Weber, K., and Osborn, M. (1979): Identification of multiple microtubule initiating sites in mouse neuroblastoma cells. Europ.J. Cell Biol. 20:45-50.
8. Marchisio, P.C., Naldini, L., and Calissano, P. (1979):Intracellular distribution of nerve growth factor in rat pheochromocytoma PC12 cells: Evidence for a peri- and intranuclear location. Proc. Natl. Acad. Sci. U.S.A. 76.
9. Marchisio, P.C., Cirillo, D., Naldini, L., and Calissano, P. (1980): Cellular distribution of nerve growth factor (NGF) in chick embryo sympathetic neurons in vitro. J. Neurocytol., submitted.
10. Partlow, L.M. and Larrabee, M.G. (1971): Effects of a nerve-growth factor, embryo age and metabolic inhibitors on growth of fibers and on synthesis of ribonucleic acid and protein in embryonic sympathetic ganglia. J. Neurochem. 18:2101-2118.
11. Spiegelman, B.M., Lopata, M.A., and Kirschner, M.W. (1979): Aggregation of microtubule initiating sites preceding neurite outgrowth in mouse neuroblastoma cells. Cell 16:253-263.

Control Mechanisms in Animal Cells,
edited by L. Jimenez de Asua et al.
Raven Press, New York © 1980.

The Effect of EGF on Cell Proliferation and Gene Expression

D. Gospodarowicz, I. Vlodavsky, N. Savion, and L. K. Johnson

Cancer Research Institute, University of California Medical Center, San Francisco, California 94143, U.S.A.

One of the central problems of cell biology is the elucidation of the factors and mechanisms which control cell proliferation and gene expression. A group of peptides called growth factors has been shown to be involved in the control of cell proliferation. Among them is the epidermal growth factor (EGF), which has been shown to be mitogenic both in vivo and in vitro for the basal cell layer of the skin and corneal epithelia as well as for fibroblasts maintained in culture (for a review, see refs. 5, 13). Although the mitogenic effect of EGF has been studied mostly with fibroblast cultures and primarily with established cell lines such as BALB/c 3T3 or Swiss 3T3 (5), the main event analyzed has been the initiation of DNA synthesis rather than active cell proliferation. This is due primarily to the poor proliferative response of high density cell cultures exposed to EGF under restrictive conditions (serum starvation) and their lack of response when maintained under sparse conditions. The experimental conditions under which one can observe a mitogenic effect of EGF on these cell types are therefore limited. Among other cell types also used to study the mitogenic effect of EGF is the A-431 carcinoma cell (5). Although these cells have an enormous number of EGF binding sites, they do not proliferate in response to EGF, thereby limiting their usefulness as a cell model for the study of the control of cell proliferation by growth factors. However, the mitogenic effect of EGF is not restricted to a few cell types but extends over a vast spectrum of cells originating from various tissues. Among these, ovarian cells have been shown to be particularly responsive to the mitogenic stimuli provided by EGF (10-12). Granulosa cells of bovine or porcine origin when maintained in culture depend on EGF for proliferation, regardless of the serum concentration or cell density at which the cultures are started. In contrast, luteal cells, which are derived via cytomorphosis from post-ovulatory granulosa cells, are no longer responsive to the mitogen. The granulosa-luteal cell system, because it involves intimately related cell types, may therefore provide the system of choice with which to study which steps are relevant or irrelevant to the initial mitogenic stimulus leading to cell proliferation. Until now, such study has been difficult due to the lack of mutant cells capable of both binding a mitogen and remaining unresponsive to it while the original cell population retains its responsiveness.

Equally important is the study of the effect of mitogens on the phenotypic or genetic expression of cell types with which they interact. Although it is likely that the main effect of growth factors in sparse cell populations which are maintained under conditions which favor growth will be to stimulate cell proliferation, it is possible that at confluence, when cells enter a resting state or lose their potential for proliferation, the same factors could directly or indirectly affect the phenotypic expression of the cells. We have explored this possibility using as a cell model the GH_3 cells. Although EGF has no apparent mitogenic effect on these cells, it can strongly affect their pattern of gene expression, as reflected by a deceleration of growth hormone synthesis and a greatly increased production of prolactin (17).

I. THE CONTROL OF CELL PROLIFERATION BY EGF

A) The effect of EGF on proliferation of granulosa, luteinized granulosa, and luteal cells maintained in culture

When the mitogenic activity of EGF is tested on bovine granulosa cell cultures originating from small- and medium-sized follicles, the results indicate that these cells are far more sensitive to EGF than are fibroblasts. When its effect on initiation of DNA synthesis was studied, as little as 10^{-3} ng/ml EGF is enough to induce initiation, and saturation is observed at 0.3 ng/ml EGF. Concentrations five- to ten-fold higher are required to obtain a maximal increase in cell number and the effect can be as high as a 10,000- to 20,000-fold increase in cell number over control, if one worked at clonal density (50 cells/6 cm dish). As with fibroblasts, the mitogenic effects of EGF were dependent on the serum concentration in which the cells were maintained. Increasing concentrations of serum (from 0.1% to 10%) enhanced the growth of cells when maintained in the presence of FGF or EGF, but a mitogenic response can be observed with cells kept in serum concentrations as low as 0.1% (10, 12).

When luteal cells maintained in culture are tested for their response to EGF, they, unlike granulosa cells, no longer respond to it. This indicated that during the luteinization process granulosa cells shift their sensitivity to mitogens. Similar results are observed when granulosa cells obtained from Graafian follicles are allowed to luteinize spontaneously in vitro. One therefore had to conclude that, although luteal and granulosa cells are interrelated cell types, the control of their proliferation in vitro is quite different, since the sensitivity of the cells to EGF is lost during luteinization (11).

Cells exposed to a mitogen are capable of binding it specifically to cell surface receptor sites, and following binding the receptor-mitogen complexes are internalized and degraded in lysosomes. Although it is not yet known which of these steps are necessary and which are sufficient to induce the mitogenic response, it is possible that the lack of response of the luteal cells to EGF could either have been due to a lack of EGF receptor sites or was secondary to a lack of internalization and/or degradation. We have therefore compared the binding, internalization, and degradation of EGF by cultured granulosa, luteinized granulosa, and luteal cells.

B) Comparison of the binding, internalization, and degradation of EGF in granulosa versus luteal and luteinized granulosa cell cultures

When the time-course of binding of ^{125}I-EGF to bovine granulosa and luteal cells was compared, maximal binding was reached after 30 min at 37°C for the granulosa cells and at 1 hour for the luteal cells (Fig. 1). The maximal amount of hormone specifically bound by the luteal cells was 8-fold higher than the amount bound at saturation by the granulosa cells. Upon continued incubation of luteal cells with the labeled hormone, the amount of cell-bound radioactivity decreased in 4 hours to 62% of the initial maximal amount bound at one hour. The same observation was made with granulosa cells; by 4 hours the amount of cell-bound radioactivity was 55% of that observed at 30 minutes (Fig. 1), indicating that in both cases exposure to EGF down-regulates its cell surface receptor sites.

EGF binding was a saturable process with a maximal binding at 2 ng/ml (3 X 10^{-10}M) and half-maximal binding at 0.75 ng/ml (1.1 X 10^{-10}M) for the granulosa cells. With luteal cells maximal binding was observed at 12 ng/ml (1.8 X 10^{-9} M) (Fig. 2). The dissociation constant (K_D) for EGF binding to granulosa cells was 2.4 X 10^{-10} M, and the binding at saturation was 1036 pg of EGF per 10^6 cells (25).

When the binding of EGF to granulosa and luteal cell cultures was compared to its mitogenic effect on these cultures, it was observed, as previously reported, that although EGF did bind to luteal cells it had no mitogenic effect, as indicated by the lack of any significant increase in cell number. In contrast, EGF was a strong mitogen for granulosa cell cultures and gave a maximal (up to 20-fold) increase in cell number at concentrations as low as 4.5 X 10^{-11} M (25).

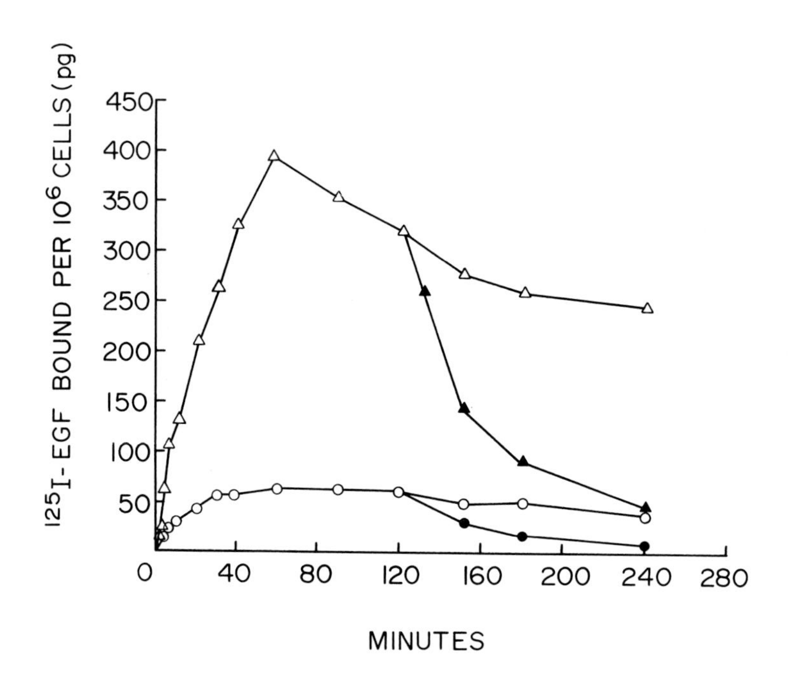

FIG. 1. Time-course of ^{125}I-EGF binding to granulosa and luteal cells. The granulosa (o) and luteal (Δ) cells densities were 31,000 cells/cm^2. Two ml of binding medium, containing 2.4 ng/ml of ^{125}I-EGF (45,000 cpm/ng), were added to each dish and the dishes were incubated at 37°C. At the indicated time intervals the specific cell-bound radioactivity (O, Δ) was determined. Non-specific binding was measured by adding 250 ng/ml of unlabeled EGF to the incubation medium at 0 min. Displacement of bound EGF from the cells was followed after the addition of 250 ng/ml of unlabeled EGF to the incubation medium at 120 min (O, Δ).

Similar results were obtained when the binding of EGF to cultured granulosa cells versus luteinized granulosa cells was compared. Although luteinized granulosa cell cultures no longer responded to EGF, they had a 6-fold higher binding capacity. These results therefore demonstrate that there is a total lack of correlation between number of receptor sites on the cell surface and proliferative response, since during the process of losing their responsiveness to EGF both luteal and luteinized granulosa cells show a net increase in the total number of EGF cell surface receptor sites. This observation is paradoxical, to say the least (25).

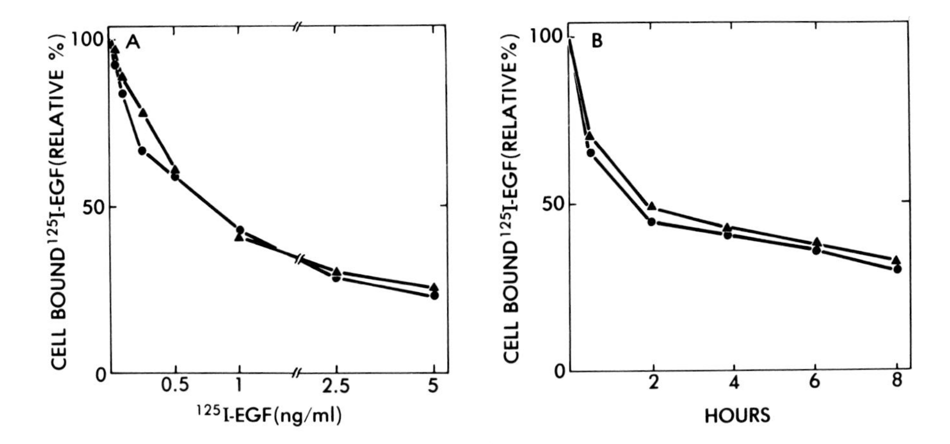

FIG. 2. Loss of EGF-binding capacity induced by preincubation with unlabeled EGF. (A) Ten hours preincubation with saturating and non-saturating concentrations of unlabeled EGF. (B) Preincubation for different time periods with unlabeled EGF (5 ng/ml). Cultures of granulosa cells that responded (▲) or lost response (●) to EGF were incubated at 37°C with unlabeled EGF in F-12 medium containing 1% calf serum. The cells were then washed free of unbound EGF and incubated in the binding medium (3 hr, 37°C) to permit degradation of the bound hormone. Cells were washed four times, transferred to 4°C, and their capacity to rebind ^{125}I-EGF (5 ng/ml, 370,000 cpm/ng) was determined (4°C, 75 min). The results are expressed as the relative percentage of cell-bound radioactivity, taking as 100% the amount of bound radioactivity in cells that were not exposed to unlabeled EGF.

Although a decrease in EGF receptor sites is not likely to be the reason for the lack of response of luteal cells or luteinized granulosa cells to EGF, a block in the internalization process or a lack of degradation could account for it, should internalization or degradation be required to mediate the mitogenic stimulus provided by EGF. That this is unlikely to be the case was concluded from comparative studies of the rates of EGF internalization and degradation in cultures of granulosa cells versus luteinized granulosa cells (14, 25).

To measure the release of EGF degradation products, monolayers were incubated at 4°C with labeled EGF, washed extensively to remove the unbound mitogen, and the amount of cell-bound radioactivity determined at various times after being transfered to 37°C. With both EGF-responsive and non-responsive granulosa cells the bound radioactivity decreased rapidly ($t_{1/2}$ = 60 min) during the subsequent incubation at 37°C. As with other cell types, 85-95% of the initial cell-bound radioactivity was not associated with the cells after two hours of incubation in these conditions. A 60-80% inhibition of this release was obtained in the presence of chloroquine (0.1 mM) or 2.4-dinitrophenol (0.2 mM). This indicates that the loss of cell-bound radioactivity is dependent on the prior internalization and degradation of the EGF molecules which require both the generation of a metabolic energy and the proteolytic activity of lysosomal enzymes. Likewise, both the loss of EGF binding sites induced by preincubation with unlabeled EGF (Fig. 2) and their recovery (Fig. 3) in EGF-responsive and non-responsive cells were exactly the same, both as a function of EGF concentration and as a function of time (14, 25).

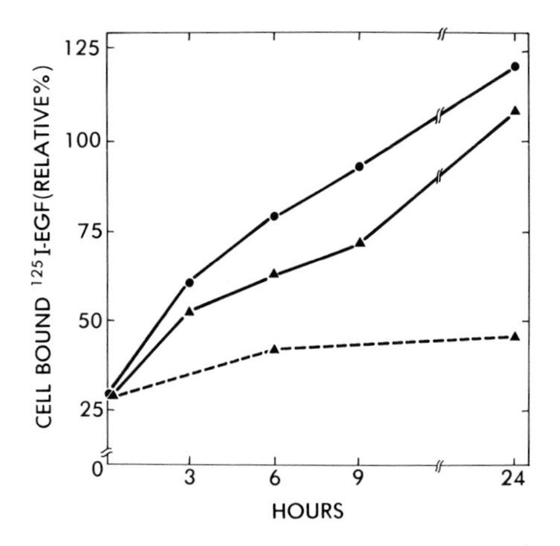

FIG. 3. Recovery of EGF-binding capacity in EGF-responsive and non-responsive cells that were first exposed to EGF. EGF-responsive (Δ--Δ) and non-responsive (●—●) granulosa cultures were preincubated with unlabeled EGF (5 ng/ml, 10 hr) in F-12 medium containing 1% calf serum. The cells were then washed free of unbound EGF and incubated in the binding medium for 3 hr at 37°C to permit degradation of the bound hormone. The cells were washed extensively, and the medium replaced with F-12 medium containing 10% calf serum (—) or 0.1% bovine serum albumin (---). At the indicated times, cells were transferred to 4°C and assayed for ^{125}I-EGF (5 ng/ml, 270,000 cpm/ng) binding. The extent of EGF-binding to cells that were similarly incubated but not exposed to unlabeled EGF was taken as 100%, and the results are expressed as the relative percentage of cell-bound radioactivity.

Our results on the binding of EGF to granulosa and luteal cells as well as to early and late luteinizing passages of granulosa cells therefore indicate a lack of correlation between EGF-binding capacity and mitogenic activity. Granulosa cells which proliferate in response to EGF bind ten times more EGF than is needed to induce a maximal proliferative response, whereas luteal cells which do not respond at all can even bind 4-6 times more EGF than granulosa cells. These observations lead us to conclude that 1) binding studies should be correlated with biological activity as expressed by the increase in cell number, since it is often implied that binding of a given agent will induce a biological effect. The example of EGF-binding to granulosa cells versus that to luteal cells demonstrates that this is not always the case; 2) the injection of a radioactively labeled mitogen in vivo and its subsequent distribution in the body can lead to false conclusions as to where the mitogen acts. For example, with ^{125}I-EGF the in vivo binding to granulosa cells will be barely detectable because of the high sensitivity of the cells, while luteal cells which do not respond to EGF will show, in comparison to granulosa cells, a detectable binding. This could lead to the wrong conclusion that luteal cells, instead of granulosa cells, are the target cells for EGF.

We have shown that during the in vitro luteinization of granulosa cells the number of surface receptors for EGF is increased and, as with EGF-responsive cells, can be down regulated by the extracellular concentration of EGF. Our experiments on the binding of EGF clearly indicate that the loss of a mitogenic response to EGF in cells that undergo luteinization is not due to a defect in the internalization and degradation of cell-bound EGF and/or in the EGF-induced loss and recovery of surface receptor sites for EGF.

The induction of cell division is accompanied by various changes in cellular physiology and properties of the cell surface. It is, however, not clear which of the induced alterations are correlative and which are both necessary and sufficient for the G_o-G_1 transition. Our results indicate that the internalization and the induced loss and recovery of EGF receptor sites are by themselves not sufficient to commit the cells to undergo cell division. These factors might, however, have a role in transmitting the mitogenic signal.

C) The role of the degradation of EGF in mediating the mitogenic response

Das and Fox (8) have proposed four transduction mechanisms to explain how the EGF-receptor interaction can lead to a production of a second messenger which drives the cell through G_1 and commits it to enter the S phase of the cell cycle. These are: 1) stimulation of second messenger production by receptor-mediated catalysis at the cell surface-- this stimulation may be the phosphorylation of a specific cell surface protein, as observed by Carpenter et al (3); 2) receptor-mediated transport of EGF, which then serves as the second mesenger either with or without further modification; 3) proteolytic processing of EGF in lysosomes, leading to the formation of second messenger; 4) proteolytic processing of receptors or some other endocytosed membrane proteins to yield new factors which can either serve as second messengers or have the capacity to produce one. However, to date no regulatory action of a protein hormone has been shown to require hormone degradation and it is not even clear whether the internalization, with or without degradation, is required in order to elicit an effect on cell proliferation and DNA synthesis or whether it is simply needed to inactivate the mitogen.

In an effort to evaluate the role of degradation in the induction of cell proliferation by EGF, we have used two proteases inhibitors, leupeptin and antipain, which have been shown to inhibit specifically the lysosomal protease cathapsin B (23, 24). Our results demonstrate that an inhibition of the lysosomal degradation of ^{125}I-labeled EGF (Table I) has no effect on the proliferative response of cultured bovine granulosa cells to that mitogen.

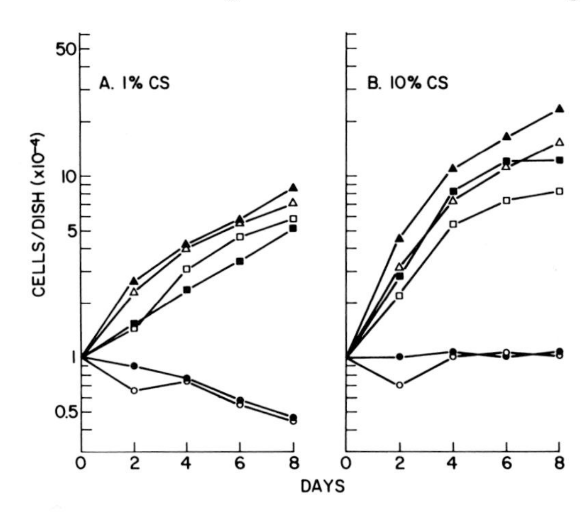

FIG. 4. Growth of granulosa cells induced by EGF or FGF in the presence or absence of leupeptin. Bovine granulosa cells were seeded at a density of 10^4 cells per 35 mm dish. 8 hours later, the medium in half of the dishes was replaced by fresh medium containing 1% calf serum (A), while the rest of the dishes contained 10% calf serum (B). At the same time, leupeptin (80 ug/ml) was added to some of the dishes. One day later, and every other day thereafter, EGF (10 ng/ml) or FGF (50 ng/ml) was added. Duplicate cultures of each system were trypsinized and the cells counted with a Coulter counter. The systems were: none (●), leupeptin (o), EGF (■), EGF + leupeptin (□), FGF (▲) and FGF + leupeptin (Δ).

At concentrations (80 ug/ml) which almost totally inhibited the degradation process (Table I), neither leupeptin nor antipain inhibited the proliferation of cells induced by EGF or FGF in cultures maintained with 1% calf serum (Fig. 4A). With cells maintained in 10% calf serum, there was a slight inhibition (10%) of cell proliferation when both leupeptin and EGF were present, as compared to cells exposed to EGF alone (Fig. 4B). That slight inhibition could be attributed to the accumulation of cytotoxic factors which are no longer degraded within the granulosa cells. It

Table I

The degradation of ^{125}I-EGF in the presence or absence of leupeptin

	12 hour incubation		24 hour incubation	
	free ^{125}I	degraded ^{125}I-EGF	free ^{125}I	degraded ^{125}I-EGF
		(cpm/dish)		
PBS	--	--	146,140	--
cells + unlabeled EGF in excess	143,160	--	147,880	--
cells	184,150	40,990	224,980	77,100
cells + leupeptin	142,800	0	155,498	7,618

Freshly prepared ^{125}I-EGF (5 ng/ml, 370,000 cpm/ng) was added to confluent cultures of bovine granulosa cells which had been preincubated for 16 hours in the presence or absence of leupeptin (80 ug/ml). The ^{125}I-EGF was also added to PBS solution to measure the spontaneous breakdown of ^{125}I-EGF and to granulosa cultures in the presence of unlabeled EGF in excess (2 ug/ml) to measure the nonspecific degradation. The dishes were then incubated for 12 and 24 hours at 37°C and the cell-free medium subjected to double immunoprecipitation (rabbit anti-EGF antiserum followed by goat anti-rabbit IgG). Alternatively, cell-free medium was submitted to gel filtration on a Bio-Gel P-10 column as described (20). The amount of free ^{125}I plus low molecular weight ^{125}I degradation products (mostly iodotyrosine) determined by both techniques was the same and the mean value is presented in the table. The amount of degraded ^{125}I-EGF is expressed as the difference between the amount of free ^{125}I obtained in the presence or absence of leupeptin and that detected in the presence of an excess of unlabeled EGF.

is also conceivable that such an inhibition by leupeptin is due to an inhibition of proteolytic activity normally present in the calf serum and which could potentiate the mitogenic effect of EGF (15). In this regard, leupeptin and antipain, in addition to being potent and specific inhibitors of cathapsin B, are known to inhibit, although to a smaller extent, the activity of plasmin and trombokinase (23, 24). Cultures maintained in the presence of EGF and leupeptin incorporated 20 to 30% less (^{3}H)thymidine than cultures maintained in the presence of EGF alone, whether the EGF concentration was saturable (10 ng/ml) or non-saturable (0.1 ng/ml).

That even the slight inhibition of cell proliferation and DNA synthesis caused by the proteases inhibitors is not due to the inhibition of EGF degradation was supported by studies on DNA synthesis induced by non-saturable concentrations of EGF (0.1 ng/ml). Under those conditions, the initiation of DNA synthesis was not inhibited to a higher extent than that induced by saturable concentrations of that mitogen (10 ng/ml), although the rate-limiting factor in the stimulation of (^{3}H)thymidine incorporation by 0.1 ng/ml EGF is the EGF concentration. If the degradation of EGF were involved in its mitogenic effect, then a 90% inhibition of EGF degradation should greatly inhibit the response to EGF (20).

The hypothesis that only a very small fraction of the EGF surface receptor sites is specifically involved in mediating the mitogenic effect of EGF on granulosa cells was suggested previously by Gospodarowicz et al (10). Even if such a small fraction of the bound EGF is internalized and has to undergo a degradation process in order to induce a proliferative response, then at a suboptimal concentration of EGF a 90% inhibition of the degradation process should extensively inhibit the response to EGF. Our observation that the mitogenic activity at suboptimal concentration of EGF was not affected by leupeptin any more than the 10% to 20% inhibition observed in the presence of leupeptin and optimal concentration of EGF supports the hypothesis that the degradation process has no role in the mitogenic effect of EGF. It is therefore more likely that the physiological function of the degradation of EGF by the lysosomal system is merely to destroy the mitogen, so that each molecule can act only once. Shechter et al (21) suggested that the processes of hormone internalization, degradation, and "down regulation" may be irrelevant to the effects of EGF on DNA synthesis. This was based on the observation that in cultured human fibroblasts occupation of only a negligable fraction of binding sites by irreversibly bound EGF might be enough to enhance DNA synthesis (21). This hypothesis was not confirmed with granulosa cells. A simple wash of the monolayer, even 4 hours after the addition of EGF, abolished the incorporation of (^{3}H)thymidine into DNA. Whether or not the internalization process of EGF is relevant to its mitogenic effects on granulosa cells therefore still remains to be studied; however, on the basis of the present results it is reasonable to conclude that the degradation process has little or no role in inducing a mitogenic response of granulosa cells to EGF. It is therefore unlikely that the proteolytic processing of EGF in lysosomes, leading to the formation of a second messenger, or that proteolytic processing of receptors or some other endocytosed membrane protein(s), yielding new factors, could play any role in the mitogenic stimulus.

In addition to allowing the determination of the role of the intracellular degradation of mitogens in the proliferative response of granulosa cells, the use of lysosomal inhibitors which caused an intracellular accumulation of intact EGF has allowed us to study the fate of the cell-associated ^{125}I-EGF within the cells.

D) The accumulation of intact EGF within granulosa cell nuclei

As previously described, the degradation process of the ^{125}I-EGF was found to be irrelevant to its mitogenic effect on granulosa cells. Moreover, the internalization of mitogens and the associated receptor down regulation have not been shown to be sufficient to commit the cells to undergo division. It is therefore necessary to look for other sites of interaction which might be involved in the induction of a mitogenic response. These might be present on the cell surface, cytosol, or nucleus. Recent studies on the binding of insulin and nerve growth factor to nuclear receptor sites led us to investigate whether an interaction between EGF and the cell nucleus can be demonstrated subsequent to surface binding and internalization in granulosa cells. In view of the rapid degradation of the internalized EGF, the average lifetime of intact EGF inside the cells might be too short to allow its detection in the cell nucleus. Our approach was therefore to induce intracellular accumulation of intact EGF by using either chloroquine or leupeptin to inhibit its degradation and under these conditions to study whether acumulation of EGF in the cell nuclei can be demonstrated. Cells were incubated (18 hr, 37°C) with ^{125}I-EGF in the absence and presence of leupeptin or chloroquine, lyzed with 0.5% NP_{40}, and tested for the amount of radioactivity associated with the cell lysate and the nuclear layer remaining on the dish. It was demonstrated that as much as 15% of the total amount of EGF that had been accumulated in the cytoplasm of chloroquine-treated cells became associated with the nuclei, as opposed to less than 2% of the amount found in leupeptin-treated cells (Fig. 5). This suggests that a certain percentage of the cell-surface-bound EGF is delivered into the nucleus, possibly *via* fusion of endocytotic vesicles with the outer nuclear membrane. This pathway is slightly or not at all detectable in the absence of chloroquine due to rapid degradation of EGF in the lysosomes. The reason for the lack of EGF accumulation in the nuclei of leupeptin-treated cells is still unclear. This result may indicate, however, that detection of ^{125}I-EGF in the nuclei of chloroquine-treated cell preparations is not due to a contamination

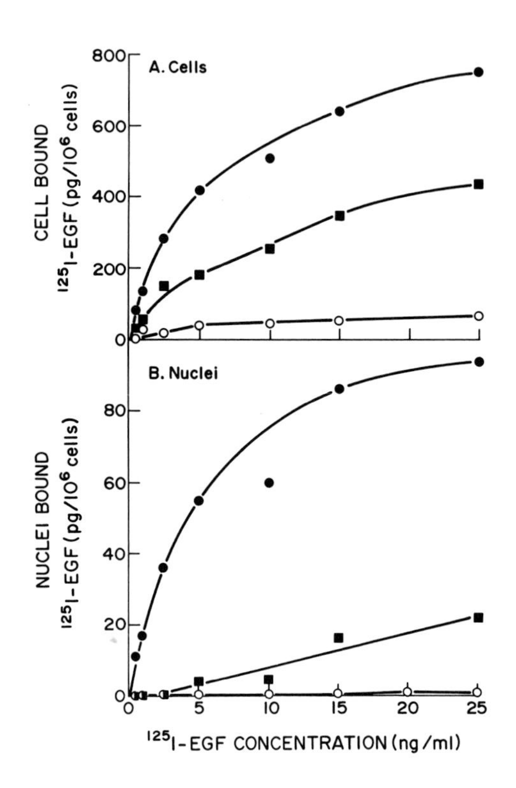

FIG. 5. Accumulation of ^{125}I-EGF in the nuclei of granulosa cells. Confluent bovine granulosa cultures were preincubated for 4 hr or 16 hr in the presence of chloroquine (25 uM; ●) or leupeptin (100 ug/ml; ■), respectively, or without any addition (o). Cultures were then incubated (18 hr 37°C) with various concentrations of ^{125}I-EGF in the presence or absence of unlabeled EGF (2 ug/ml) and unbound radioactivity removed by 10 washes with cold (4°C) phosphate-buffered saline (PBS, pH 7.4) containing 0.1% albumin. The cells were lysed with NP_{40} containing lysis buffer (10 mmM Tris, pH 7.5, 10 mM NaCl, 3mM $MgCl_2$, 15 mM β-mercaptoethanol, 0.5% NP_{40}), and the radioactivity present in the lysate fraction was counted and is referred to as "cell-bound ^{125}I-EGF." The remaining nuclear monolayers were washed ten times with PBS containing albumin, solubilized in NaOH (0.1 N), and counted to determine the amount of nuclear associated ^{125}I-EGF.

with plasma membrane or cytoplasmic-bound EGF. An examination of the subnuclear localization of EGF using micrococcal nuclease for enzymatic fractionation of the nuclear chromatin suggested that the bound EGF is not localized in the chromatin fraction of the nucleus but is rather associated with the inner nuclear membrane or nuclear matrix fraction.

Our present results indicate that the nuclear accumulation of EGF is mainly due to the inhibition of its degradation by chloroquine and not to other possible effects of the drug. Whether the nuclear pathway functions in cells that are actively degrading the mitogen and whether specific EGF binding can be demonstrated with isolated nuclei are questions which remain to be studied. By using cells (granulosa) which bind and respond to EGF and closely related cells (luteal) which bind but no longer respond to the mitogen, we hope to find out whether an interaction with the cell nucleus is essential for the induction of cell proliferation by EGF.

II. THE CONTROL OF GENE EXPRESSION BY EGF

Although EGF can control the proliferation *in vivo* of a variety of cell types and thereby act as a mitogen, it can also influence morphogenetic events. This is best demonstrated by the precocious eyelid opening and tooth eruption observed in neonatal mice treated with EGF (3). Both in organ culture and *in vivo*, EGF will either delay or prevent the fusion of the palate shelves and may therefore be considered to be teratogenic under special circumstances (1, 16). Although most of these morphogenetic effects could be secondary to the proliferation of the basal cell layer of the epithelium, EGF may also directly regulate the phenotypic expression of the cells. Evidence that EGF could to some extent control the phenotypic expression of a given cell type has already been provided. In mouse 3T3 cells, for example, EGF induces the appearance of a cell surface protein named fibronectin at concentrations which are lower than those required to induce cell proliferation (4). Likewise, the addition of EGF to cultured human choriocarcinoma cells stimulates the secretion of biologically active human chorionic gonadotropin (hCG) and, to a lesser extent, the secretion of free hCG- units (2).

EGF has been shown to be a potent inhibitor of gastric acid secretion in the gastric mucosa (15). All these effects are clearly unrelated to cell proliferation and indicate that, in addition to influencing morphogenesis, EGF can also influence gene expression, as reflected by its effect on the synthesis of differentiated cell products. In earlier studies, we have investigated the effect of EGF *in vivo* on the development and maturation of ovarian follicles in neonate rats (12). We observed that animals subjected to a daily injection of 5 ug of EGF between the ages of 5 to 15 days had a body size which was $^1/_3$ to $^1/_2$ that of their littermates injected with saline alone. This suggested that EGF could adversely affect, either directly or indirectly, the growth of the body. Particularly attractive was the possibility that EGF could modulate the synthesis and release of growth hormone (GH) by pituitary cells. Since this hormone is responsible for the growth of the body, any inhibition of its synthesis which will result in a lower plasma concentration of GH will soon be reflected in a lower rate of bodily growth. Although the effect of EGF on the synthesis of GH in the pituitary of 5-day-old rats may be difficult to study, such an effect can be studied easily using established pituitary cell lines which produce GH and prolactin and which have been shown to retain their hormonal response insofar as the control of synthesis and secretion of both GH and prolactin by thyroxine and cortisol is concerned.

A) The effect of EGF on the synthesis of GH versus prolactin by GH_3 cells

GH_3 cells synthesize both GH and prolactin. In addition, the production of these differentiated gene products and their mRNA's are under multi-hormonal control. Growth hormone synthesis is controlled by both thyroxine and glucocorticoid (19), while prolactin expression is induced by estradiol and thyrotropin-releasing hormone (TRH) (7) and inhibited by glucocorticoids (6). To test for an effect of EGF on either of these two gene products, cells were exposed to EGF (20 ng/ml, 0.5% serum) for 48 hours. The pattern of protein synthesis analyzed on two-dimensional gels was then compared to that of parallel cultures exposed to thyroxine (10 nM) and dexamethasone (1 uM) (Fig. 6). Using a short labeling pulse (30 min), the relative incorporation of label into growth hormone and prolactin directly reflects the quantity of translatable mRNA for each protein (9, 19). As already reported (19), after a 48 hour exposure to thyroxine and dexamethasone, growth hormone synthesis is dramatically induced while prolactin is inhibited and hardly detectable. In contrast, a 48 hour incubation with EGF (25 ng/ml) (Fig. 6C) produces an entirely opposite result. Growth hormone synthesis is totally inhibited by EGF, while prolactin production is stimulated.

The opposing effects of these two hormone treatments were also investigated in more detail using one dimensional SDS gel electrophoresis of proteins synthesized after exposure of cells to various hormone combinations (Fig. 7). In control cultures exposed to EGF alone, the patterns of proteins synthesized by the cells during the labeling period are remarkably similar except for the changes in growth hormone and prolactin synthesis (Fig. 7A, B, anc C). In both cases, 25 ng/ml EGF (C) was much more effective than 0.25 ng/ml (B), although prolactin is clearly induced above control values even at this subsaturating dose of the growth factor.

In confirmation of previous reports (19), thyroxine alone induces GH synthesis after 48 hours (Fig. 7D). However, EGF, when added with thyroxine (Fig. 7E, 0.25 ng/ml and F, 25 ng/ml) reduces growth hormone synthesis almost to its basal level. Thus, EGF also blocked the thyroid hormone response in GH_3 cells. As in control cells, prolactin was again induced by EGF, although not as dramatically.

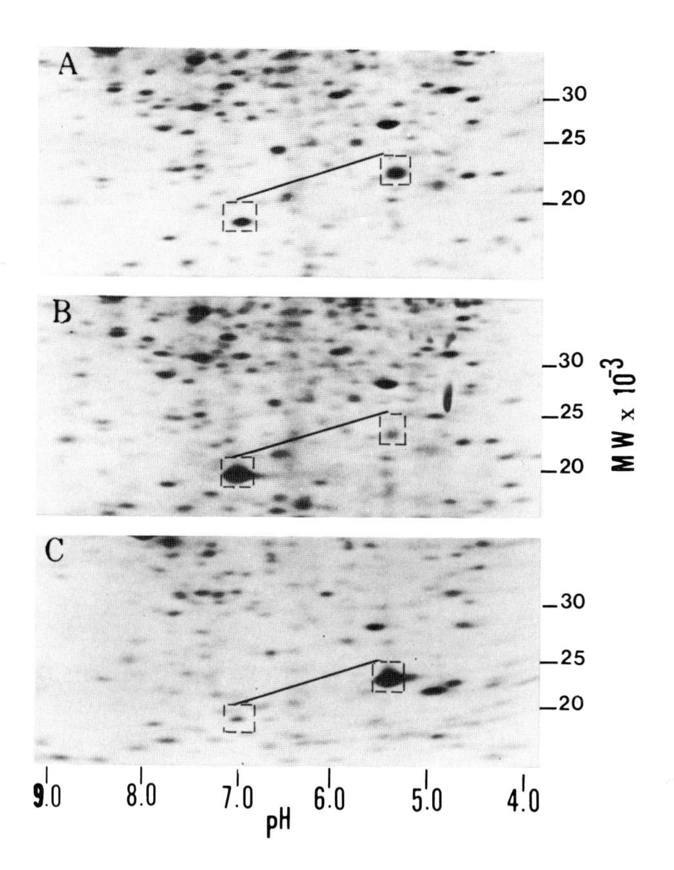

FIG. 6. Two-dimensional gel electrophoresis of newly synthesized GH_3 cell proteins after treatment with EGF or thyroid hormone and dexamethasone. (A) Newly synthesized proteins were labeled in DMEM containing 0.5% serum, 10 uM methionine, and 500 uCi/ml (^{35}S)methionine for 30 min at 37°C. Control cells. (B) T_3 (10 nM), dexamethasone (1 uM); (C) EGF (20 ng/ml). All inductions were for 48 hr in the presence of 0.5% serum. Boxes marked the position of growth hormone and prolactin. 500,000 cpm were applied to the first dimension and the second dimension was a 10-16% exponential gradient of polyacrylamide. Exposure time was ten days.

The synergistic effects of thryoxine and dexamethasone on stimulating GH production are clearly apparent (G). In this case, GH synthesis is dramatically stimulated when compared to either thyroxine-treated (D) or untreated cells (A). However, even in the presence of both thyroxine and dexamethasone, EGF was an effective dose-dependent inhibitor of the induced levels of GH synthesis (Fig. 7H, 0.25 ng/ml and I, 25 ng/ml). In contrast to control or

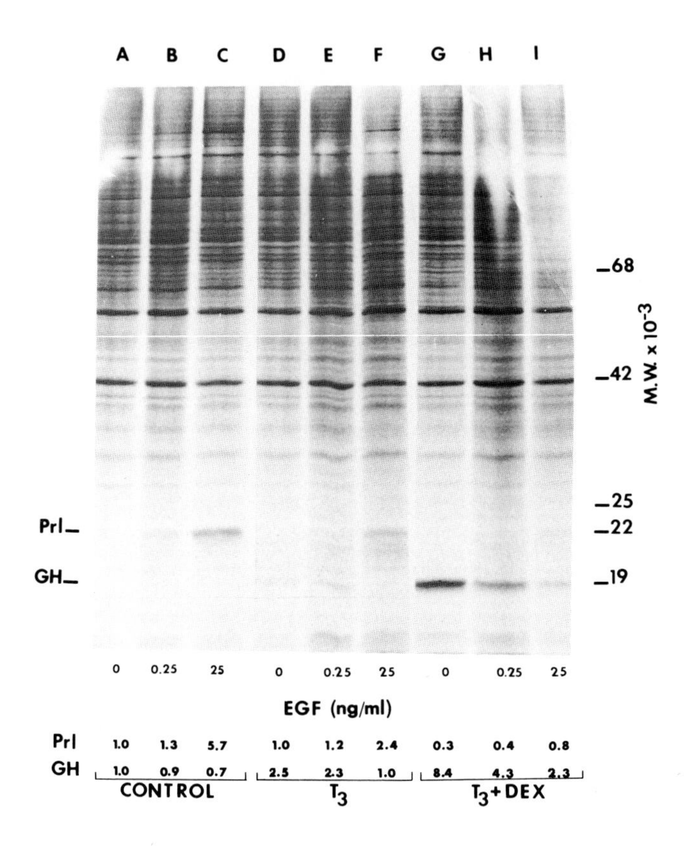

FIG. 7. Effect of EGF on the hormonal control of growth hormone and prolactin expression. Cells were induced for 48 hr in the presence of 0.5% serum with EGF; T_3 (10 nM) alone and with EGF; T_3 (10 nM) and dexamethasone (1 uM) alone and with EGF. Newly synthesized proteins were labeled with (^{35}S)methionine (400 uCi/ml, 30 min, 37°C) and fractionated on 10-16% exponential gradient of polyacrylamide in SDS. 50,000 cpm were applied to each lane.

thyroxine-treated cells, the addition of dexamethasone dramatically inhibits the EGF induction of prolactin synthesis (compare C and I). Thus, the glucocorticoid has an overriding inhibitory influence on EGF-stimulated prolactin expression, but a submissive role in the EGF inhibition of growth hormone synthesis. These results demonstrate that EGF can induce changes in the expression and hormonal control of two differentiated gene products, all of which are independent of a mitogenic effect of EGF. Indeed, it has been shown that, although GH_3 cells have high affinity cell surface binding sites specific for EGF (34,000 per cell, Kd - 1nM) and degrade EGF within the cells through a chloroquine-sensitive pathway, prolonged treatment of the cells with EGF does not stimulate cell division but instead results in an inhibition of the proliferation induced by thyroid hormone (17).

B) Alteration of chromatin structure induced by EGF

Because the actions of thyroid hormone on GH synthesis are thought to be mediated by chromatin receptor proteins (18) and are inhibited by EGF, it was of interest to determine if EGF itself might elicit some nuclear response. To test this, GH_3 cells maintained in 0.5% serum were exposed for 48 hours to 20 ng/ml EGF. The nuclei were then isolated from treated and control cells and the number of rifampicin-resistant initiation sites for E. coli RNA polymerase were examined in both nuclei preparations. When the number of initiation sites was compared in control nuclei versus nuclei from EGF-treated cells, EGF increased by 40% the number of initiation sites. At saturation, nuclei from EGF-treated cells bound 0.22 pmol of RNA polymerase per ug DNA, while untreated nuclei bound 0.17 pmol of RNA polymerase per ug of DNA (17). In both cases, however, the average chain lengths of the in vitro synthesized transcripts were similar (1124 bases vs. 1062 bases for EGF and control chromatin respectively), as determined by sucrose density gradient centrifugation (17). This result demonstrates that EGF treatment affects chromatin structure in some manner which allows for the exposure of additional initiation sites detected by the non-specific bacterial polymerase. Because such effects on chromatin have been shown to occur concomitantly with changes in gene expression within the rat pituitary cells (17), it is possible that they reflect the increased prolactin synthesis induced by EGF. It can therefore be concluded that EGF induced alteration in the structure of chromatin isolated from the cells, as reflected by an increase in the number of rifampicin-resistant sites for E. coli RNA polymerase assayed in isolated nuclei, an alteration correlative with alterations in the pattern of gene expression (18). Growth hormone synthesis was repressed while prolactin production was increased. Since these changes were detected during a short pulse of the cells with (^{35}S)methionine, it is likely that they represent EGF-induced alterations in the quantity of growth hormone and prolactin mRNA as has been observed for the thyroxine, dexamethasone, and TRH (9, 19) regulation of these gene products.

C) The accumulation of EGF within GH_3 cell nuclei

The effect of EGF on chromatin structure, as well as on the synthesis of two differentiated gene products, prolactin and growth hormone, suggest that various nuclear functions could be modulated by EGF in GH_3 cells despite its failure to have an effect on DNA synthesis or cell proliferation. Recently, Yanker and Shooter have reported on the nuclear accummulation of nerve growth factor (NGF) in PC12 cells (26). As with EGF and GH_3 cells, NGF does not promote cell division but does influence RNA and protein synthesis while inducing neurite outgrowth. The similarities between the two systems and the various theories regarding the mechanism by which mitogens exert their growth-promoting and other effects led us to investigate whether an interaction between EGF and the cell nucleus can be demonstrated subsequent to surface binding and internalization in GH_3 cells.

Previous studies with GH_3 and other cell types have demonstrated that the cell-bound ^{125}I-EGF is rapidly degraded (t½ = 20') to mono(^{125}I)iodotyrosine by lysosomal proteases following endocytic internalization. Due to this process, the intracellular amount of intact EGF reaches, within 1-2 hr of exposure to EGF, a steady state concentration which is much lower than the actual amount of internalized EGF and which later even decreases, due to a loss (down-regulation) of the appropriate cell surface receptor sites (17). If translocation of EGF into the nucleus does occur, this might require a time longer than the average half-life of the internalized EGF and thereby reduce the probability of detecting EGF accumulation within the nuclei. Therefore, to increase the half-life of internalized EGF, we use the lysosomal inhibitor chloroquine, which has been shown to inhibit the degradation of EGF and thus to allow the intracellular accumulation of intact EGF molecules. Cells were incubated with ^{125}I-EGF in the absence and presence of chloroquine, lyzed with 0.5% NP_{40}, and tested for the amount of radioactivity associated with the nuclear pellet and cell lysate. Nuclei isolated from detergent-lysed cells which were first exposed to 12.5 ng/ml ^{125}I-EGF for 18 hours contained about 1% of the total internalized label. In contrast, when cells were incubated with EGF in the presence of increasing concentrations of chloroquine, the amount of nuclear-associated radioactivity increased and at saturation represented 14% of the internalized EGF (Fig. 8). Under these conditions, the total amount of cell-associated ^{125}I-EGF (bound plus internalized) was increased 6-fold. The nuclear accumulation of EGF in chloroquine-treated cells lags significantly behind the rate of internalization of the growth factor. Thus, after a 4 hour incubation with EGF, only about 2% of the internalized EGF (measured in the NP_{40} lysate) was associated with the isolated nuclei, whereas a 24 hour incubation resulted in a nuclear accumulation of as much as 14% of the internalized EGF (Fig. 8). Such a difference in the kinetics of cytoplasmic accumulation versus nuclear binding also suggests that the observed nuclear accumulation of EGF is not due to a contamination of the nuclei by membrane or cytoplasmic bound EGF.

Exposure of the cells to chloroquine (75 uM) had no effect on, or increased by at most 2-fold, the amount of surface-bound, and hence trypsin-releasable ^{125}I-EGF. However, the cell lysate from

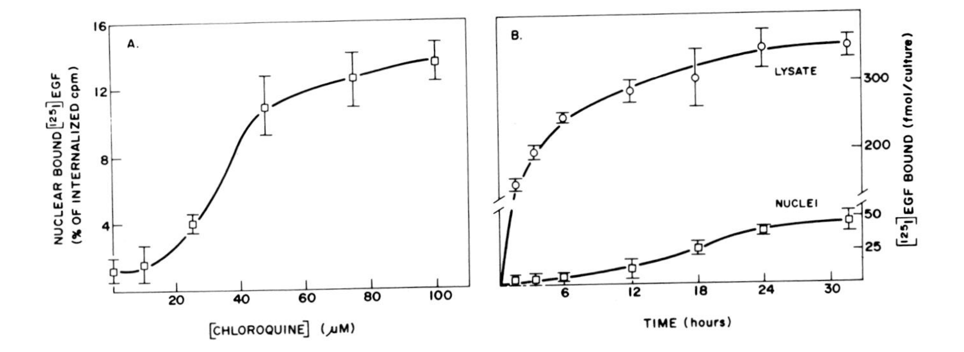

FIG. 8. Nuclear accumulation of ^{125}I-EGF in GH_3 cells. (A) Effect of chloroquine concentration on the percent of total cell-associated ^{125}I-EGF found in the nuclear fraction after 18 hr. (B) Kinetics of ^{125}I-EGF accumulation in GH_3 cell lysates and nuclei. 5 X 10^6 GH_3 cells were seeded into 35 mm culture dishes and 24 hours later the medium was replaced by 1.0 ml of medium containing 0.5% fetal calf serum and the indicated chloroquine concentrations. One hour later 12.5 ng/ml of ^{125}I-EGF (460,000 cpm/ng) was added in the presence and absence of 2.5 ug/ml unlabeled EGF. After incubation at 37°C cultures were washed ten times with phosphate-buffered saline containing 1 mg/ml BSA and lysed in buffer containing 0.5% NP-40 (17). The initial supernatant after removal of the nuclei (1,000 X g, 5 min) was designated the lysate fraction and the nuclei were purified further as described (17) before counting of radioactivity. The mean and range of triplicate determinations are shown for each experiment.

chloroquine-treated cultures contained as much as 19.2 times more immunoprecipitable and, hence, presumably intact ^{125}I-EGF than control cells. This increase was directly paralleled by a 19-fold increase in the amount of nuclear-associated ^{125}I-EGF. As in the lysate of chloroquine-treated cells, 75% of the nuclear-bound ^{125}I-EGF could be precipitated with anti-EGF after solubilization from the nuclei. The results of these experiments suggest that in control cells, due to a rapid degradation rate of the internalized

EGF, only a small percentage (less than 1%) of the internalized growth factor could be detected in the nucleus. In contrast, in chloroquine-treated cultures EGF degradation is greatly diminished, thus allowing a cytoplasmic accumulation of EGF and making a greater proportion of the internalized, intact EGF available for nuclear binding. The presence of unlabeled EGF (10 ug/ml) in the lysis buffer had no effect on the amount of ^{125}I-EGF associated with the nuclear pellet, thereby suggesting that no binding of ^{125}I-EGF occurs during the lysis and nuclear isolation and that, once the internalized EGF is tranlocated into the nuclei, it becomes either tightly bound or inaccessible to exchange with free EGF.

The nature of the nuclear sites which interact with ^{125}I-EGF was examined. Two distinct classes of binding affinity are apparent: a high affinity component with a Kd of 0.28 nM and a lower affinity class of sites possessing a Kd of 2.7 nM. The binding capacity of each class of sites is 10^3 and 1.2×10^4 sites/nucleus respectively. Interestingly, this observation is similar to findings regarding the nuclear binding of nerve growth factor in PC12 cells. In that system, two classes of binding sites which differ in affinity by an order of magnitude were also observed (26), although the number of high-affinity sites was 2- to 5-fold greater than that seen for EGF in GH_3 cells. The NGF nuclear receptor sites in PC_{12} cells have been shown to be resistant to extraction with triton, suggesting that they may be localized at the inner nuclear membrane. We therefore examined the subnuclear localization of EGF by subjecting isolated nuclei with bound ^{125}I-EGF to enzymatic fractionation of (^{14}C)thymidine-labeled chromatin (Fig. 9). Micrococcal nuclease solubilized approximately 90% of the labelled DNA within 10 min. The solubilized DNA fragments showed the characteristic nucleosome repeat pattern of chromatin and contained DNA fragment sizes from monomers (approx. 180 base pairs) to pentamers (Fig. 9 inset). However, no (^{125}I)EGF was released by the nuclease above that which dissociated during the course of the experiment (approx. 19% in the zero time controls), nor was any release observed when the EGF-loaded nuclei were lysed with 1 mM EDTA. This suggests that the bound EGF is not localized in the chromatin fraction of the nucleus but is rather associated with the inner nuclear membrane or nuclear matrix fraction. The latter is known to contain the so-called "nuclear scaffolding," structures thought to participate in the higher order condensation and packaging of chromatin into metaphase chromosomes prior to cytokinesis. This is in contrast to the subnuclear location of two other regulators of growth hormone and prolactin m RNA synthesis, namely the thyroid and glucocorticoid hormone receptors (17). Both of these receptors are liberated from GH cell chromatin upon DNAse digestion, and in the case of T_3 there may be a preferential localization of the receptor in nuclear fractions more active in RNA synthesis. Direct EGF binding to receptors in the inner nuclear membrane seems unlikely, since nuclei isolated by detergent showed no binding despite their retention of the inner membrane.

In the case of GH_3 cells, although binding to isolated nuclei has not yet been observed, a translocation pathway leading to a nuclear accumulation of EGF could be described in accordance with the present observations. This seems to involve internalization of EGF by adsorptive endocytosis, fusion of endocytotic vesicles with the

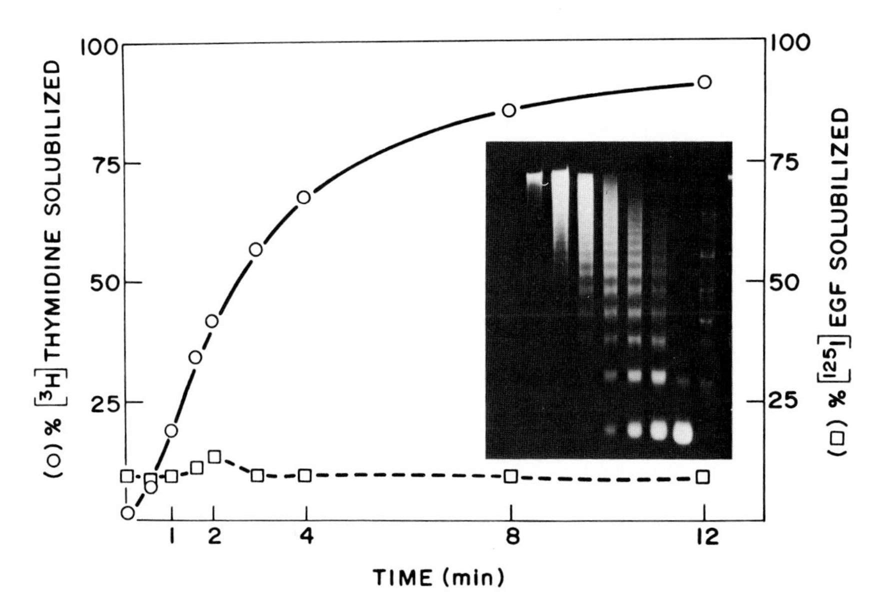

FIG. 9. Influence of micrococcal nuclease on release of ^{125}I-EGF and (^{14}C)thymidine-labeled DNA from isolated GH_3 nuclei. Cells (1.5×10^7/dish) were labeled 24 hr with 25 uCi (5 uCi/ml) (^{14}C)thymidine, followed by an additional 18 hr incubation with ^{125}I-EGF (12.5 ng/ml) in the presence of 75 uM chloroquine. After isolation of the nuclei, aliquots were incubated at 22°C with 40 units/ml micrococcal nuclease (Worthington) for the indicated times. The reaction was stopped, nuclei were lysed by the addition of a 10-fold volume of 1 mM EDTA, and unreleased materials were removed by centrifugation. ^{125}I-EGF in the two fractions (pellet and supernatant) was counted first. Each was subsequently brought to 1.0 M PCA, heated to 70°C for 30 min, and cooled on ice. Precipitated protein was removed by centrifugation and (^{14}C) detected by scintillation counting in a Triton-toluene-omnifluor cocktail. The percent of each label released was determined by dividing the amount of label present in the supernatant by the total (supernatant plus pellet) for each time point.

outer nuclear membrane followed by release of EGF, either bound or unbound to its surface receptor sites, into the perinuclear space, and interaction with the inner nuclear membrane or nuclear scaffolding structures. Cytoplasmic vesicles containing fluorescein-conjugated EGF were observed to form a perinuclear ring in A-431 cells, and fusion of lysosomal vesicles with the nuclear membrane has been considered as a possible means of introducing estrogens into the nucleus (22). In the present study, a nuclear accumulation of EGF in GH_3 cells was unequivocally demonstrated only when its lysosomal degradation was inhibited. To date, no regulatory action of protein hormone has been shown to require hormone degradation. On the contrary, recent results from our laboratory demonstrate a full induction of thymidine incorporation and cell proliferation by EGF, despite the inhibition of its degradation by either leupeptin or antipain (20). Whether or not the nuclear accumulation of either intact or modified EGF is in fact responsible for the various nuclear effects exerted by EGF on GH_3 cells requires a further investigation.

REFERENCES

1. Bedrick, A.D. and Ladda, R.L. (1978):Teratology, 17:13-18.
2. Benveniste, R., Speeg, K.V., Jr., Carpener, G., Cohen, S., Lindner, J., and Rabinowitz, D. (1978):J. Clin. Endocrinol. Metab., 46:169-172.
3. Carpener, G. and Cohen, S. (1978): In:Biochemical Actions of Hormones, edited by G. Litwack, vol. 5, pp. 203-247, Academic Press, New York.
4. Chen, L.B., Gudor, R.C., Sun, T.T., Chen, A.B., and Mosesson, M.W. (1977):Science, 197:776-778.
5. Cohen, S. And Carpenter, G. (1979):Ann. Rev. of Biochem., 48:193-216.
6. Dannies, P.S. and Tashjian, A.H. (1973):J. Biol. Chem., 248:6174-6179.
7. Dannies, P.S. and Tashjian, A.H. (1973): In:Tissue Culture, Methods and Applications, edited by P. Kruse and M. Patterson, pp. 561-569, Academic Press, New York.
8. Das, M. and Fox, F. (1978):Proc. Natl. Acad. Sci. USA, 75:2644-2648.
9. Evans, G.A., David, D.N., and Rosenfeld, M.G. (1978):Proc. Natl. Acad. Sci. USA, 75:1294-1298.
10. Gospodarowicz, D., Ill, C.R., and Birdwell, C.R. (1977):Endocrinology, 100:1108.
11. Gospodarowicz, D., Ill, C.R., and Birdwell, C.R. (1977):Endocrinology, 100:1121.

12. Gospodarowicz, D., Ill, C.R., Mescher, A.L., and Moran, J. (1976): In: Fifth Int. Cong. of Endocrinology, Hamburg, vol. 2, pp. 196-205, Excerpta Medica, Elsevier, New York.
13. Gospodarowicz, D. and Moran, J. (1976):Ann. Rev. Biochem., 45:531-556.
14. Gospodarowicz, D., Vlodavsky, I., Bialecki, H., and Brown, K. (1978): In: Fifth Brook Lodge Meeting on Novel Aspects of Reproductive Biology, edited by J. Wilkes, pp. 107-178, Plenum Press, New York.
15. Gregory, H. (1975):Nature, 257:325-327.
16. Hassell, J.R. (1975):Develop. Biol., 45:90-92.
17. Johnson, L.K., Baxter, J.D., Vlodavsky, I, and Gospodarowicz, D. (1979):Proc. Natl. Acad. Sci. USA, in press.
18. Latham, K.R., MacLeod, K.M., Papavasiliou, S.S, Martial, J.A., Seeburg, P.H., Goodman, H.M., and Baxter, J.D. (1978): In: Receptors and Hormone Action, vol. 3, edited by B.W. O'Malley and L. Birnbaumer, pp. 76-100, Academic Press, New York.
19. Martial, J.A., Seeburg, P.H., Guenzi, D., Goodman, H., and Baxter, J.D. (1977):Proc. Natl. Acad. Sci. USA, 74:42934295.
20. Savion, N., Vlodavsky, I., and Gospodarowicz, D. (1979):Proc. Natl. Acad. Sci. USA, submitted.
21. Schechter, Y., Hernaez, L., and Cuatecasas, P. (1978):Proc. Natl. Acad. Sci., 75:5788-5791.
22. Szego, C.M. (1974):Rec. Prog. Horm. Res., 30:171-233.
23 Umezawa, H. (1972): In: Enzyme Inhibitors of Microbial Origin, edited by H. Umezawa, pp. 15-52, University Park Press, Pennsylvania.
24. Umezawa, H. (1976):Methods in Enzymology, 55:678-695.
25. Vlodavsky, I., Brown, K., and Gospodarowicz, D. (1978): J. Biol. Chem., 253:3744-3750.
26. Yanker, B.A. and Shooter, E. (1979):Proc. Natl. Acad. Sci. USA, 76:1269-1273.

Control Mechanisms in Animal Cells,
edited by L. Jimenez de Asua et al.
Raven Press, New York © 1980.

The Biological Activity of Early Simian Virus 40 Proteins

A. Graessmann, M. Graessmann, and C. Mueller

Institute for Molecular Biology and Biochemistry, Free University of Berlin, 1 Berlin 33, West Germany

The oncogenic potency of simian virus 40 (SV40) is linked to the expression of the early viral genome region (map position: 0.655-0.17) (17). This genome part codes for two known proteins, the large T-antigen and the small t-antigen (2,3,11). The large T-antigen (94k) is a nuclear DNA binding protein, encoded by two discontinuous DNA segments (0.655-0.60; 0.533-0.17). The coding region for the cytoplasmic small t-antigen (17k) is located between map coordinates 0.655 and 0.53 (12).

However, more than two functions can be attributed to the early SV40 proteins:

i. stimulation of cell DNA synthesis and T-antigenicity (19);
ii. helper function for Adenoviruses in monkey cells and U-antigenicity (17);
iii. induction of viral DNA replication (14);
iv. regulation of late viral gene expression(1);
v. loss of actin cables (13);
vi. initiation and maintenance of cell transformation (10,15).

Our approach to further study these functions involves microinjection of SV40 DNA, DNA fragments, early SV40 RNA or purified T-antigens into mammalian tissue culture cells.

i. STIMULATION OF CELL DNA SYNTHESIS AND T-ANTIGENICITY

The first experimental evidence that T-antigen is a virus-coded protein inducing cellular DNA synthesis was obtained by microinjection of early SV40 RNA transcribed in vitro by E.coli DNA dependent RNA polymerase. After microinjection of this RNA intranuclear T-antigen formation also occurred when cellular RNA synthesis was blocked by actinomycin D. Additionally, incorporation of (^{3}H)-thymidine into quiescent primary mouse kidney cells or TC7 cells was found after cRNA injection as in cells injected with SV40 DNA or infected with the virus (3,7). Table 1 demonstrates the temporal correlation between intranuclear T-antigen appearance and stimulation for cell DNA synthesis after microinjection of either the purified large T-antigen, isolated from SV80 cells(16), or of superhelical SV40 DNAI.

In order to test which part of the early SV40 genome region is

TABLE 1. Stimulation of DNA synthesis in primary mouse kidney cells microinjected with SV40 T-antigen or viral DNA[a]

Labeling period (hours post injection)	Injection of SV40 T-antigen		Injection of SV40 DNA I	
	% of T-antigen positive cells	% of T-antigen positive cells stimulated for DNA synthesis	% of T-antigen positive cells	% of T-antigen positive cells stimulated for DNA synthesis
0 - 5	95	0	0	0
5 - 10	95	15	60	0
10 - 15	85	80	99	35
15 - 20	75	75	99	92
20 - 25	42	25	99	50

[a]Primary mouse kidney cells grown to confluence on glass slides were transferred into serum free medium 24 hours before microinjection. (^3H)-Thymidine was added at a final concentration of 0.5μCi/ml for the time intervals indicated. Thereafter, cells were washed with phosphate buffered saline, fixed and stained for T-antigen and finally processed for autoradiography.

essential for this function, SV40 DNA fragments of different size were microinjected into primary mouse kidney cells. The fragments used were obtained by digestion of SV40 DNA with restriction endonucleases (HpaI, HpaII, PstI, BamHI). The relevant cleavage sites of these enzymes are shown in Figure 1. The conjunction of immunofluorescent staining with autoradiography used in these studies allows a direct correlation between T-antigen formation and DNA synthesis. These experiments have proven that the information between map coordinates 0.27-0.17 is not necessary to stimulate host cell DNA synthesis. Cells microinjected with the HpaII/PstI-A DNA fragment (0.735-0.27) incorporated thymidine with the same efficiency as cells injected with full size DNA. DNA sequences between 0.655-0.375 are not sufficient to induce cell DNA synthesis. Cells microinjected with the HpaII/HpaI-B fragment (0.735-0.375) stained positive for T-antigen but did not incorporate thymidine (9).

ii. HELPER FUNCTION OF SV40 FOR ADENOVIRUS IN MONKEY CELLS AND U-ANTIGENICITY

Monkey cells (e.g. TC7, CV1) are non permissive for adenovirus (17). TC7 cells infected with adenovirus type 2 (50 PFU/cell) syn-

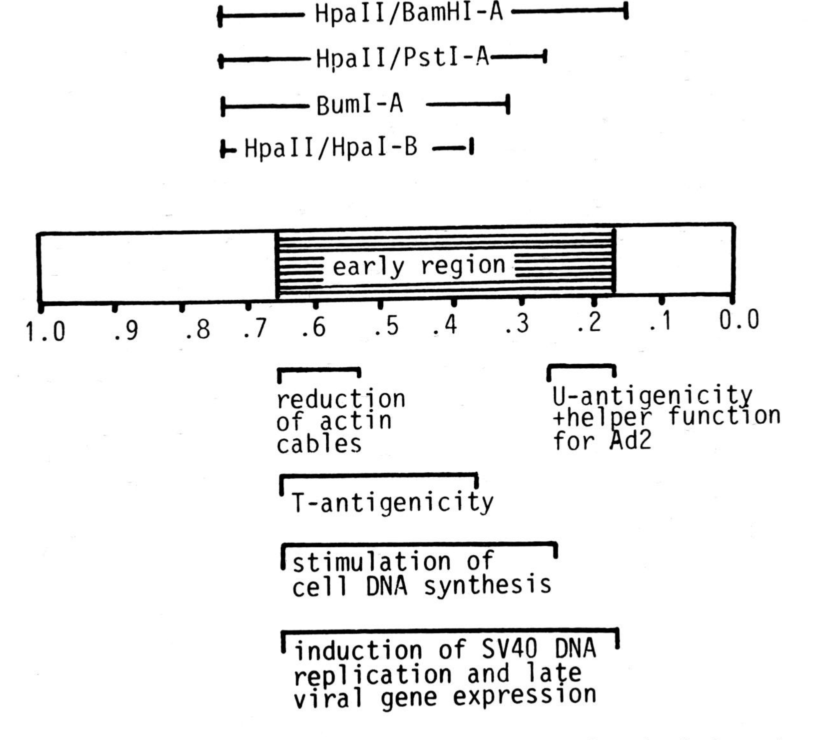

FIG.1. Assignment of SV40 DNA fragments and early viral functions to the physical map of the SV40 genome.

thesize hexon but not fiber protein (both are virus capsid proteins). Following coinfection with SV40 (10PFU/cell), fiber protein synthesis and virus maturation take place. This SV40 mediated effect is a large T-antigen function, since microinjected purified large T-antigen (D2-protein, SV80 T-antigen) allows fiber formation and virus multiplication as SV40 virus. Experimental evidence that only the 50 C-terminal amino acids of the large T-antigen are involved in the helper function was obtained by microinjection of the SV40 DNA fragments and of a 23k protein. The 23k protein is a hybrid protein partly coded by the adeno fiber gene and partly by the early SV40 gene. The SV40 portion of this protein are 50 amino acids of the C-terminal part of large T-antigen. This fusion protein too exerts the helper function (16).

Intranuclear U-antigen synthesis is a further event associated with early SV40 gene expression (17). This antigenicity is part of the large T-antigen since anti-U serum precipitates purified T-antigen. SV40 DNA fragments lacking sequences downstream map position 0.27 do not exert helper function for adenovirus (9). These results implicate that U-antigenicity and helper function are closely related to the C-terminus of the large T-antigen (Fig.1).

iii. INDUCTION OF SV40 DNA REPLICATION

SV40 DNA replication requires efficient synthesis of T-antigen in terms of quality and quantity (4,5). SV40 capsid protein synthesis (V-antigen) can be used as an indicator for viral DNA replication since expression of the late genome part occurs only after the onset of SV40 DNA replication (17). In our studies complementation (V-antigen synthesis) of early temperature-sensitive mutants (tsA) at the non permissive temperature of 41.5°C was used as a test system for viral DNA replication. So far, complementation of tsA7 or tsA58 virus with wild-type early DNA fragments was only obtained with the HpaII/BamHI-A fragment. The HpaII/PstI-A fragment which induced cell DNA synthesis failed to stimulate viral DNA replication (9).

Induction of viral DNA replication depends not only on the quality but also on the quantity of T-antigen molecules synthesized per cell. The correlation between the amount of T-antigen molecules present present per cell and the onset of viral DNA replication becomes obvious through double-staining experiments using fluoresceine conjugated anti-T and rhodamine-B conjugated anti-V sera (5). The time course of intranuclear T-antigen accumulation in SV40 infected CV1 cells is shown in Figure 2. In these experiments intranuclear T-antigen concentration was measured in arbitrary units (AU) with a microscope fluorescence photometer. In order to test whether the onset of intranuclear V-antigen accumulation can be correlated to a threshold amount of T-antigen, T-antigen concentration in randomly chosen V-antigen positive cells was measured at different hours after infection. Independent of the multiplicity of infection (100, 10 or 1 PFU/cell), the minimal concentration of T-antigen in V-antigen positive cells was always 28 AU (5).

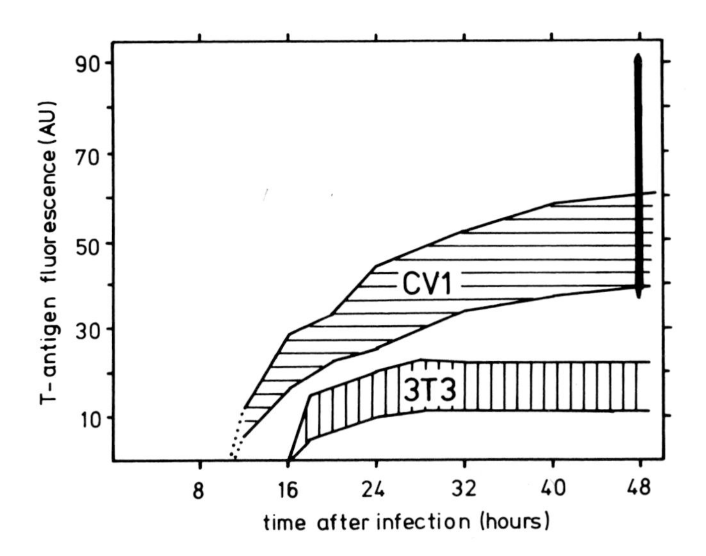

FIG.2. Time course of T-antigen accumulation in CV1 monkey and 3T3 mouse cells. The black bar represents T-antigen concentration in 3T3 cells 48 hours after microinjection with 2,000-4,000 SV40 DNA molecules.

To learn how many T-antigen molecules are equivalent to 28 AU, purified T-antigen was microinjected at different concentrations into the nuclei of CV1 cells. Recipient cells were fixed and stained under the same conditions as for the above experiments. Using the direct immunofluorescence technique, the AU measured are linearly correlated to the amount of T-antigen molecules transferred per nucleus (Fig.3). The threshold concentration of 28 AU is equivalent to $1\text{-}2 \times 10^6$ T-antigen molecules. This high amount of T-antigen molecules required for the onset of viral DNA replication implicates that their function is stoichiometric and not catalytic.

Late SV40 gene expression in non permissive mouse cells

Mouse cells, infected by the conventional virus adsorption method, are non permissive for SV40. 3T3 cells (a permanent mouse cell line) infected with 500 PFU/cell synthesize T-antigen but viral DNA replication or capsid protein synthesis cannot be demonstrated (5,17). The T-antigen concentration of the infected 3T3 cells is always lower than that in permissive monkey cells (Fig. 2). However mouse cells become permissive following microinjection of a high number of SV40 DNA form I molecules. The correlation between the multiplicity of DNA injection and the number of cells synthesizing V-antigen is summarized in Table 2. Table 3 shows the virus yield at different hours after SV40 DNA injection in CV1

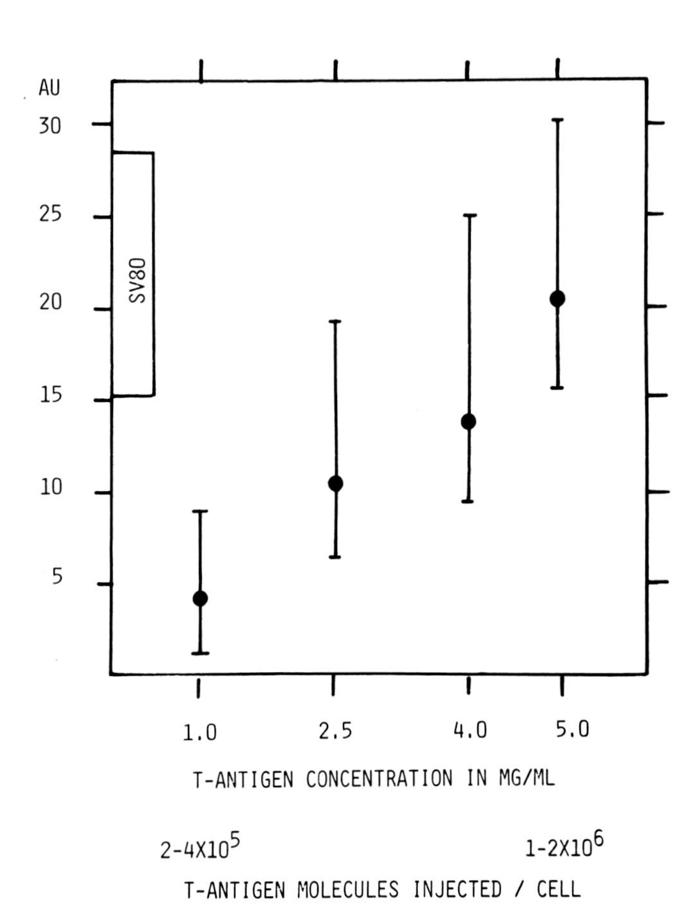

FIG.3. Correlation between AU and amount of T-antigen molecules. In general, identically stained SV80 cells served as a biological standard.

or 3T3 cells. Late SV40 gene expression is also demonstrable after microinjection of early SV40 RNA together with a small number of SV40 DNA molecules in 3T3 cells. Generally, mouse cells injected with SV40 DNA (2,000-4,000 molecules/cell) synthesized significantly higher amounts of T-antigen than cells infected by conventional virus adsorption with 500 PFU/cell. The threshold concentration of T-antigen in V-antigen positive 3T3 cells was found to be about 52 AU (5).

It remains to be determined why SV40 DNA (or virus) injected but not infected 3T3 cells gain the T-antigen concentration requi-

TABLE 2. SV40 T- and V-antigen formation in 3T3 cells microinjected with SV40 nucleic acids[a]

concentration of material injected		no. of injected DNA molecules per cell	% of injected cells positive for	
SV40 cRNA (μg/ml)	SV40 DNA I (μg/ml)		T-antigen	V-antigen
0	1,000	2,000-4,000	100	38
0	100	200- 400	100	4
0	10	20- 40	100	0
500	0	0	72	0
500	10	20- 40	100	50

[a]Cells were fixed and stained 48 or 24 (cRNA experiments) hours after injection.

TABLE 3.

Production of SV40 in 3T3 and TC7 cells microinjected with 2000-4000 DNAI molecules				
	3T3 cells			TC7
hours post injection	cell number injected	total plaques	plaques/ injected cell	plaques/ injected cell
24	300	0	0	0
48	300	0	0	2×10^2
72	100	5×10^4	5×10^2	2×10^3
96	100	1×10^5	1×10^3	N.D.

red for late viral gene expression. SV40 T-antigen injection experiments indicate that this is not a question of higher T-antigen turnover rate in 3T3 cells, as shown in Figure 4.

iv. REGULATION OF LATE VIRAL GENE EXPRESSION

Using tsA mutants of SV40, Tegtmeyer and coworkers demonstrated that viral DNA replication is essential for efficient late viral gene expression (1). We investigated two of the prominent consequences of DNA replication as possible regulatory steps for capsid protein synthesis:
- an increase in the number of SV40 DNA molecules;
- conformational changes of the viral DNA molecules.
For this high amounts of SV40 DNA I molecules or low numbers of

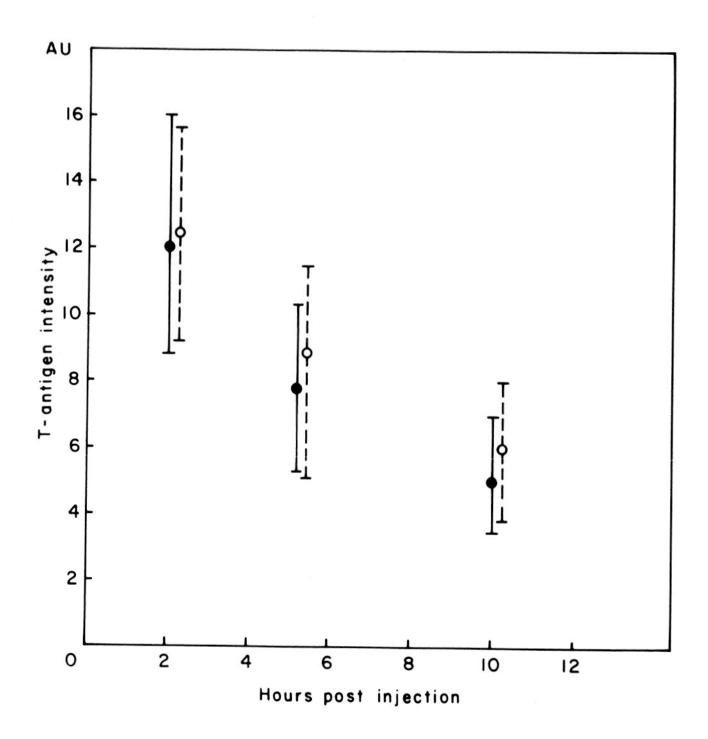

FIG.4. Turnover of T-antigen in CV1 and 3T3 cells microinjected with SV40 T-antigen. CV1----; 3T3 ——.

replicative intermediate DNA (RI-DNA), nicked SV40 DNA (DNA II), or partially denatured DNA I ($DNAI_{den}$) were microinjected into cells at conditions excluding viral DNA replication. Three different methods were used to prevent SV40 DNA replication:

- chemical inhibitors (araC, FUdR);
- multinucleated rat myotubes (postmitotic cells) as recipients;
- incubation at the nonpermissive temperature for cells injected with tsA DNA.

As summarized in Table 4, capsid protein synthesis was not demonstrable at restricted conditions following microinjection of 2,000-4,000 SV40 DNA I molecules but T-antigen synthesis proceeded unaffected. In contrast, microinjection of 200-400 RI-DNA molecules induced in about 30% of the recipient cells intranuclear V-antigen formation. However, this number indicates that the RI-DNA per se cannot induce capsid protein synthesis, since at permissive conditions injection of 1-2 DNA molecules causes T- and V-antigen formation. More likely, a small proportion of the molecules present in the RI-DNA preparation is used for late gene expression. Since randomly nicked SV40 DNA II partially denatured DNA molecu-

TABLE 4. SV40 gene expression at conditions restrictive for viral DNA replication in cells microinjected with various forms of SV40 DNA[a]

Type of SV40 DNA injected	No. of SV40 DNA molecules transferred per cell	% of antigen-positive cells							
		TC7 cells and inhibitors				rat myotubes		TC7 cells at 41.5°C	
		araC		FUdR					
		T	V	T	V	T	V	T	V
DNA I	2,000-4,000	99	0	99	0	80	0	99	80
RI-DNA	200- 400	99	32	99	30	80	30	ND	ND
DNA II	200- 400	90	2	ND	ND	ND	ND	ND	ND
DNA I_{den}	200- 400	95	16	95	16	ND	ND	ND	ND
tsA7 DNA I	2,000-4,000	99	0	ND	ND	ND	ND	99	0
tsA7 RI-DNA	200- 400	ND	ND	ND	ND	ND	ND	99	16

[a]SV40 DNA was extracted from plaque purified virus and further purified by CsCl/ethidium bromide equilibrium centrifugation, yielding SV40 DNA I and DNA II. Form II contained on the average one random nick per molecule. Replicative intermediate DNA (RI-DNA) was isolated 32 hours after infection of TC7 cells with either wt777 or tsA7 virus. Purification of this DNA included digestions with RNase, RNase H, and Pronase, CsCl/ethidium bromide equilibrium centrifugation and either velocity centrifugation through sucrose gradients or chromatography on benzoylated-naphthoylated DEAE-cellulose. Neutral denatured SV40 DNA I_{den} was obtained by a 3 min exposure of DNA I to pH13.2, followed by neutralisation. Cells were fixed and stained 24 hours after microinjection. ND, not done.

les also induced V-antigen synthesis we may conclude that introduction of single-stranded regions or nicks at specific sites initiates late viral gene expression. Generation of these features is enhanced during viral DNA replication (4).

v. LOSS OF CYTOPLASMIC ACTIN CABLE STRUCTURE

The loss of the cytoplasmic actin cable structure is a common feature of many transformed cells (18). In order to test which part of the early SV40 genome region is required for this change, SV40 DNA I, DNA fragment HpaII/HpaI-B or purified large T-antigen were microinjected into rat cells (rat-1, REF). Reduction of actin cables was observed after microinjection of the HpaII/HpaI-B fragment. About 80% of the T-antigen positive cells became actin cable negative after microinjection. The same proportion of actin negative cells was obtained after microinjection of SV40 DNA I. In contrast, large T-antigen (D2-protein, SV80 T-antigen) did not reduce this structure above background (6).

This observation indicates that small t-antigen causes the change of the cytoskeleton. However the HpaII/HpaI-B fragment contains not only the coding region for the small t-antigen but also additional sequences for the large T-antigen (a 30k polypeptide). Therefore it may be that this 30k protein, related to large T-antigen acts coordinately with the small t-antigen.

vi. INDUCTION AND MAINTENANCE OF CELL TRANSFORMATION

The ultimate goal of the fragment injection experiments is to provide some clue of the early SV40 specific functions involved in the process of cell transformation. Experiments with tsA virus transformed cell lines revealed that continuous synthesis of large T-antigen is required for the maintenance of the transformed state (10,15).

So far, cell lines permanently positive for T-antigen were obtained only from permissive monkey cells and non permissive rat 1 cells microinjected with DNA fragments containing the entire early information. These cells are maximally transformed (8). Cells microinjected with the HpaII/PstI-A fragment exhibited T-antigen synthesis only for 100-120 hours after injection (Figure 5). This period of time is too short to test anchorage independent growth in soft agar or methylcellulose, which is the most reliable in vitro test for tumorigenicity.

If the stabilisation of viral information (for example via enhanced integration of the viral DNA into the host DNA) transiently requires large T-antigen, permanently tumor antigen positive cell lines should be obtained by coinjection of the smaller early SV40 DNA fragments together with either early SV40 cRNA or purified large T-antigen.

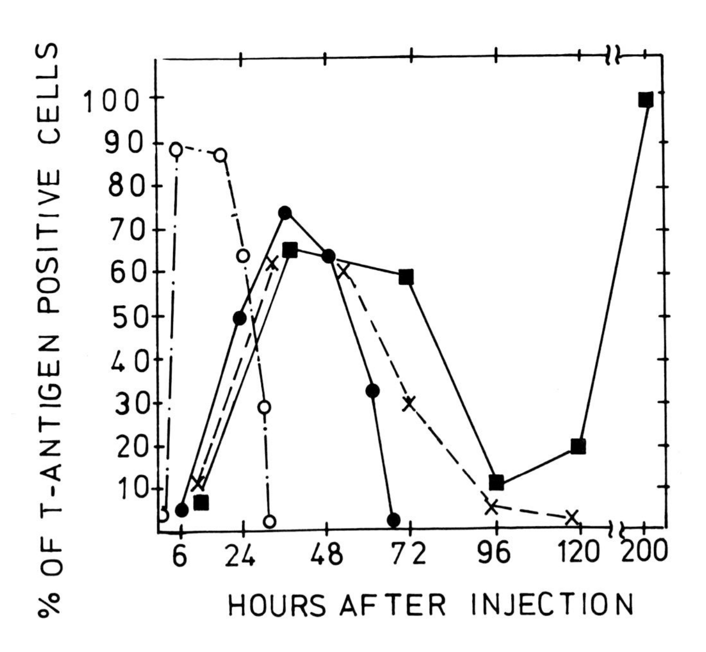

FIG.5. Time course of T-antigen synthesis in rat 1 cells microinjected with: cRNA o-·-·-o; T-antigen •——•; HpaII/PstI-A fragment X----X; HpaII/BamHI-A fragment □——□ .

ACKNOWLEDGMENTS

We are grateful to Eva Guhl for skillful technical assistance. This work was supported by the Deutsche Forschungsgemeinschaft (Gr 599/2, Gr 384/7).

REFERENCES

1. Cowan,K.,Tegtmeyer,P.,and Anthony,D.D.(1973):Proc.Natl.Acad.Sci. USA, 70:1927-1930.
2. Crawford,L.V.,Cole,C.N.,Smith,A.E.,Tegtmeyer,P.,Rundell,K.,and Berg,P.(1978):Proc.Natl.Acad.Sci.USA,75:117-121.
3. Graessmann,A.,Graessmann,M.,Hoffmann,E.,Niebel,J.,Brandner,G., and Mueller,N.(1974):FEBS Lett.,39:249-251.
4. Graessmann,A.,Graessmann,M.,andMueller,C.(1977):Proc.Natl.Acad. Sci.USA,74:4831-4834.
5. Graessmann,A.,Graessmann,M.,Guhl,E.,and Mueller,C.(1978):J.Cell Biol.,77:R1-R8.
6. Graessmann,A.,Graessmann,M.,Tjian,R.,and Topp,W.(1979):J.Virol., in press.

7. Graessmann,M.,and Graessmann,A.(1976):Proc.Natl.Acad.Sci.USA, 73:366-370.
8. Graessmann,M.,Graessmann,A.,and Mueller,C.(1979):Cold Spring Harbor Symp.Quant.Biol.,44: in press.
9. Mueller,C.,Graessmann,M.,and Graessmann,A.(1978):Cell,15:579-585.
10. Osborn,M.,and Weber,K.(1975):J.Virol.,15:636-644.
11. Prives,C.,Gilboa,E.,Revel,M.,andWinocour,E.(1977):Proc.Natl. Acad.Sci.USA,74:457-461.
12. Reddy,V.B.,Thimmappaya,B.,Dhar,R.,Subramanian,K.N.,Zain,B.S., Pan,J.,Gosh,P.K.,Celma,M.L.,and Weissman,S.M.(1978):Science, 200:494-502.
13. Risser,R.,and Pollack,R.(1974):Virology,59:477-489.
14. Tegtmeyer,P.(1972):J.Virol.,10:591-598.
15. Tegtmeyer,P.(1975):J.Virol.,15:613-618.
16. Tjian,R.,Fey,G.,and Graessmann,A.(1978):Proc.Natl.Acad.Sci.USA, 75:1279-1283.
17. Tooze,J.,editor(1973):The Molecular Biology of Tumor Viruses. Cold Spring Harbor Laboratory, Cold Spring Harbor, New York.
18. Topp,W.C.,Rifkin,D.,Graessmann,A.,Chang,C.M.,and Sleigh,M.J. (1979):In:Hormones and Cell Culture,edited by G.Sato,and R. Ross.Cold Spring Harbor Laboratory,Cold Spring Harbor,New York.
19. Weil,R.,Salomon,C.,May,E.,and May,P.(1974):Cold Spring Harbor Symp.Quant.Biol.,39:381-395.

Control Mechanisms in Animal Cells,
edited by L. Jimenez de Asua et al.
Raven Press, New York © 1980.

Specific Increase in the Phosporylation of a Protein in Mitogen-Stimulated 3T3 Cells Compared with SV40-Transformed 3T3 Cells: Comparison of *In Vivo* and *In Vitro* Assays

Marit Nilsen-Hamilton and Richard T. Hamilton

Cell Biology Laboratory, Salk Institute for Biological Studies, San Diego, California 92138, U.S.A.

The phosphorylation of a 33,000 dalton protein (33K protein), tentatively identified as the ribosomal protein S6, is increased within five minutes after the stimulation of ^{32}P-labeled 3T3 cells with serum or fibroblast growth factor from brain or pituitary. In ^{32}P-labeled SV40-transformed 3T3 cells the phosphorylation of this protein is not increased by serum. We have identified an endogenous phosphorylation in microsomes from 3T3 cells that is stimulated by cyclic AMP and high concentrations of cyclic GMP. This activity is not present in microsomes from SV40-transformed 3T3 cells. We are investigating the possibility that this cyclic nucleotide-dependent phosphorylation is responsible for the increased incorporation of ^{32}P into the 33K protein in mitogen-stimulated 3T3 cells.

Diverse compounds such as plant lectins, prostaglandins, and peptides such as epidermal growth factor and fibroblast growth factor (FGF) are classified as "mitogens" because of their common effect on DNA synthesis and cell division. As well as initiating DNA synthesis, these mitogens initiate a number of early metabolic events in the cell such as changes in transport of ions (10,11,24, 26,34) and nutrients (7,11,15), alterations in phospholipid turnover (6,22), phospholipase activity (8) and phosphorylation (5,20,33). Apparently beginning immediately after addition of the mitogen, these early alterations generally have two phases, an early increase, peaking within the first half hour, followed by a decline or a plateau and then a subsequent steady rise. The initial increase in these activities is independent of protein synthesis (10,12). The second steady rise is regulated at the level of translation (10,12). DNA synthesis does not begin to increase until eight to fourteen hours after addition of the mitogen, depending upon the cell type. The metabolic pathway leading from the early events that occur directly after mitogen addition to DNA synthesis and cell division is unknown. So far there is no evidence that these early events do lead to DNA synthesis and are not branch pathways diverging from the initial stimulatory signal. In fact, some alterations in transport have been shown to be unnecessary for stimulation of DNA synthesis (2,18). It is also possible that there is no main pathway to DNA synthesis and that each mitogen stimulates growth via an independent pathway. Even so, it is clear that eventually, if only at the level of DNA synthesis, these pathways must converge. The point of convergence is probably before DNA synthesis, as no mitogen has been shown capable of stimulating a cell to proceed from G_1 into the S phase of the cell cycle in the absence of protein synthesis. Most likely, the synthesis of specific proteins is important to stimulation of DNA synthesis.

Alterations in protein turnover - both increased synthesis (25) and decreased degradation (36) - occur early after mitogen addition. The effect of intracellular magnesium ion concentration on protein synthesis is paralleled by similar effects on subsequent DNA synthesis (27). Specific proteins are made during this period, some found within the cell (9,23,25), and others that are secreted into the medium by mitogen-stimulated cells (14,21). The appearance of newly synthesized, specific protein occurs later after mitogen stimulation than the early changes discussed above.

The mechanism by which protein synthesis is regulated during the shift from the quiescent to the growing state is unknown, but it is known that quiescent 3T3 cells contain a large number of ribonucleoprotein particles and free ribosomal subunits (16,29,30) that form polysomes immediately after mitogen stimulation (29,35). These polysomes would be expected to participate in the formation of those proteins specifically induced by mitogenic stimuli.

Our ultimate goal is to define the metabolic events that, when activated by mitogens, lead to the initiation of DNA synthesis in mouse fibroblasts. As many alterations occur in 3T3 cells after the addition of mitogens, it is necessary to develop criteria that will distinguish those alterations important to the regulation of DNA synthesis. There is no identifying characteristic that can be assigned to those alterations necessary for the stimulation of DNA synthesis by mitogens. It is therefore necessary to determine whether the effect of a mitogen on a particular metabolic event, in this case phosphorylation, can be correlated with DNA synthesis in a variety of cells under different environmental conditions. In addition, if this event, which is sensitive to the presence of mitogens, is involved in the mechanisms allowing 3T3 cells to regulate the length of time they remain in the G_1 (G_0) phase of the cell cycle, then cells, such as SV40-transformed 3T3 (SV3T3) cells, that are unable to regulate the length of G_1 (G_0) independently of the remainder of the cell cycle, might lack or be altered in this particular mitogen-sensitive event.

METHODS

3T3 and SV3T3 cells were grown in Dulbecco-Vogt modified Eagle's medium (DME) in plastic, tissue culture dishes. The cells were kept in a 37°C incubator containing a constant, water-saturated atmosphere of 15% CO_2, 85% air. Cell stocks were periodically tested for mycoplasma contamination by autoradiography.

Procedures for labeling monolayer cultures with ^{32}P are described in the legend to the appropriate figure.

Phosphorylation *in vitro* was achieved by incubating microsomal samples for two minutes at 37°C in 10 mM $MgCl_2$, and 10 mM Hepes, pH 7.3, followed by the addition of 10 μM γ-[^{32}P]-ATP (3-6 μC/mmole) and a further incubation of one minute. The phosphorylation reaction was stopped by adding an equal volume of gel buffer, 4% (w/v) SDS, 20% glycerol, 4% (v/v) β-mercaptoethanol, 118 mM Tris-PO_4, pH 6.9, 0.008% bromophenol blue). Samples were heated at 98°C for two minutes before resolution by SDS polyacrylamide gel electrophoresis on 7.5-15% gradient gels according to procedures previously described (20). Gels were stained, destained, dried and exposed to Kodak NS5T film to determine the position and intensity of radiolabeled bands. Once localized, bands were cut from the gels and counted, or the films were scanned with a densitometer to determine the amount of ^{32}P incorporated per band.

Fibroblast growth factor (FGF) was isolated from bovine brain according to the method of Gospodarowicz *et al.* (3).

RESULTS

1. Specific Increase in Phosphorylation Following Mitogen Addition to Quiescent ^{32}P-labeled 3T3 Cells

We find that addition of serum or fibroblast growth factor from pituitary or from brain to ^{32}P-labeled quiescent 3T3 cells stimulates a rapid increase in the phosphorylation of a protein with a molecular weight of 33,000 daltons (20; Fig. 1). Although this protein is present in phosphorylated form in all subcellular fractions, the stimulated phosphorylation is found mainly in the particulate fraction (100,000 x g 60 min pellet). In several cell types a mitogenic stimulus is rapidly followed by an increase in the phosphorylation of a ribosomal protein designated S6 (4,5,35). S6 is the major phosphorylated component of the 40S ribosomal subunit. The addition of serum to quiescent 3T3 cells results in a rapid increase in the amount of TCA-precipitable ^{32}P associated with particles that travel in the position of polyribosomes on sucrose gradients (Fig. 2). Experiments with cells labeled for 16 hours with [^{3}H]uridine show that the peak of ^{32}P incorporation that occurs after serum addition sediments further into the sucrose gradient than the 80S monoribosomes. When sucrose gradient fractions are analyzed by SDS polyacrylamide gel electrophoresis, the 33K protein is found in the polyribosome region of the gradient, indicating that this protein is the ribosomal protein S6. We are currently analyzing this protein by two-dimensional electrophoresis to unambiguously identify it as S6. Recently, another group has reported an increase in phosphorylation of S6 after the addition of serum to quiescent 3T3 cells (35).

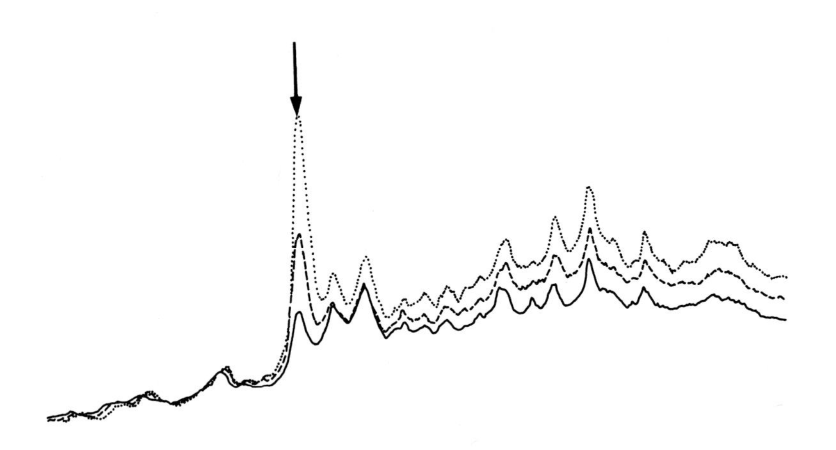

FIG. 1. *The effect of FGF and serum on phosphorylation of a 33,000 dalton protein in quiescent 3T3 cells.* Quiescent 3T3 cells were labeled with ^{32}P as described previously (20). The ^{32}P-labeling medium was removed and replaced by the medium in which the cells had been grown with the addition of either 0.005% BSA (——), 50 ng/ml brain FGF plus 0.005% BSA (---), or 10% serum (···). Twenty minutes later, the medium was removed and the cells scraped into buffer (0.15 M NaCl, 0.05 M NaF, 0.01 M KPi, pH 7.2) pelleted, and microsomal preparations isolated as previously described (20). The membranes were resuspended to equal protein concentrations and analyzed by SDS polyacrylamide gel electrophoresis. Densitometric tracings of an autoradiogram of the gel are shown. The position of the 33K protein is indicated by the arrow. The top of the gel is to the right of the figure.

2. Cyclic Nucleotide Dependent Phosphorylation of the 33K Protein

The amount of ^{32}P associated with the 33K protein increases rapidly after the addition of mitogen (20). As with S6 phosphorylation in other systems the increased phosphorylation can be observed within five minutes of mitogen addition (5,35). There are several mechanisms by which mitogens could specifically stimulate phosphorylation of the 33K protein. Mitogens

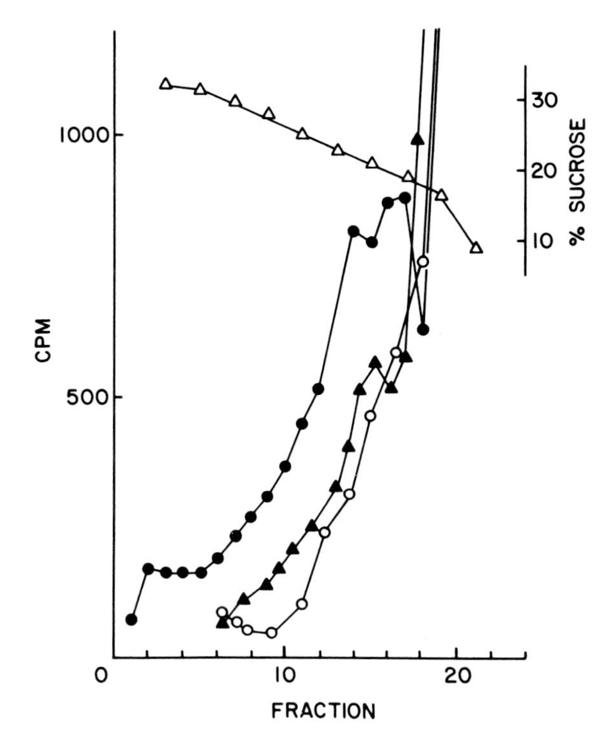

FIG. 2. *Fractionation on sucrose gradients of the post nuclear supernatants of ^{32}P-labeled 3T3 cells after stimulation with serum.* 3T3 cells were labeled for two hours in DME lacking Pi but with 330 μC/ml ^{32}P plus 0.2% serum. Serum was added to a final concentration of 10% for the last 10 seconds (O), 30 minutes (▲) or 60 minutes (●) of the ^{32}P-labeling period. The cells were harvested and lysed in 0.5% NP40 according to the method of Singer and Penman (32). The lysate was spun for 10 minutes at 12,000 x g and the supernatant from this spin was spun through a gradient of 15-33% sucrose at 32,000 rpm for 90 minutes in an SW41 rotor. TCA precipitable counts in 50 μl of each fraction were determined in the presence of 2 mM ATP, 9 mM KPi, 0.14% Cohn fraction V proteins, and 10% TCA.

could 1) alter the conformation of the protein making it a better kinase substrate or poorer phosphatase substrate; 2) alter the subcellular localization of the protein such that it is now available to protein kinases or inaccessible to phosphatases; 3) alter the activities of specific protein kinase(s) or phosphatase(s) by, for example, regulating the levels of cofactors or endogenous regulatory molecules. We began by investigating the third mechanism.

Using microsomal preparations, we investigated the properties of the endogenous enzyme(s) that phosphorylate the 33K protein. We found that the phosphorylation is stimulated fairly well by cyclic AMP and poorly by cyclic GMP (Fig. 3). The stimulation is selective for the 33K protein and a 15,000 dalton protein. The 15,000 dalton protein is not phosphorylated to any appreciable degree *in vivo*, although *in vitro* it is a very good substrate for the cyclic nucleotide-dependent enzymes.

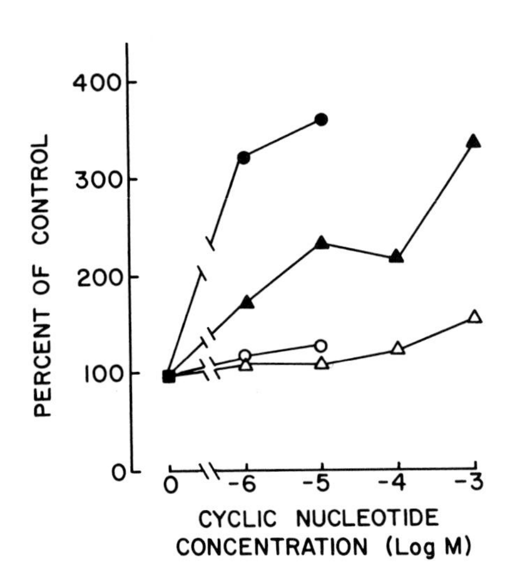

FIG. 3. *Cyclic nucleotide dependent phosphorylation of the 33K protein in microsomes from quiescent 3T3 cells.* Microsomes were prepared and phosphorylated as described in *Methods*. Cyclic nucleotides were present during both the two-minute preincubation and the one-minute phosphorylation periods. Cyclic AMP (●,O); Cyclic GMP (▲,△). Phosphorylated samples were resolved by electrophoresis and autoradiograms of the dried gels were scanned with a densitometer. The relative area under the peak is plotted against cyclic nucleotide concentration for the 33K protein (●,▲) and a protein of about 76,000 daltons (O,△). The 76,000 dalton protein is representative of most of the proteins in the microsomal fraction, whose phosphorylation are little affected by the presence of cyclic nucleotides. The error in estimating the areas was ± 10% (S.D.).

3. Cation Requirements for Phosphorylation of the 33K Protein

Phosphorylation of the 33K protein depends on the presence of magnesium - requiring 10 to 20 mM $MgCl_2$ for maximal activity. We find that most endogenous phosphorylation reactions in microsomal preparations from 3T3 cells have a similar high Mg^{++} requirement. Some endogenous phosphorylations are particularly sensitive to inhibition by calcium. This is not so for phosphorylation of the 33K protein. Under our conditions, we do not observe any calcium-stimulated phosphorylations in the microsomal fraction from 3T3 cells, although we do observe a calcium-dependent phosphorylation in the cytoplasmic fraction of these cells (19).

4. Comparison of the Endogenous Phosphorylation of the 33K Protein in 3T3 and SV3T3 Microsomes

The cyclic nucleotide dependent, phosphorylating activity specific for the 33K protein is present in microsomes from both growing and quiescent 3T3 cells. The sensitivity to cyclic nucleotide stimulation appears to be the same in both growth states. SV3T3 cells do not possess the endogenous activity that phosphorylates the 33K protein (Fig. 4). That they possess the 33K protein can be shown by the fact that an exogenously added cyclic AMP-dependent protein kinase or its catalytic subunit will phosphorylate the protein (Fig. 4) to an equal extent in both SV3T3 and 3T3 microsomal preparations (Fig. 5). Mixing experiments show that in SV3T3 cells the cyclic nucleotide-dependent activity cannot be found in other subcellular fractions.

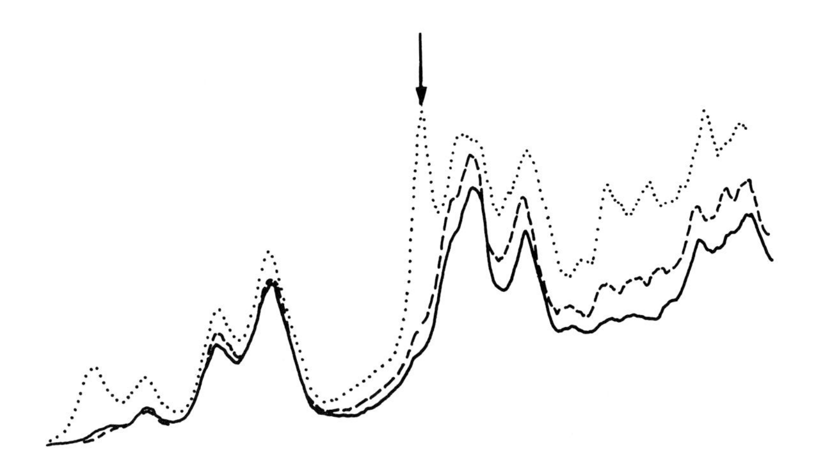

FIG. 4. *Phosphorylation of the 33K protein in microsomes from SV40-transformed 3T3 cells.* Microsomes from SV3T3 cells were phosphorylated, as described in *Methods*, in the presence of 10^{-5} M cyclic AMP (---), the catalytic subunit of bovine lung cyclic AMP dependent protein kinase (···), or with no addition (——). The position of the 33K protein is indicated by the arrow. The top of the gel is to the right of the figure.

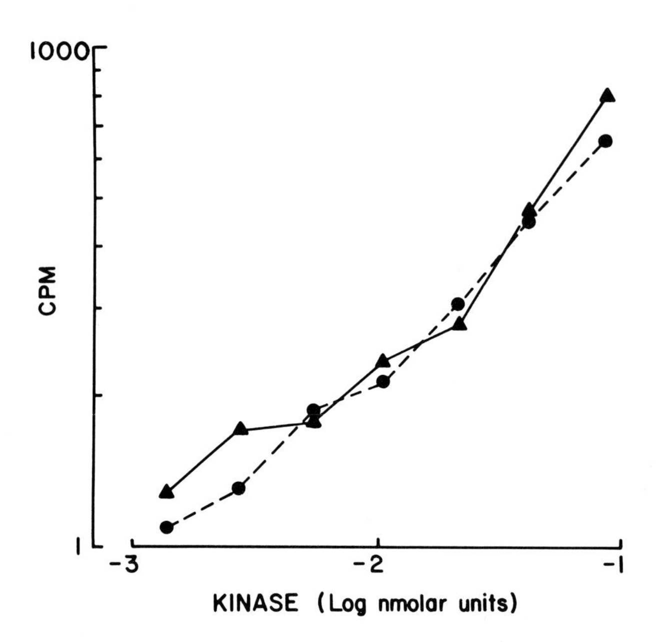

FIG. 5. *Phosphorylation of microsomes from SV3T3 and 3T3 cells by cyclic AMP dependent protein kinase.* As described in *Methods*, microsomes from 3T3 (O) and SV3T3 (Δ) cells were phosphorylated in the presence of 10^{-6} M cyclic AMP by a cyclic AMP-dependent protein kinase isolated from bovine heart according to the method of Rubin *et al.* (28). Phosphorylated samples were resolved by electrophoresis and the 33K protein was identified on autoradiograms. The band was cut from the gel and the radioactivity per band measured in a scintillation counter. ^{32}P cpm per band are plotted against log nmolar units of enzyme activity. The protein kinase was assayed according to Rubin *et al.* (28) and one nmolar unit of enzyme activity represents the amount of enzyme that will incorporate one nanomole of ^{32}P into one μg of protamine sulphate per minute at 30°C. The concentration of microsomal protein per assay was 0.6 mg/ml.

DISCUSSION

Stimulation of phosphorylation of the 33K protein, which we have tentatively identified as S6, occurs rapidly after addition of mitogen. It appears to occur simultaneously with other rapid changes that are independent of protein synthesis. To determine how phosphorylation is regulated *in vivo*, we have developed an *in vitro* assay to analyze the factors affecting phosphorylation of the 33K protein. By this means, we identified cyclic nucleotides, particulary cyclic AMP, as stimulatory for phosphorylation of the 33K protein. But published evidence does not support the hypothesis that increased concentrations of intracellular cyclic AMP could stimulate a protein kinase specific for the 33K protein. Serum and FGF have been reported to decrease the intracellular concentration of cyclic AMP (9,31). Although intracellular cyclic GMP has been reported to increase after serum stimulation (31,37), a decrease was reported by others (17). Also, our data suggests that the intracellular cyclic GMP concentrations must reach high, nonphysiological levels before appreciable enhancement of 33K protein phosphorylation is observed. Our results can be reconciled with published evidence, by assuming a local increase in intracellular concentration of cyclic nucleotide(s) increases the kinase activity for the 33K protein, or that phosphorylation of the 33K protein is regulated by a specific cyclic AMP-dependent phosphatase.

We have examined the possibility that divalent cations influence the phosphorylation of the 33K protein. Magnesium is required at concentrations much greater than necessary for formation of the Mg-ATP substrate complex. At 10 to 20 mM $MgCl_2$, incorporation of ^{32}P is maximal for the 33K protein and for most proteins in the microsomal preparations. It has been suggested that alterations in the intracellular concentration of magnesium ions between 15 and 20 mM could regulate protein synthesis and consequently, DNA synthesis (27). If this were so, we expect that, unless the alteration in magnesium concentration was local, mitogens would increase the phosphorylation of all proteins nonspecifically. Although we observe an increase in the phosphorylation of many cellular proteins after addition of mitogens, the increase in phosphorylation of the 33K protein is quantitatively greater than for other proteins (20).

Alterations in calcium ion fluxes have also been suggested as taking part in the initiation of DNA synthesis. We find that, although calcium inhibits the phosphorylation of several microsomal proteins,it does not affect cyclic AMP-dependent phosphorylation of the 33K protein.

In growing and quiescent 3T3 cells, the 33K protein is phosphorylated in a cyclic nucleotide dependent manner and to similar degrees relative to other phosphorylated proteins. We have, as yet, observed no difference in cyclic nucleotide-dependent phosphorylation in microsomes isolated from serum-stimulated cells, compared with the controls. If the specific protein phosphorylation measured *in vitro* is the activity responsible for phosphorylating the 33K protein in the intact cell after mitogen-stimulation, then our results suggest that the alteration in phosphorylation *in vivo* is due to a regulatory event that is destroyed by breaking open the cell. It is possible that FGF induces alterations in intracellular compartmentalization of the kinase, a phosphatase, the 33K protein, or a dissociable inhibitor or activator of protein kinase activity. Alternatively, during preparation of the microsomal fraction, there could be loss of the serum-stimulated activity or a stimulation of the activity in quiescent cells.

SV3T3 cells are unable to arrest their growth in the G_1 (G_0) phase of the cell cycle and also do not possess the cyclic nucleotide-dependent phosphorylation of the 33K protein. It is possible that the regulatory mechanism for cell arrest that is absent in SV3T3 cells involves the cyclic nucleotide-dependent regulation of phosphorylation of the 33K protein. To test this possibility, it is necessary to correlate, in a series of cell lines, the presence of cyclic-nucleotide dependent phosphorylation of the 33K protein observed *in vitro* with the ability of mitogens to influence phosphorylation of the 33K protein *in vivo*. We are doing these experiments now with spontaneously transformed cell lines and benzpyrene-transformed cells. The latter cell lines, although they have the morphological characteristics of transformed cells and grow to high densities in tissue culture, can be arrested in G_1 (G_0) of the cell cycle and respond to serum with increased growth. The results of these experiments will answer two questions: The first is whether the specific cyclic nucleotide-dependent phosphorylation observed *in vitro* is the enzyme responsible for mitogen-stimulated phosphorylation of the 33K protein *in vivo*. The second is whether phosphorylation of the 33K protein is stimulated in all cell lines and under all conditions that lead to stimulation of DNA synthesis.

ACKNOWLEDGMENT

This investigation was supported by Grant Number 1 RO1 CA24395, awarded by the National Cancer Institute, DHEW.

REFERENCES

1. Allen, W. R., Nilsen-Hamilton, M., Hamilton, R. T., and Gospodarowicz, D. (1979): J. Cell. Physiol., 98:491-502.
2. Barsh, G. S., Greenberg, D. B., and Cunningham, D. D. (1977): J. Cell. Physiol., 92:115-128.
3. Gospodarowicz, D., Bialecki, H., and Greenburg, G. (1978): J. Biol. Chem., 253:3736-3743.
4. Gressner, A. M., and Wool, I. G. (1974): J. Biol. Chem., 249: 6917-6925.
5. Haselbacher, G., Humbel, R., and Thomas, G. (1979): FEBS Lett., 100:185-190.
6. Hoffmann, R., Ristow, H.-J., Pachowsky, H., and Frank, W. (1974): Eur. J. Biochem., 49:317-324.
7. Hollenberg, M. D., and Cuatrecasas, P. (1975): J. Biol. Chem., 250:3845-3853.
8. Hong, S. L., and Levine, L. (1976): Proc. Natl. Acad. Sci. USA, 73:1730-1734.
9. Jimenez de Asua, L., Richmond, K. M. V., Otto, A. M., Kubler, A. M., O'Farrel, M. K., and Rudland, P. S. (1979): In: Hormones and Cell Culture - Cold Spring Harbor Conferences on Cell Proliferation, Vol. 6, pp. 403-424. CSH, New York.

10. Jimenez de Asua, L., and Rozengurt, E. (1974): Nature, 251: 624-626.
11. Jimenez de Asua, L., Rozengurt, E., and Dulbecco, R. (1974): Proc. Natl. Acad. Sci. USA, 71:96-98.
12. Kletzien, R. F., and Perdue, J. F. (1974): J. Biol. Chem., 249:3383.
13. Lastick, S. M., Nielsen, P. J., and McConkey, E. H. (1977): Mol. Gen. Genetics, 152:223-230.
14. Lee, L.-S., and Weinstein, I. B. (1978): Nature, 274:696-697.
15. Lever, J. E., Clingan, D., and Jimenez de Asua, L. (1976): Biochem. Biophys. Res. Commun., 71:136-143.
16. Levine, E. M., Becker, Y., Boone, C. W., and Eagle, H. (1965): Proc. Natl. Acad. Sci. USA, 53:350-356.
17. Miller, Z., Lovelance, E., Gallo, M., and Pastan, I. (1975): Science, 190:1213-1215.
18. Naiditch, W. P., and Cunningham, D. D. (1977): J. Cell. Physiol., 92:319-332.
19. Nilsen-Hamilton, M., Massoglia, S. L., and Hamilton, R. T., in preparation.
20. Nilsen-Hamilton, M., and Hamilton, R. T. (1979): Nature, 279: 444-446.
21. Nilsen-Hamilton, M., Shapiro, J. M., Massoglia, S. L., and Hamilton, R. T. (1979): Cell, in press.
22. Northoff, H., Dörken, B., and Resch, K. (1978): Exp. Cell Res., 113:189-195.
23. O'Farrell, M. K. (1977): Biochem. Soc. Trans., 5:940.
24. Quastel, M. R., and J. G. Kaplan (1970): Exp. Cell Res., 63:230.
25. Riddle, V. G. H., Durrow, R., and Pardee, A. B. (1979): Proc. Natl. Acad. Sci. USA, 76:1298-1302.
26. Rozengurt, E., and Heppel, L. (1975): Proc. Natl. Acad. Sci. USA, 72:4492-4495.
27. Rubin, A. H., Terasaki, M., and Sanui, H. (1979): Proc. Natl. Acad. Sci. USA, 76:3917-3921.
28. Rubin, C. S., Erlichman, J., and Rosen, O. (1972): J. Biol. Chem., 247:36-44.
29. Rudland, P. S. (1974): Proc. Natl. Acad. Sci. USA, 71:750-754.
30. Rudland, P. S., Weil, S., and Hunter, A. R. (1975): J. Mol. Biol., 96:745-766.
31. Seifert, W. E., and Rudland, P. S. (1974): Nature, 248:138-140.
32. Singer, R. H., and Penman, S. (1972): Nature, 240:100-102.
33. Smith, C. J., Wejksnora, P. J., Warner, J. R., Rubin, C. S., and Rosen, O. M. (1979): Proc. Natl. Acad. Sci. USA, 76: 2725-2729.
34. Smith, J. B., and Rozengurt, E. (1978): J. Cell. Physiol., 97:441-450.
35. Thomas, G., Siegmann, M., and Gordon, J. (1979): Proc. Natl. Acad. Sci. USA, 76:3952-3956.
36. Warburton, M. J., and Poole, B. (1977): Proc. Natl. Acad. Sci. USA, 74:2427-2431.
37. Yasuda, H., Hanai, N., Kurata, M., and Yamada, M. (1978): Exp. Cell Res., 114:111-116.

Control Mechanisms in Animal Cells,
edited by L. Jimenez de Asua et al.
Raven Press, New York © 1980.

Intracellular Signals of Proliferation Control: Diadenosine Tetraphosphate (Ap_4A)–A Trigger of DNA Replication

F. Grummt

Max-Planck-Institute for Biochemistry, D-8033 Munich, West Germany

Stimulation of growth in quiescent G_1-arrested cells comprises extra- and intracellular events: After the mitogen-receptor interaction at the outer cell membrane the mitogenic signal has to be transmitted into intracellular metabolic changes which eventually lead to the onset of DNA replication. It is generally accepted that the initiation of DNA synthesis at the G_1/S phase boundary is under the control of cytoplasmic factors (10, 19). Those factors were described either as heat-labile, non-dialysable proteins of high molecular weight (3, 17, 18, 21, 23) or as heat-stable low molecular weight compounds (9, 28). The controversial results demonstrate that the nature of the intracellular molecule(s) involved in triggering DNA replication after mitogenic stimulation remains to be elucidated. Our recent results suggest that diadenosine tetraphosphate, Ap_4A, (structural formula see FIG. 4) is involved in the induction of nuclear DNA replication of quiescent mammalian cells. This purine nucleotide was described by Zamecnik et al. to be a byproduct of the aminoacylation step of protein biosynthesis (30). Ap_4A was found to be distributed in eukaryotic as well as in prokaryotic cells (24, 25, 30). In mammalian cells the concentration of this odd nucleotide fluctuates drastically in response to the proliferation rate, i.e. between 10 nM in resting or slowly growing cells and 1 µM in rapidly proliferating normal or in malignant transformed cells (25). Therefore, Ap_4A could play the role of a signal molecule that accumulates during the G_1 phase of the cell cycle and triggers the onset of DNA synthesis when a critical threshold concentration is reached at the G_1/S phase boundary (25). In order to prove this assumption, we have carried out experiments which indeed revealed that Ap_4A is able to induce DNA replication in G_1-arrested baby hamster kidney (BHK) cells in vitro (12).

RESULTS

G_1-arrested BHK cells synthesize DNA during incubation with Ap_4A

When G_1-arrested BHK cells were rendered permeable by hypotonic shock and then incubated in vitro in the presence of both the four ribonucleoside triphosphates and the four deoxyribonucleoside triphosphates, they incorporate [^{3}H]dTTP into DNA, provided that Ap_4A is present in the reaction mixture (12, 13). This Ap_4A-stimulated DNA synthesis is dose-dependent with regard to the amount of Ap_4A and the number of cells added (12). Only structurally intact Ap_4A is able to induce [^{3}H]dTTP incorporation, neither the putative degradation products of Ap_4A, i.e. ATP, ADP, AMP, adenosine and pyrophosphate nor Ap_4A pretreated with snake venom phosphodiesterase are able to substitute for Ap_4A. The stimulatory effect is very specific for Ap_4A.The structural analogs Ap_2A, Ap_3A, Ap_5A, Ap_6A, Ap_4T and Gp_4G cannot replace Ap_4A in the stimulatory action (13).

In subsequent studies we could demonstrate that the Ap_4A-stimulated DNA synthesis reflects induction of replicative DNA synthesis rather than enhanced repair synthesis by three different sets of experiments (1) by using specific inhibitors of replication, (2) by demonstrating that the Ap_4A-induced DNA synthesis occurs discontinuously, and (3) by scoring replication eyes induced by Ap_4A (12, 13). These results provided evidences that Ap_4A plays indeed a crucial role as a signal molecule triggering DNA replication in quiescent BHK cells.

Ap_4A is a ligand of DNA polymerase α

To approach an understanding of the molecular mode of action of Ap_4A in DNA replication we carried out experiments to characterize the intracellular target of Ap_4A. Since DNA polymerase α was shown to be involved in the replication of nuclear and viral DNA in mammalian cells (2, 4, 5, 16, 22, 27, 29) we investigated whether Ap_4A binds to this enzyme. FIG. 1 shows an equilibrium dialysis experiment which demonstrates the binding of [^{3}H]AP_4A to DNA polymerase α in a dose-dependent manner with regard to the amount of DNA poly-

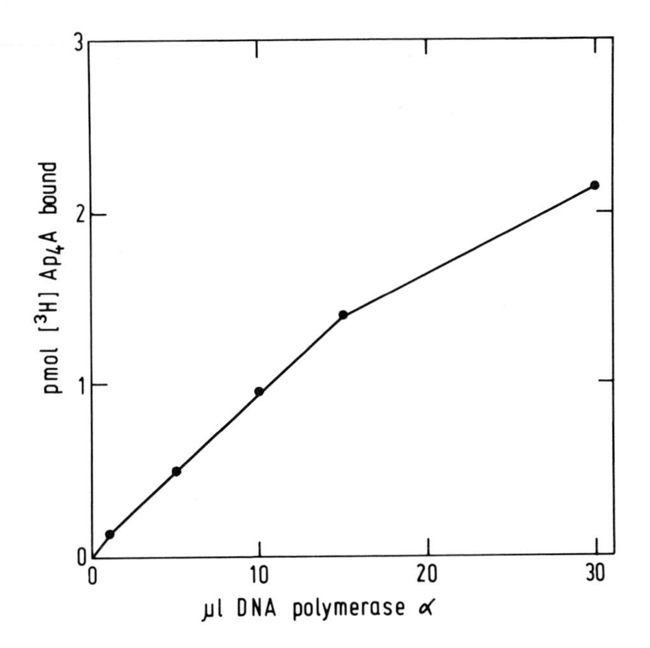

FIG. 1. Dose dependence of ^3H-Ap_4A-binding to DNA-polymerase α.

merase α present in one of the dialysis chambers. Since the DNA polymerase α preparation used in this experiment was not purified to molecular homogeneity, it remained unclear whether DNA polymerase α by itself or a contaminating protein was the Ap_4A-binding component. Conclusive evidence for Ap_4A being a specific ligand of DNA polymerase α were gained by purification of DNA polymerase α to homogeneity and studying both the catalytic and the Ap_4A-binding activity of the enzyme. DNA polymerase α was purified from calf thymus extracts by means of chromatography on phospho-, DEAE-cellulose, gel filtration and electrophoresis in non-denaturing polyacrylamide gels (14). The comparison of catalytic activity and the Ap_4A-binding activity revealed a co-purification of both [^3H]dTTP incorporation activity and the Ap_4A-binding activity during all the isolation procedures applied. DNA polymerase α of apparent homogeneity was isolated by this purification scheme. This DNA polymerase α preparation reveals seven protein bands of different M_r (64.000, 63.000, 62.000, 60.000, 57.000, 55.000 and 52.000) if separated under denaturating conditions in sodium dodecylsulfate polyacrylamide gels (TABLE 1). Thus, DNA polymerase α isolated from calf thymus is of a complex nature and is in this respect similar to DNA polymerase III holoenzyme isolated from E.coli as described by Kornberg and collaborators (20).

TABLE 1. M_r of eu- and prokaryotic DNA polymerases

E.coli DNA polymerase III (holoenzyme)	calf thymus DNA polymerase α
α = 140 kdal	64 kdal
β = 40 "	63 "
γ = 52 "	62 "
δ = 32 "	60 "
ε = 27 "	Ap_4A ----→ 57 "
τ = 83 "	55 "
θ = 2 × 9 "	52 "
Σ = 392 kdal	Σ = 413 kdal (SDS gel)
	404 kdal (gel filtration)

In order to decide which one of the protein constituents of DNA polymerase α represents the Ap_4A-binding subunit we carried out affinity labeling experiments with periodate-oxidized [^{3}H]Ap_4A which demonstrated that predominantly the subunit with M_r 57.000 can be labeled with Ap_4A (14). The covalent binding of oxidized Ap_4A to DNA polymerase α inactivates the enzymes (FIG. 2).

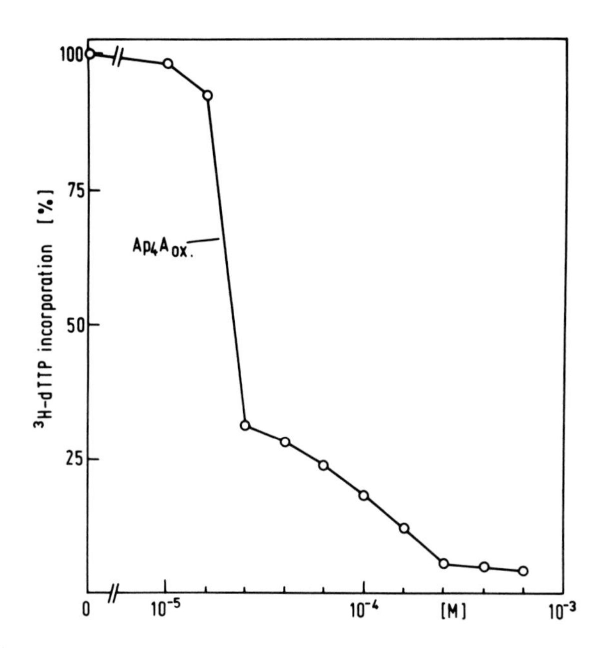

FIG. 2. Inhibition of DNA polymerase α activity by covalently binding of oxidized Ap_4A to the enzyme. Ap_4A was oxidized by preincubation with a 10-fold excess of sodium periodate for 30 min at 0°C.

Ap_4A-binding activity is lost concomitantly with DNA polymerase α in cerebral neurons of rats

If solely DNA polymerase α is the intracellular target of Ap_4A in animal cells the Ap_4A-binding activity should be diminished in those cell types having lost their polymerase α activity. We have studied whether this proves right by using neuronal cells of different developmental stages. Cerebral rat neurones develop from actively proliferating precursor cells at late fetal stages *via* non-deviding immature neurons thereafter. These changes are accompanied by a specific and eventually complete loss of DNA polymerase α correlating with the decline of the *in vivo* rate of mitotic activity (15). In order to find out whether the cellular-Ap_4A-binding capacity declines concomitantly with the DNA polymerase α activity the binding of [^3H]Ap_4A by lysates of rat neuronal cells was analyzed. FIG. 3 demonstrates that neurones from rat embryos at day 5 before birth have a significant Ap_4A-binding activity. The capability to bind Ap_4A decreases sharply at the end of the fetal period resulting eventually in a com-

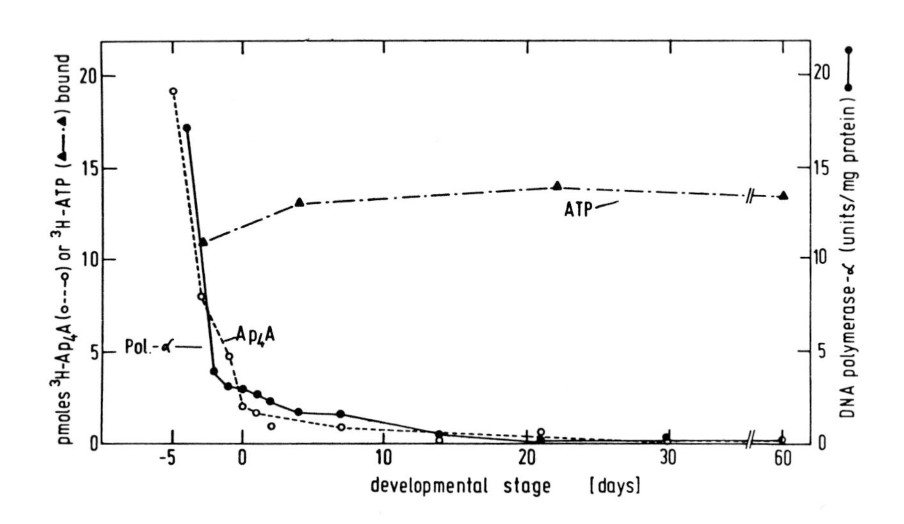

FIG. 3. Loss of Ap_4A-binding activity during development of rat brain neurones. (o), ^3H-Ap_4A-binding activity, (▲), ^3H-ATP-binding activity in neuronal extracts. The extracts were prepared as described elsewhere (14). (●), DNA polymerase α activity in neuronal extracts according to Huebscher (15).

plete loss in the early period after birth. FIG. 3 also shows the decrease of the DNA polymerase α activity during the development of rat neurones. The results clearly show that the capacity of neuronal lysates to bind Ap_4A declines with a similar rate as the activity of the replicating enzyme in neuronal cells. In contrast to the Ap_4A-binding capacity, no loss of ATP-binding activity was observed in rat neuronal cells during the pre- and postnatal development (FIG. 3). Therefore, a specific correlation exists between the level of replicating activity and the Ap_4A-binding capacity in neuronal cells.

Methylene-bis-ADP as antagonist of Ap_4A

Condensation of formaldehyde with adenosine or adenylates results in the formation of the respective methylene-bridged dimers (1, 6, 8) (structural formula see FIG. 4). In those dimers the base moieties appear to be in a "stacking" conformation (7) like that postulated for Ap_4A (26). Therefore, methylene-bis-ADP is a structural analog the special conformation of which is very similar to that of Ap_4A. We synthesized this analog by a method described by Feldman (6) and studied its competition with Ap_4A for the binding site of DNA polymerase α, its effects on DNA synthesis as well as its influence on cell proliferation in tissue cultures.

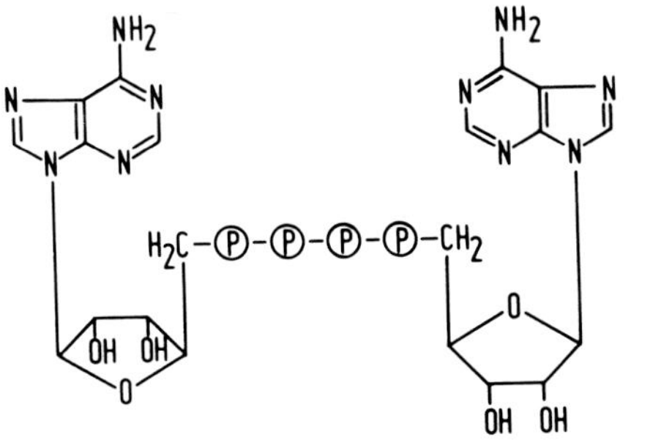

(Ap_4A) Diadenosin 5',5'''-P^1,P^4-tetraphosphat

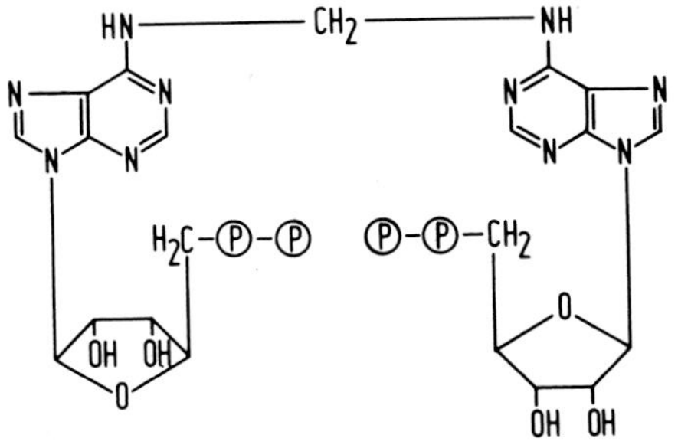

Methylen-bis-ADP (MB-ADP)

FIG. 4. Structural formula of diadenosine tetraphosphate (top) methylene-bis-ADP (bottom).

FIG. 5 shows that methylene-bis-ADP competes at a 1:1 ratio with Ap_4A for its binding site at DNA polymerase α. Methylene-bis-AMP is inactive in this respect,where-

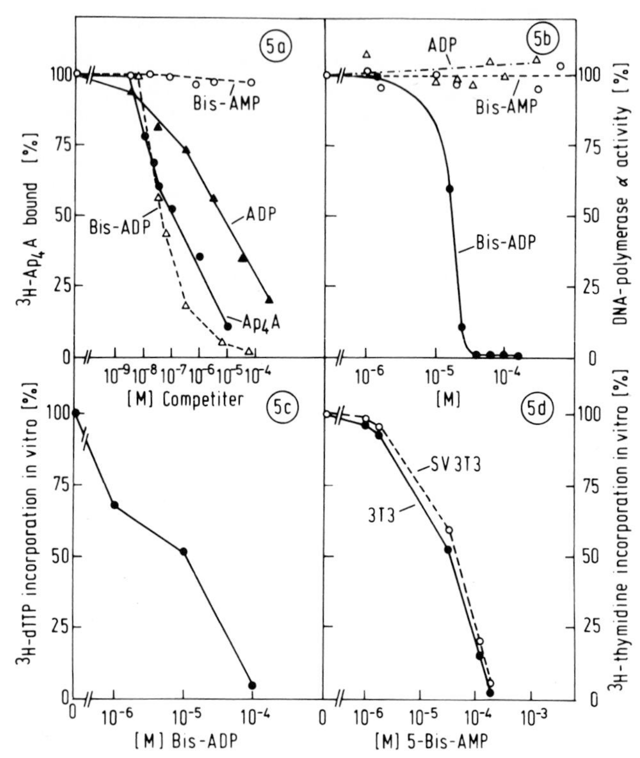

FIG. 5. a) Binding of ^{3}H-AP_4A to DNA polymerase α from calf thymus and its inhibition by unlabeled Ap_4A, methylene-bis-ADP, methylene-bis-AMP,and ADP. 10 µl of calf thymus DNA polymerase α were used in equilibrium dialysis experiments as described previously (14).
b) Effects on DNA polymerase α activity of methylene-bis-ADP in the in vitro assay for isolated DNA polymerase α as described previously (13).
c) Effects on DNA polymerase α of methylene-bis-ADP assayed in the complex lysate system containing permeabilized baby hamster kidney (BHK) cells as described (12).
d) Effects of methylene-bis-(5')DMP on in vitro DNA synthesis of mouse 3T3 and SV 3T3 fibroblasts. Cells were grown and ^{3}H-thymidine incorporation assayed as described (11).

as ADP competes only at an 100-fold excess with Ap_4A for its binding site. DNA synthesis *in vitro* either in a system with purified DNA polymerase α (FIG. 5b) or in a complex system containing cell lysates (FIG. 5c) is strongly inhibited by methylene-bis-ADP. At this concentration neither methylene-bis-AMP nor ADP have any effect on DNA replication *in vitro* (FIG. 5b).

To sutdy whether methylene-bis-adenylate derivatives can inhibit DNA replication *in vivo* we synthesized methylene-bis-adenosine and -AMP, added these compounds to 3T3 and Simian virus 40-transformed 3T3 (SV 3T3) cells and measured the incorporation of [^{3}H]-thymidine into DNA. FIG. 5d shows that methyline-bis-(5')AMP inhibits *in vivo* DNA synthesis in both 3T3 and SV 3T3 cells. Since *in vitro* the DNA synthesis was exclusively inhibited by the ADP dimer we assume that *in vivo* the dimers are transported as adenosine derivatives through the cell membrane and are then phosphorylated intracellularily to the eventually active methylene-bis-ADP. Addition of methylene-bis-adenosine to the cultures does not only completely inhibit DNA replication but also the proliferation of those cells. BHK cells stop growing about 24 h after addition of 0.2 mM methylene-bis-adenosine. However, most of the cells survive at least four days in the presence of the drug and remain attached to the plastics of the tissue culture dishes. These results let us hope that these Ap_4A analogs might prove useful tools for the elucidation of the molecular mechanism of Ap_4A action as well as potential cytostatic drugs.

REFERENCES

1. Alderson, T. (1973): Nature New Biol., 244: 3-6
2. Arens, M., Yamashita, T., Padamanabhan, R., Tsuro, T., and Green, M. (1977): J. Biol. Chem., 252: 7947-7954.
3. Benbow, R.M., and Ford, C.C. (1975): Proc. Natl. Acad. Sci. USA, 72: 2437-2441.
4. Bertazzoni, U., Stefanini, M., Pedrali-Noy, G., Giulotto, E., Nuzzo, F., Falaschi, A., and Spadari, S. (1976): Proc. Natl. Acad. Sci. USA, 73: 785-789.
5. Edenberg, H.J., Anderson, S., and DePamphilis,M.L. (1978): J. Biol. Chem., 253: 3273-3280.
6. Feldman, M.Y. (1967): Biochim. Biophys. Acta, 149: 20-34.
7. Feldman, M.Y. (1973): Progr. Nucleic Res. and Mol. Biol., 13: 1-49.
8. Feldman, M.Y., Balabanova, H., Bachrach, U. and Pyshnov, M. (1977): Cancer Res., 37, 501-506.
9. Friedman, D.L., and Mueller, G.C. (1968): Biochim. Biophys. Acta, 161: 455-468.
10. Graham, C.R., Arms, K., and Gurdon, J.B. (1966): Dev. Biol., 14: 349-381.
11. Grummt, F., Grummt, I., and Mayer, E. (1969): Europ. J. Biochem., 97: 37-42.
12. Grummt, F. (1978): Proc. Natl. Acad. Sci. USA, 75: 371-375.
13. Grummt, F. (1978): Cold Spring Harbor Symp. on Quant. Biol., 43: 649-653.
14. Grummt, F., Waltl, G., Jantzen, H.-M.,Hamprecht,K., Huebscher, U., and Kuenzle, C.C. (1979): Proc. Natl. Acad. Sci. USA, 76: (12), in press
15. Huebscher, U., Kuenzle, C.C., and Spadari, S. (1977): Nucleic Acids Res., 8: 2917-2929.
16. Huebscher, U., Kuenzle, C.C., Limacher, W.,Scherrer, P., and Spadari, S. (1979): Cold Spring Harbor Symp. on Quant. Biol., 43: 625-630.
17. Jazwinski, S.M., Wang, J.L., and Edelman, G.M. (1976): Proc. Natl. Acad. Sci. USA, 73: 2231-2235.
18. Jazwinski, S.M., and Edelman, G.M. (1976): Proc. Natl. Acad. Sci. USA, 73: 3933-3936.
19. Johnson, R.T.,and Harris, H. (1969): J.Cell Sci., 5: 625-644.
20. Kornberg, A. (1978): Cold Spring Harbor Symp. on Quant. Biol., 43: 1-9.
21. Murakami-Murofushi, K., and Mano, Y. (1977): Biochim. Biophys. Acta, 475: 254-266.
22. Otto, B., and Fanning, E. (1978): Nucleic Acids Res., 5: 1715-1728.

23. DePamphilis, M.L., and Berg, P. (1975): J. Biol. Chem., 250: 4348-4354.
24. Plesner, P., Stephenson, M.L., Zamecnik, P.C., and Bucher, N.L.R. (1979): In: Alfed Benzon Sympos., edited by J. Engenberg, H. Klenow, and V. Leick, Vol.13: in press
25. Rapaport, E., and Zamecnik, P.C. (1976): Proc. Natl. Acad. Sci. USA, 73: 3984-3988.
26. Scott, J.F., and Zamecnik, P.C. (1969): Proc. Natl. Acad. Sci. USA, 64: 1308-1314.
27. Spadari, S., and Weissbach, A. (1975): Proc. Natl. Acad. Sci. USA, 72: 503-507.
28. Thompson, L.R., and McCarthy, B.J. (1968): Biochem. Biophys. Res. Commun., 30: 166-172.
29. Waqar, M.A., Evans, M.J., and Huberman,J.A.(1978): Nucleic Acids Res., 5: 1933-1946.
30. Zamecnik, P.C. and Stephenson, M.L. (1969): In: Alfed Benzon Symp., edited by H.M. Kalckar, H.Klenow, A. Munch-Peterson, M. Ottesen, and J. Hess Thaysen, Vol. 1: 276-289. Munksgaard, Copenhagen.

Control Mechanisms in Animal Cells,
edited by L. Jimenez de Asua et al.
Raven Press, New York © 1980.

Glucocorticoid Receptors in Resting and Growing 3T3 Cells

S. Iacobelli, F. O. Ranelletti, C. Natoli, and P. Longo

Laboratory of Molecular Endocrinology, Catholic University, 00168 Rome, Italy

Glucocorticoids (GCs) are known to have different effects on a large variety of functions in target cells. With regard to the growth effects of GCs on fibroblastic cell lines exclusively, paradoxical results have been shown. GCs seem to have little or no effect on the growth regulation of mouse fibroblastic cell lines not density-inhibited such as 3T6, polyoma-transformed 3T3, and SV40-transformed 3T3 cells (26). On the contrary, they stimulate DNA synthesis and cell division in density-inhibited 3T3 cells, but at steroid concentrations well above the physiological range (26), while at physiological concentrations they have been reported to have no effect when used alone (11). Furthermore, in the same cell line when GCs are used in combination with growth factors, they can inhibit the initiation of DNA synthesis induced by Prostaglandin $F_{2\alpha}$ ($PGF_{2\alpha}$)(11) and Epidermal Growth Factor (EGF)(A. Otto et al., manuscript in preparation). At the same time they have a synergistic effect with Fibroblast Growth Factor (FGF) (8, 21) on DNA synthesis and cell division. GCs can also stimulate growth of density-inhibited early passages of human foreskin fibroblasts (HF cells) (26). It is interesting to note that GCs enhance the mitogenic action of EGF on these HF cells, probably modulating cell surface receptors for EGF (2). On the other hand, it is also well known that GCs inhibit the growth of L929 mouse fibroblasts (L929 cells) (19) as well as that of normal chick embryo cells (7).

It is now generally accepted that most steroid hormone actions are necessarily mediated through the initial binding of the hormone to a specific cytoplasmic receptor and the subsequent translocation of the "activated" steroid-receptor complex into the nucleus, where it can bind to chromatin and/or DNA to induce characteristic effects. Specific GC receptors (GCRs) have been reported in L929 cells (1). The GC effect on this particular cell line is strictly inhibitory while in 3T3 cells it can be either stimulatory or inhibitory depending on which growth factor is present. The evocability of opposite responses by the same hormone renders the 3T3 cell line particularly interesting for the study of the relationship between hormone receptors and responsiveness.

MATERIALS AND METHODS

Cell culture. Swiss 3T3 fibroblasts (27) were maintained in Dulbecco's modified Eagle medium (DEM) supplemented with 10% fetal calf serum, 100 units of penicillin/ml, and 100 μg streptomycin/ml. Subconfluent cultures were grown in 100-mm Falcon petri dishes at 37°C under 10% CO_2 in air.

Steroid binding assays. The binding assays were performed on quiescent cell cultures (when no mitotic cells were visible) and on cultures exposed to the stimulating agents for 24 hr. Three different methods were used. The first method was the cytoplasmic competitive binding assay described by Baxter and Tomkins (3) with certain modifications. The buffer for homogenization (0.01 M Tris, 0.0015 M EDTA, 0.0012 M monothioglycerol, pH 7.4) was supplemented with 20% glycerol which is thought to increase the stability of the GCRs (23). The cells were detached from monolayers with a rubber policeman, collected by centrifugation, and washed twice in 50 ml of phosphate-buffered saline (PBS). After homogenization at 4°C with a teflon/glass homogenizer, the cytosol was obtained by centrifugation at 145,000 x g for 45 min in a refrigerated Spinco Model L centrifuge. In the standard binding assays, 50 μl aliquots of cytosol were pipetted in Beckman microfuge tubes containing increasing concentrations (0.7-33 nM) of (^{3}H) triamcinolone acetonide ((^{3}H)TA) (22 Ci/mM, The Radiochemical Centre, Amersham, England). Following incubation (4°C for 16-18 hr), 100 μl of dextran-coated charcoal (1% Norit A charcoal-0.1% Dextran T70 in homogenizing buffer) were added to adsorb free steroid. Radioactivity in the supernatant was measured with 6 ml of scintillation fluid.

The second method, whole-cell binding, was carried out by a modification of the Sibley and Tomkins method (24) as follows: Cells were harvested with trypsin-EDTA and washed twice with PBS at 25°C. For measurement of steroid uptake, cells suspended in DEM (4-6 x 10^6 cells/ml) were incubated with (^{3}H)TA (0.7-33 nM) at 37°C for 60 min. Cells were washed 3 times with cold PBS and extracted with 1 ml of 80% ethanol. The ethanol extracts were counted with 8 ml of scintillation fluid.

The third method was steroid binding to cells adhering to plastic dishes (also referred to as cell-culture binding). For these studies cells were grown in 33-mm petri dishes. To measure steroid binding, cell monolayers were first incubated at 37°C for 30 min in serum-free medium (to remove most endogenous hormone) and then incubated with (^{3}H)TA (1-40 mM) in 700 µl DEM per dish at 37°C for 60 min. After incubation, the medium was removed and the cells washed 3 times with PBS at 25°C and then extracted as above with 1 ml of 80% ethanol which was then counted.

Nonspecific binding was measured by the amount of bound radioactivity when binding was carried out in the presence of a 1,000-fold excess of unlabeled TA. For determination of the concentration and apparent equilibrium dissociation constant (K_D) of receptor sites, specific binding data were analyzed by the Scatchard equation (21) using an Olivetti P6060 desktop computer. In some cases the number of binding sites per cell was assayed after incubating cell cultures with a single saturating concentration of (^{3}H)TA (40 mM) with and without 4 µM unlabeled TA. The number of saturable binding sites per cell determined by this procedure did not differ significantly from the value observed using a complete saturation curve.

Nuclear translocation assay. This was done by incubating cells (harvested with trypsin-EDTA) for 2 hr at 2°C with 30 nM (^{3}H)TA, taking aliquots for determining cytoplasmic and nuclear binding, then warming the cells at 37°C for 15 min and again determining cytoplasmic and nuclear radioactivity (20).

GCR specificity. To determine the stereospecificity of binding sites, cells were incubated as described above for "whole-cell binding" except that incubations contained a fixed concentration of (^{3}H)TA (20 nM) along with increasing concentrations of unlabeled competing steroids.

Sucrose gradient. Sucrose density gradient centrifugation was carried out by layering 100 µl of cytosol, labeled with (^{3}H)TA, on 5-20% sucrose gradients made up in homogenizing buffer. The gradients were centrifuged at 1°C for 16 hr at 48,000 rpm in a SW 50.1 rotor. Fractions (5 drops) were collected and counted in 6 ml of scintillation fluid (20).

PRESENCE OF GCRs IN 3T3 CELLS

Using either cytosol, whole-cell, or cell-culture binding assays saturable GCRs were detected in 3T3 cells. Results of the cytosol assay indicating that quiescent cells contain high affinity, limited capacity binding sites for (^{3}H)TA are given in Fig.1 which shows a binding curve for increasing concentrations of steroid. At 4°C these cytoplasmic receptors became saturated at a hormone concentration of approximately 20 nM. Scatchard analysis of the binding curve (Fig. 1, inset) is consistent with TA binding to a single class of receptor sites of uniform affinity with a

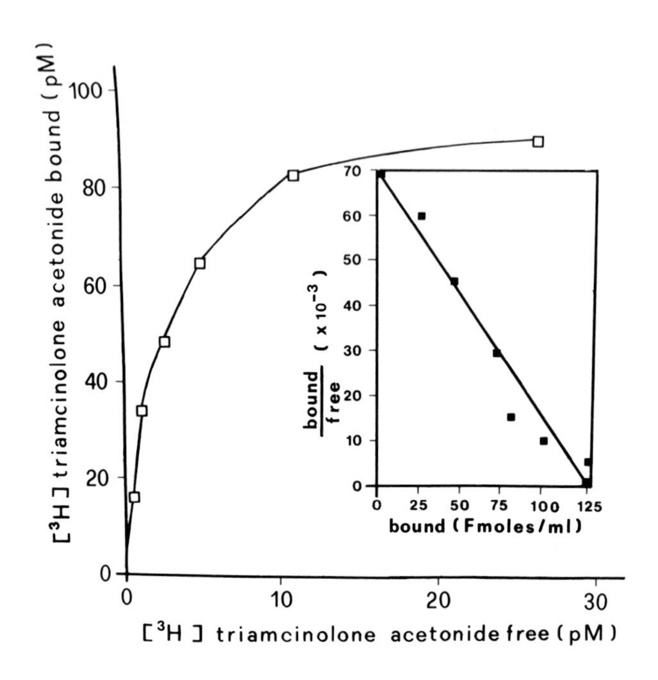

FIG. 1. Binding of (^{3}H)TA to 3T3 cell cytosol. The inset shows a Scatchard plot of the specific binding curve.

K_D of about 2.0 nM. The mean number of binding sites (4 experiments) was 801 Fmoles/mg cytosol protein.

Results of whole-cell binding assay, which measures both cytoplasmic and nuclear sites, are illustrated in Fig. 2. Using this whole-cell assay, the receptor was saturated at approximately 15 nM (^{3}H)TA. Scatchard analysis of the binding data (Fig. 2, inset) again resulted in a straight line, suggesting a single class of homogenous receptors of uniform affinity. For 6 experiments the mean number of receptor sites/cell was 34,000. Approximately 20% more receptor sites/cell were detected by "cell-culture" binding assay than by whole-cell assay. The reason for this difference is not clear and as yet has not been further investigated.

Studies have shown that upon addition of GCs and steroids, in general, to cells at 37°C, the hormone binds to cytoplasmic receptors, which are rapidly translocated to nuclear bound form (17). The formation of this nuclear receptor-hormone complex in 3T3 cells incubated with (^{3}H)TA is illustrated in Fig. 3. After 15 min at 37°C, significant nuclear translocation occurred while after 2 hr at 2°C translocation was minimal. This temperature-dependent nuclear translocation of receptor agrees well with the generally accepted model of steroid hormone action, and in particular with the role of cytoplasmic receptor as an obligatory intermediate in the formation of the nuclear complex.

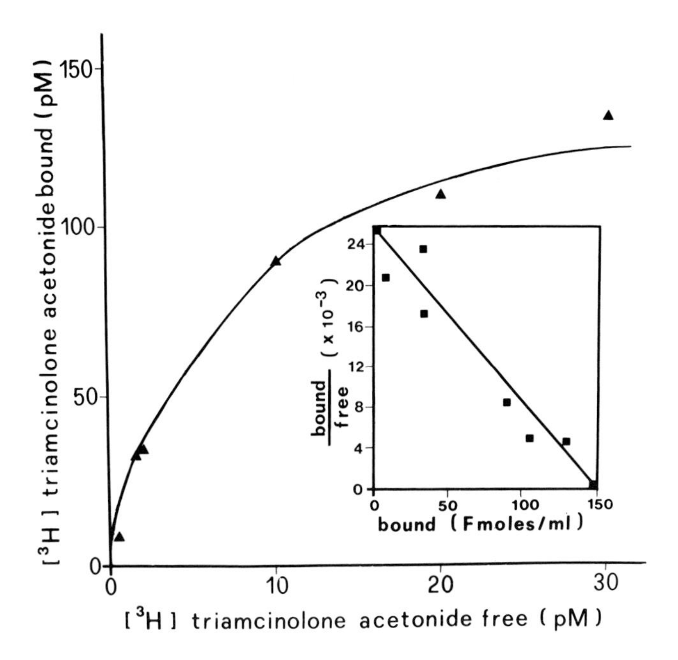

FIG. 2. Binding of (^{3}H)TA to whole 3T3 cells. The inset shows a Scatchard plot of the specific binding curve.

The specificity of GCR was demonstrated by measuring the ability of competing steroid analogues to displace (^{3}H)TA in the binding reaction. The ability of TA, dexamethasone,corticosterone and cortisol at about 100-fold molar excess to displace 50% of the bound (^{3}H)TA (Fig. 4A) agrees with the known biological activity of these steroids in other fibroblastic cell lines (6, 7) as well as in lymphoid systems (14, 18). Cortexolone, which in rodent thymocytes is an anti-GC (15), also competed for binding when present at higher molar excess (Fig. 4B). Similarly, R 5020 (17, 21-dimethyl-19-nor-4,9-pregnadiene-3,20-dione) (kindly supplied by Dr. J.P. Raynaud, Roussel Corp., Romainville, France) and aldosterone also competed but only at high concentrations, while estradiol-17β and testosterone showed weak competition (Fig. 4B).

Sucrose density gradient analysis of 3T3 cell extracts incubated with (^{3}H)TA revealed a single peak of bound radioactivity (Fig. 5). The sedimentation coefficient of the complex is 7-8S when determined in reference to bovine serum albumin (4.6S) as a marker. The addition of a 100-fold excess of unlabeled TA completely suppressed the peak. Similar results for GCRs in various target tissues have been previously reported (9, 10, 12).

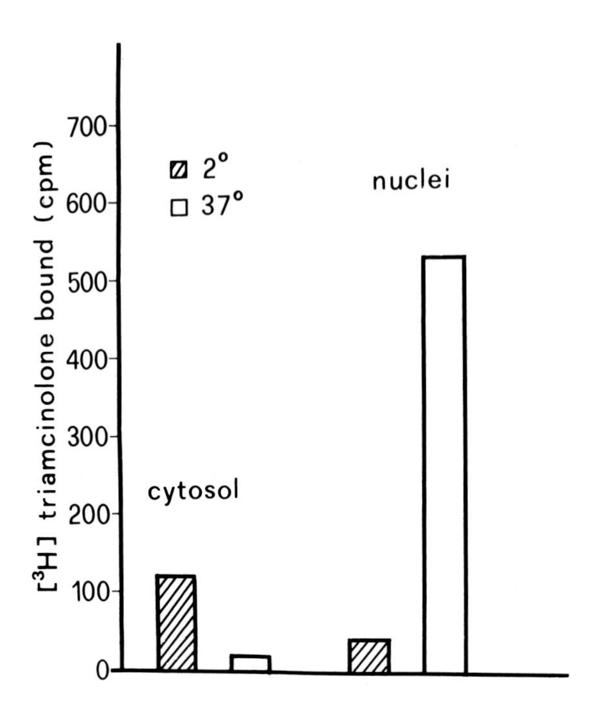

FIG. 3. Cytoplasmic-nuclear translocation of (^{3}H)TA-receptor complex at 2°C and 37°C.

INDUCTION OF GCRs BY GROWTH FACTORS AND SERUM

As mentioned above, GCs are able to produce stimulatory or inhibitory effects in 3T3 cells, depending on which growth factor initiates DNA synthesis. This observation suggests that GCs can modulate the sequence of events triggered by growth factors (21).

The mechanism through which GCs interact with growth factors in the initiation of DNA synthesis is unknown. It is not even known if GCs alone have any effect on quiescent 3T3 cells. Nor is it actually known whether or not the observed effect of GCs on cell proliferation is dependent on the modifications of cell metabolism produced by growth factors. In fact, it is possible that the metabolic state of the cells may be relevant in determining the response to GCs. A striking example is offered by human skin fibroblasts exposed to dexamethasone in nutrient-rich or poor environments (13). In the first case, GCs stimulate cell growth, while in the second case (under step-down conditions) their effect is inhibitory. Thus it may be argued that growth factors, depending on their chemical structure, set up a particular metabolic state in the cells which determines the quality of response to hormones.

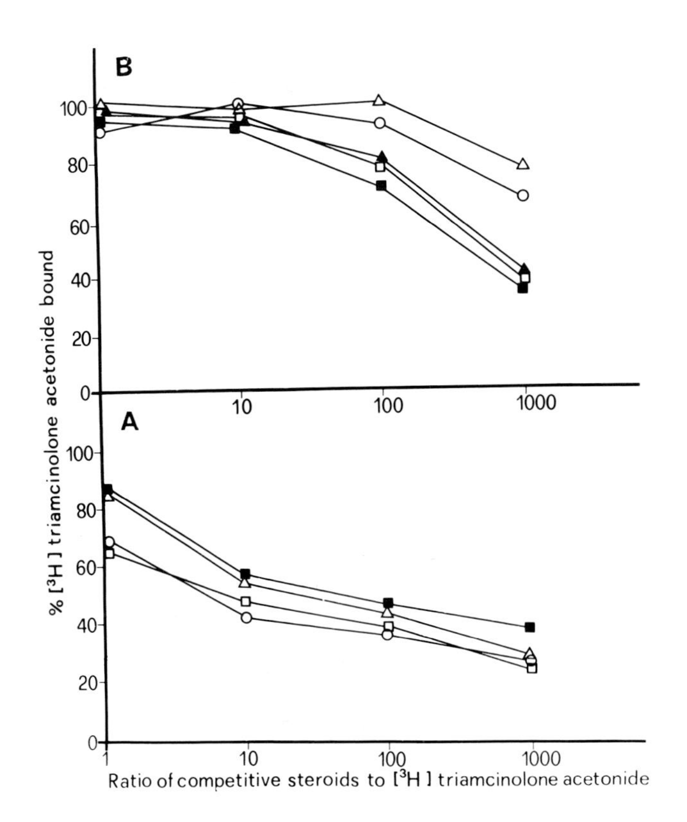

FIG. 4. Displacement of (^{3}H)TA binding by various steroids. A: (○) = TA, (□) = dexamethasone, (△) = corticosterone, (■) = cortisol; B: (■) = cortexolone, (□) = R 5020, (▲) = aldosterone, (○) = estradiol-17β , (△) = testosterone.

Among the various biochemical parameters affected by growth factors, we have focused on GCRs because of their central role in hormone action. As shown in Fig. 6, quiescent 3T3 cells exposed for 24 hr to saturating concentrations of $PGF_{2\alpha}$ (400 ng/ml) or FGF (50 ng/ml) for 24 hr contain approximately twice the amount of GCR sites/cell as compared with nonstimulated cells. These concentrations of growth factors were capable of stimulating a maximum thymidine incorporation (data not shown).

The affinity of the steroid-receptor complex and stereospecificity of the GCR sites did not vary significantly after stimulation with either $PGF_{2\alpha}$ or FGF (Fig. 6, Table 1).

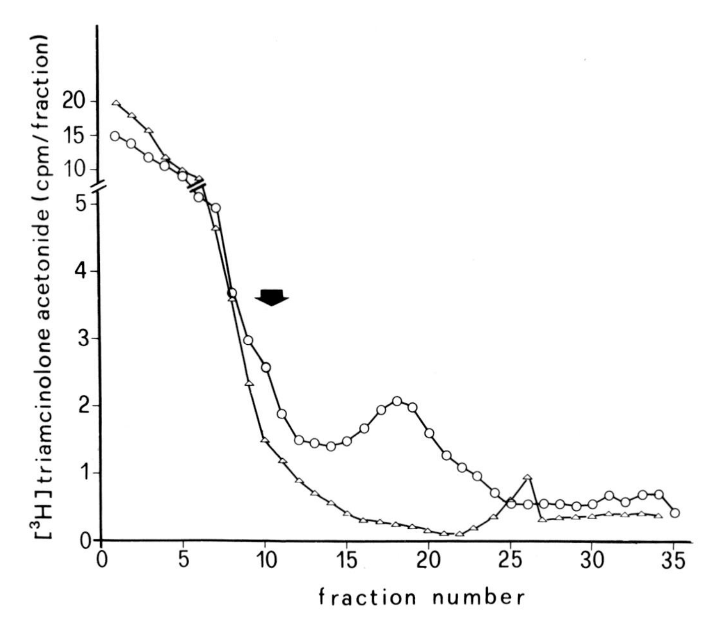

FIG. 5. Sedimentation profiles of bound (^{3}H)TA on sucrose gradients. Aliquots of cytosol were incubated with (^{3}H)TA (1 nM) in the absence (○) or presence (△) of 100 nM unlabeled TA. The arrow indicates the position of bovine serum albumin (4.6S) which was used as protein marker. Sedimentation was from left to right.

TABLE 1. Specificity of GCRs in 3T3 cells exposed to $PGF_{2\alpha}$ and FGF

Competing Steroid (100-fold molar excess)	% inhibition Control	$PGF_{2\alpha}$	FGF
TA	96	92	93
Cortisol	69	72	82
R 5020	52	51	61
Cortexolone	45	42	47
Progesterone	43	43	50
17α-hydroxyprogesterone	35	40	46
Cortisone	0	4	9
Tetrahydrocortisol	0	11	2
Estradiol- 17	0	9	13
5α - dihydrotestosterone	0	0	15
21- deoxycortisone	8	0	16

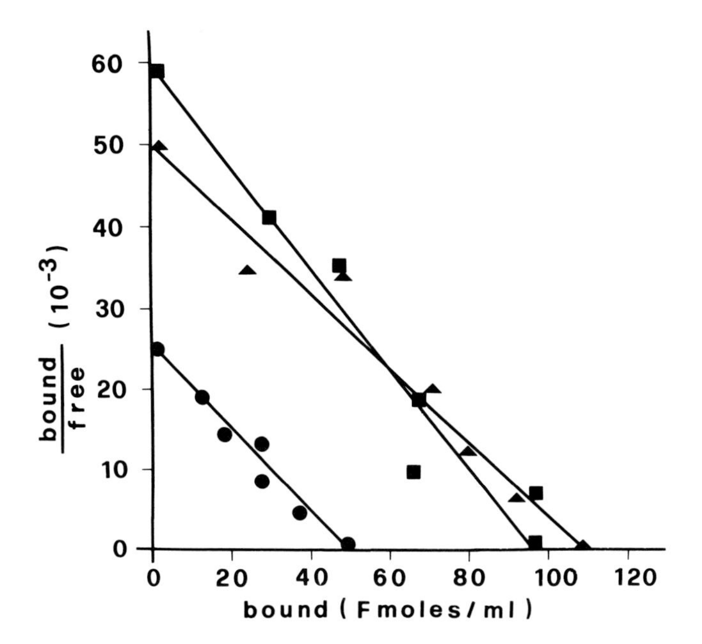

FIG. 6. Scatchard analysis of (^{3}H)TA binding to whole 3T3 cells. (●), no additions; (▲), $PGF_{2\alpha}$ (400 ng/ml); (■), FGF (50 ng/ml).

GCRs AS INDICATORS OF CELL GROWTH?

The above observations raise some important questions on the role of GCRs in 3T3 cells and in cells in general. Is the increase of the number of GCRs triggered only by $PGF_{2\alpha}$ and FGF or also by other growth stimulating agents? Does the increase account for the variable pattern of interactions between GCs and $PGF_{2\alpha}$ and FGF? If not, can this increase be attributed to other parameters of cell function?

The answer to the first question is given by the results illustrated in Table 2 which shows an appreciable increase of GCR sites/ cell after the cells were exposed for 24 hr to either EGF alone or EGF plus insulin or serum. Therefore it seems that the increase of GCRs is not brought about by a specific growth factor, but instead appears to be a more general phenomenon of cell activation.

Regarding the second question, since $PGF_{2\alpha}$ and FGF both cause a similar increase of receptor number with the same affinity and specificity, it seems unlikely that the differential effects of GCs can be attributed to receptor modifications either quantitatively or qualitatively. This concept is further substantiated by the observation that both EGF (alone or plus insulin) and serum significantly increase the number of GCRs (Table 2) despite the fact that GCs have no effect on growth stimulation by serum (11)

TABLE 2. Binding of (^{3}H)TA in 3T3 cells exposed to EGF (alone or plus insulin) and serum

Culture	GCR sites/cell[a] (%)
Control (no additions)	100
EGF (25 ng/ml)	140
EGF (25 ng/ml) + Insulin (50 ng/ml)	160
Serum (10%)	210

(a) The specific binding of (^{3}H)TA to receptors was measured by cell-culture assay with a single saturating concentration of (^{3}H)TA (40 nM). For details see Materials and Methods. Values given are the means of 3 experiments. Differences between control and EGF-treated cultures significant ($p < 0.01$; t test).

while they inhibit the stimulus given by EGF (A. Otto et al., manuscript in preparation).

The third question, whereby the increase of receptors is attributed to other parameters of cell function, can be answered in light of recent experimental data which show that in synchronized HeLa cells there is a doubling of GCR content as the cells pass through late G_1 phase followed by the return to the original number after mitosis (5). Other studies on peripheral blood lymphocytes have clearly shown that the number of GCR sites/cell increases markedly following mitogen stimulation of the cells with plant lectins (4, 16, 25). This "cell-cycle dependency" of GCR content seems to be a nonspecific event but it is generally interpreted as an indication of an overall activation of cell metabolism in preparation for mitosis. Our findings could be explained in terms of this hypothesis since quiescent 3T3 cells achieve a relative degree of synchronization after exposure to growth factors or serum. Finally, the increase of GCRs in stimulated 3T3 cells might shed further insight on the molecular mechanism of the induction process.

SUMMARY AND CONCLUSIONS

1. Specific GCRs are present in 3T3 cells.

2. Growth factors such as $PGF_{2\alpha}$ and EGF, whose stimulatory effects on DNA synthesis are known to be decreased by GCs, and FGF, whose stimulatory effect on DNA synthesis is known to be increased by GCs, are all capable of producing a similar increase in the number of GCR sites/cell without an apparent qualitative difference in the induced sites.

3. Serum, whose stimulatory effect on DNA synthesis is not affected by GCs, is likewise capable of producing a marked increase in the number of GCR sites/cell.

These observations, taken together, suggest that the increase in GCR sites might be related to cell activation by growth factors or serum.

ACKNOWLEDGMENTS

We wish to thank L. Jimenez de Asua for his stimulating discussions and L.R. Oliphant and M. Schiavoncini for their help in preparing the manuscript. This work was carried out in part at the Friedrich-Miescher Institute, Basel (Switzerland) where C. Natoli was supported by an E.M.B.O. fellowship.

REFERENCES

1. Aronow, L. (1979): In: Glucocorticoid Hormone Action, edited by J.D. Baxter, and G.G. Rousseau, pp. 327-340. Springer-Verlag, Berlin, Heidelberg, New York.
2. Baker, J.B., Barsh, G.S., Carney, D.H., and Cunningham, D.D. (1978): Proc. Natl. Acad. Sci. USA, 75: 1882-1886.
3. Baxter, J.D., and Tomkins, G.M. (1971): Proc. Natl. Acad. Sci. USA, 68: 932-937.
4. Bonnard, G.D., Lippman, M.E. (1978): Schweiz med. Wschr., 108: 1603.
5. Cidlowsky, J.A., and Michaels, G.A. (1977): Nature, 266: 643-645.
6. Cristofalo, V.J., and Rosner, B.A. (1979): Fed. Proc., 38:1851-1856.
7. Fodge, D.W., and Rubin, H. (1975): Nature, 257: 804-806.
8. Gospodarowicz, D. (1974): Nature, 249: 123-127.
9. Groyer, A., Picard-Groyer, M.T., and Robel, P. (1979): Mol. Cell. Endocrinol., 13: 255-267.
10. Iacobelli, S., Ranelletti, F.O., Sica, G., and Barile, G. (1978): J. Endocrinol. Invest. 1: 209-213
11. Jimenez de Asua, L., Carr, B., Clingan, D., and Rudland, P. (1976): Nature, 265: 450-452.
12. Kaiser, N., Milholland, R.J., and Rosen, F. (1973): J. Biol. Chem., 248: 478-483.
13. Lorin, K.J., Lan, N.C., and Baxter, J.D. (1979): J. Biol. Chem. 254: 7785-7794.
14. Moscona, A.A., and Peddington, R. (1967): Science, 158: 496-497.
15. Munck, A., and Brinck-Johnsen, T. (1968): J. Biol. Chem., 243: 5556-5565.

16. Munck, A., Crabtree, G.R., and Smith, K.A. (1978): J. Toxicol. Environ. Health, 4: 409-425.
17. Munck, A., and Leung, K. (1977): In: Receptors and Mechanism of action of Steroid Hormones, edited by J.R. Pasqualini, part 2, pp. 311-397. Marcel Dekker, New York.
18. Nicholson, M.L., and Young, D.A. (1978): Cancer Res., 38: 3673-3680.
19. Pratt, W.B., and Aronow, L. (1966): J. Biol. Chem., 241: 5244-5250.
20. Ranelletti, F.O., Carmignani, M., Iacobelli, S., and Tonali, P. (1978): Cancer Res., 38: 516-520.
21. Rudland, P.S., Jimenez de Asua, L. (1979): Biochim. Biophys. Acta, 560: 91-133.
22. Scatchard, G. (1949): Ann. N.Y. Acad. Sci., 51: 660-672.
23. Schmid, H., Grote, H., and Sekeris, C.E. (1976): Mol. Cell. Endocrinol., 5: 223-241.
24. Sibley, C.H., and Tomkins, G.M. (1974): Cell, 2: 221-227.
25. Smith, K.A., Crabtree, G.R., Kennedy S.J., and Munck, A.V. (1977): Nature, 267: 523-525.
26. Thrash, C.R., and Cunningham, D.D. (1973): Nature, 242: 399-401.
27. Todaro G.J., Lazar, G.K., and Greene, H. (1965): J. Cell. Physiol., 66: 325-334.

Control Mechanisms in Animal Cells,
edited by L. Jimenez de Asua et al.
Raven Press, New York © 1980.

Hormonal Regulation of Net DNA Synthesis in Human Breast Cancer Cells in Tissue Culture

Susan C. Aitken and Marc E. Lippman

Medicine Branch, National Cancer Institute, National Institutes of Health, Bethesda, Maryland 20205, U.S.A.

SUMMARY

Extrapolation of radioactive precursor incorporation data to rates of DNA synthesis is not straightforward. In this paper an experimental approach permitting rigorous assessment of rates of net DNA synthesis in tissue culture systems is reviewed. The conditions under which incorporation of [^{32}P] orthophosphate into DNA can be utilized as a marker for quantitation of DNA synthesis are developed. Conventional methods for isolation and measurement of DNA are adapted to minimize sample size requirements and processing time. Measurements of precursor pool sizes and specific activities are obtained. Scheduled versus non replicative DNA synthesis is assessed. Specific effects of estrogens and antiestrogens on true rates of DNA synthesis are described. The possible mechanism of action of estrogens and anti-estrogens in regulation of rates of net DNA synthesis in MCF-7 cells, an hormonally responsive breast cancer cell line, is analyzed.

INTRODUCTION

The stimulatory effects of estrogenic hormones on cell growth, protein synthesis and DNA synthesis in hormonally responsive breast cancer cells in tissue culture have been previously documented (1,2). MCF-7 cells have been shown to have specific receptors not only for estrogens but for androgens, glucocorticoids and progestins (3,4). Despite these evidences of hormonal responsiveness, many problems exist in the use of radiolabelled precursors to measure DNA synthesis.

The experiments discussed below are directed toward establishing a set of techniques which will permit an accurate determination of the net rate of DNA synthesis in tissue culture systems. It is our intention to introduce a set of methods which will allow quantification of net DNA synthesis, discuss their technical application to a particular tissue culture system, and to employ this investigative technique in a detailed study of the kinetics of estrogen and anti-estrogen effects on DNA synthesis. In this investigation $[^{32}P]\ P_i$ is employed to monitor net DNA synthesis. Specific problems which are associated with the use of such a radioactive tracer in estimating rates of net DNA synthesis include:

1. Equilibrium conditions must exist with respect to precursor pools of deoxynucleotides.
2. The effective specific activity $[^{32}P]\ P_i$ in precursor pools directly serving DNA synthesis must be known.
3. Repair versus replicative DNA synthesis must be quantified. It is possible that a part of the incorporation of labeled precursors into DNA represents repair activity rather than net DNA synthesis. Although repair activity is low in most in vitro systems as compared with proliferative activity (5), the MCF-7 human breast cancer cell line represented an unknown quantity.
4. The biochemical techniques employed must accurately reflect incorporation into what is intended authentic DNA alone.

Each of the above areas will be addressed and a number of attendant issues in the analysis of estrogen mediated effects on DNA synthesis will by explored: 1. Does the effect of estrogen represent a synchronization of responsive cells with respect to cell cycle kinetics and can the effect be localized to a shortening of S or release of cells from G1 arrest? 2. Does estrogen administration affect labeling kinetics and to what extent do transport processes and pool sizes influence the appearance of label in DNA and intracellular precursor pools for DNA synthesis? 3. The response of MCF-7 cells to tamoxifen (ICI 46, 474), a known anti-estrogen, is also discussed.

MATERIAL AND METHODS

A. Tissue Culture Techniques

MCF-7 cells, a line of human breast cancer cells obtained from the Michigan Center Foundation, were maintained in continuous tissue culture (6). This line has been demonstrated to contain estrogen receptors and to be hormonally responsive in terms of precursor incorporation into DNA, RNA and protein (2).

Cells were maintained in monolayer culture in Improved Eagle's Minimal Essential Medium (7), (IMEM), supplemented with twice the usual concentrations of glutamine, penicillin and streptomycin and with 10% fetal calf serum. Two passages prior to a given experiment cells were placed in IMEM supplemented with 10^{-7} mol/l insulin and 2.5% charcoal-treated calf serum shown to be essentially free of estradiol (8). Cells were plated replicately in plastic multiwell tissue culture dishes in the same medium. When plates became subconfluent, medium was replaced with IMEM containing a reduced concentration of phosphate (10^{-5} mol/l) and lacking asparagine. After 4 to 12 hours the medium was again replaced with low phosphate medium plus or minus 17β-estradiol (5×10^{-9} mol/l) or the antiestrogen tamoxifen (10^{-6} mol/l) (9). Labeled precursors were added at varying time intervals as described in Results. Cells were harvested, washed with 0.9% NaCl and stored at -20° pending further processing.

B. Analytical Methods

The fractionation of a given sample and the analyses performed on each aliquot are outlined in Figure 1. The acid-precipitable thymidine label is of particular interest as it will provide estimates of DNA recovery on further processing.

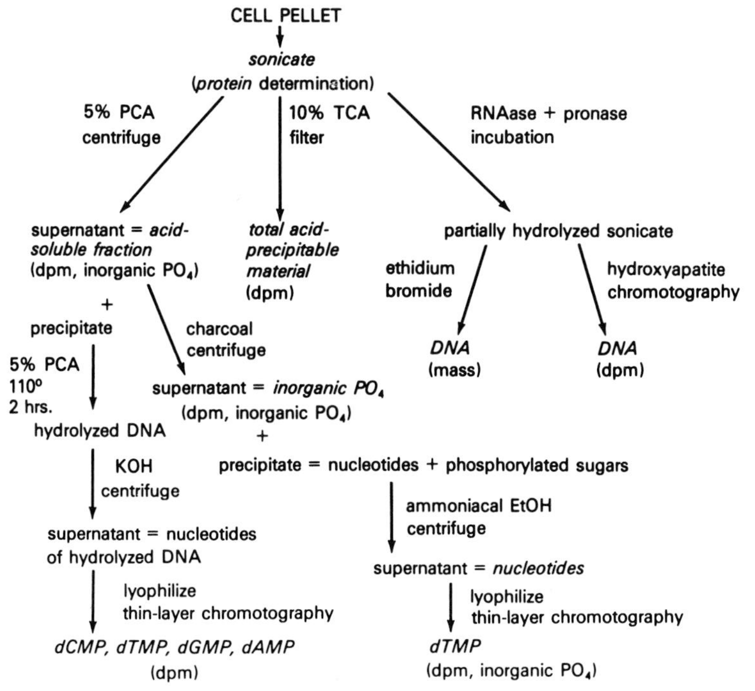

FIGURE 1. Fractionation procedure used for examination of DNA syntheis in MCF-7 cells.

Protein determination (10) and total acid precipitable material (11) are determined as previously described. An additional aliquot of the sonicate is treated with 5% perchloric acid (PCA) and centrifuged at 1500g for 10 minutes. The acid-soluble fraction is analyzed for total phosphate content by a modification of the method of Fiske and Subbarow (12). A portion of this fraction is also taken for determination of incorporation of labeled ^{32}P. These numbers provide the basis for establishing the specific activity of the phosphate pool.

Occasionally the acid-soluble fraction is analyzed further, employing dextran-coated charcoal to differentially adsorb nucleotide components (13) and thin-layer chromatography (14,15) to isolate particular nucleotides. In some cases the precipitate resulting from 5% PCA treatment is subjected to differential acid hydrolysis for DNA (16), neutralized with KOH, lyophilized, and similarly separated by thin-layer chromatography into constituent nucleotides.

Still another aliquot of the sonicate is incubated with pancreatic RNAase (type I), T1 RNAase, and pronase. Following incubation, this material is fluorometrically analyzed for DNA content employing ethidium bromide (17, 18). An additional aliquot is subjected to hydroxylapatite column chromatography (19,20) modified for very small amounts of material. The DNA eluate is precipitated on millipore filters and counted. On occasion the eluate is acid-precipitated, hydrolyzed and analyzed by thin-layer chromatography as described above.

RESULTS

In order to obtain information on rates of DNA synthesis from data derived from the incorporation of labeled precursors, information on transport and pool size is critical. Consequently the kinetics of phosphate transport and utilization in MCF-7 cells under relevant experimental conditions were investigated prior to consideration of hormonal regulation of phosphate incorporation into DNA.

^{32}P P_i has frequently been employed as a precursor in measurements of DNA synthesis (21). In order to generate substantial incorporation of [^{32}P] P_i into DNA in our tissue culture system, the ambient phosphate concentration was reduced in the culture medium to 10^{-5} mol/l. Under these circumstances MCF-7 cells showed continued growth (unlike other cell lines [22]). DNA and protein content per dish increased steadily for 60 hours after transfer of cells to 10^{-5} mol/l phosphate medium. Rates of incorporation of several labeled precursors into acid-precipitable material were examined after cells were placed in serum-free conditions The ratios of incorporation of [^{3}H] dThd [^{3}H] Urd, [^{14}C] Leu and [^{14}C] acetate in low phosphate medium to the incorporation observed in standard medium are presented in Table 1. Little or no effect on rates of incorporation was apparent. At the time periods examined it was also evident that intracellular acid-soluble phosphate (column 2) and cell protein (column 3) were similarly unaffected by reduction in extracellular phosphate to 10^{-5} mol/l. Therefore, data obtained under these circumstances can validly be compared with previous information on MCF-7 cells obtained in standard media.

TABLE 1. Effect of lowered phosphate concentration on incorporation of label into acid-precipitable material in MCF-7 cells.

Time	Protein	P_i	[^{3}H]dThd	[^{14}C]Acetate	[^{3}H]Urd	[^{14}C]Leu
12	0.97± 0.05	0.93± 0.12	1.10± 0.05	1.10± 0.22 (0.79 ± 0.24)	1.04	1.37
30	0.97± 0.07	1.18± 0.27	0.97± 0.01	0.95± 0.16 (0.77 ± 0.05)	0.99	1.18
48	1.04± 0.03	1.24± 0.18	1.27± 0.12	1.03± 0.22 (1.40 ± 0.03)	1.45	0.91

Cells were labeled with [^{14}C]acetate (5 μCi/ml, 8 hours), [^{3}H]dThd (1 μCi/ml, 2 hours) or [^{3}H]Urd (1 μCi/ml, 2 hours) plus [^{14}C] Leu (1 μCi/ml, 2 hours) and harvested at the times indicated after transfer to either 10^{-5} mol/l phosphate IMEM or standard IMEM (10^{-3} mol/l phosphate). Incorporation of label was normalized per unit protein. The incorporation of [^{14}C]acetate into DNA was also determined (values in parentheses). Numbers are presented as the ratio of values observed in 10^{-5} mol/l phosphate IMEM to values obtained in standard IMEM. Determination of protein and P_i was based on 5 independent samples, determination of [^{3}H]dThd and [^{14}C]acetate incorporation on 2 independent samples, and determination of [^{3}H]Urd + [^{14}C]Leu incorporation on a single sample. Values reported are the mean of these observations ± 1 SD (where applicable).

Analysis of the kinetics of phosphate incorporation into the acid-soluble pool of MCF-7 cells is critical to validate the experimental circumstances of labeling. The period chosen to investigate incorporation must be one in which labeled precursor has equilibrated with competing intracellular pools under all experimental conditions. It can be seen (Figure 2) that approximately 6 hours were minimally required. Hormonal treatment (control, tamoxifen, estradiol) did not appreciably alter the time required for equilibration of $[^{32}P]$ P_i with intracellular acid-soluble phosphate pools. However, the amount of label accumulating in estrogen-treated cells was significantly greater than that in controls ($p < .005$). The particular experiment illustrated represents a time between 20 and 32 hours following hormone treatment. Similar results were obtained 48 to 60 hours after hormone addition.

Since the deoxynucleotide pool which is the immediate precursor pool for DNA synthesis may represent less than 1% of the total acid-soluble pool (23) equilibration of this larger pool may not reflect conditions actually relating to

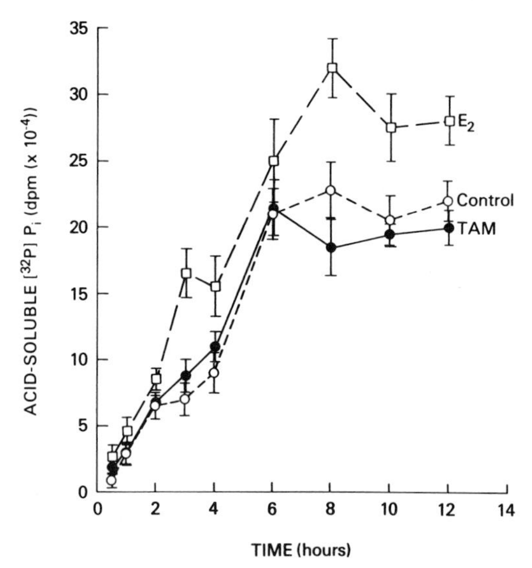

FIGURE 2. Incorporation of $[^{32}P]$ P_i into the acid-soluble pool of MCF-7 cells.

Cells were transferred to 10^{-5} mol/l phosphate medium 32 hours before time 0. 24 hours before time 0 cells were again transferred to fresh low phosphate medium ± estradiol (5×10^{-9} mol/l) or tamoxifen (2×10^{-6} mol/l). $[^{32}P]$ P_i was added at time 0 and cells were harvested at the times indicated. Values are normalized per unit protein and represent the average of 3 determinations ± 1 SD. control (0......0); tamoxifen (0———0); estradiol (□ -- □).

DNA synthesis. Examination of the time course of $[^{32}P]$ P_i incorporation into DNA however demonstrated that linearity was also achieved in all experimental circumstances within 4 to 6 hours of exposure to radioactive trace (Figure 3). The experiment illustrated refers to the period between 20 and 32 hours after hormonal administration.

In spite of the similar equilibration time requirements of the total acid-soluble pool and the precursor pool for DNA synthesis (the deoxynucleotide pool) there remains some question as to whether the specific activity of the acid-soluble pool is truly an accurate reflection of the specific activity of the deoxynucleotide pool. The former can be directly determined as previously described. The dNTP pool on the other hand is relatively small and routine determination of specific activity of $[^{32}P]$ P_i isolated in deoxynucleotides would be impossible when dealing with small numbers of cells. For extrapolation from units of incorporation to units of mass, an accurate determination of phosphate specific activity is essential. To investigate this question approximately 10^8 to 10^9 MCF-7 cells were pulsed with $[^{32}P]$ P_i and $[^3H]$ dThd and the acid-soluble

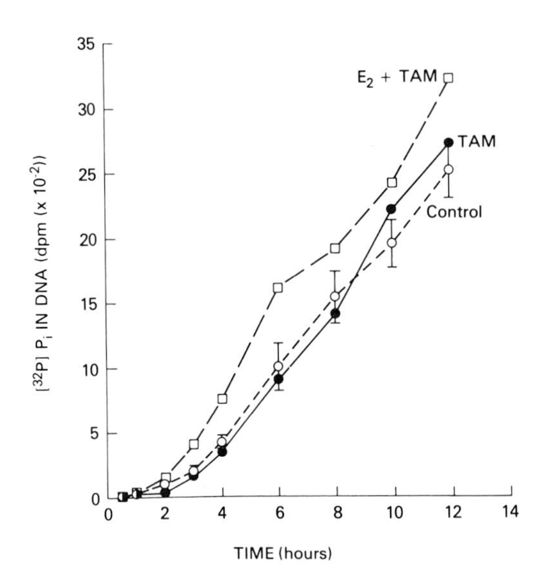

FIGURE 3. Incorporation of $[^{32}P]$ P_i into DNA of MCF-7 Cells. DNA was isolated by treatment of sonicates with RNAase plus pronase and hydroxylapatite chromatography as previously described. Standard deviations are depicted only for control samples although variation among all experimental groups was similar and in no case exceeded 15% of the reported value. control (0......0); tamoxifen (0———0); estradiol (□-----□).

fraction was obtained as previously described. Table 2 shows the subsequent fractionation of this pool and the specific activity of ^{32}P in various components of the pool. It can be seen from column 2 in which [^{3}H] dThd is utilized as a trace for the distribution and recovery of nucleotides that charcoal treatment of the acid-soluble fraction resulted in adsorption of virtually all nucleotides to the charcoal pellet. The supernatant remaining after charcoal treatment represents true inorganic phosphate (row 2). Extraction of nucleotides from the charcoal pellet with ammoniacal ethanol is represented in row 3. This extraction procedure is relatively specific for nucleotides (13). Isolation of the dTMP component employing thin-layer chromatography as previously described (14,15) was additionally performed. The results of three separate experiments (A,B,C) are presented in column 3. Examination of these data reveals that the specific activity of [^{32}P] P_i in all components of the acid-soluble pool is identical.

The above experiments strongly suggest that estimates of the specific activity of the DNA precursor pool based on data derived from the acid-soluble pool are accurate if equilibrium conditions are established. True estimates of the rate of net DNA synthesis then become possible. Equilibrium must be verified for each type of experimental manipulation performed since equilibration of the DNA precursor pool could conceivably be greatly delayed in systems not actively synthesizing DNA (24).

Examination of the relationship between incubation time and the specific activity of acid-soluble phosphate pools reveals that both time and experimental treatment constitute significant experimental variables (Figure 4). The effective specific activity of acid-soluble phosphate rose steadily over time. All points represent values at equilibrium (8 hours after addition of [^{32}P] P_i) but varying times after transfer to 10^{-5} mol/l phosphate medium ± hormones. Estradiol increased the specific activity of acid-soluble phosphate over controls at all times greater than 16 hours with greater differences at later times. Tamoxifen resulted in increased specific activities between 20 and 40 hours. This would be anticipated if extracellular phosphate (which has been substantially reduced to 10^{-5} mol/l) were progessively depleted by MCF-7 cells at rates which are additionally influenced by hormonal treatment. The specific activity of label in the medium would vary depending on the incubation time at which trace was added. Intracellular specific activities would tend to reflect extracellular values. This hypothesis was confirmed by direct analysis of the phosphate content and radioactivity in medium at varying times after transfer of cells to low phosphate medium. There was a progessive depletion in mass of phosphate with time and a consequent increase

TABLE 2. Distribution of $[^{32}P]P_i$ among components of the acid-soluble pool in MCF-7 cells.

Fraction	% Recovery ($[^3H]$dThd)	Specific Activity (dpm $[^{32}P]$/pmol P_i)
acid-soluble	100 ± 10	A. 12.8 B. 4.67 ± 1.80 C. 10.51 ± 2.37
charcoal-treated supernatant	8.7 ± 2.8	A. 11.0 B. 4.54 ± 0.23 C. 9.48 ± 2.47
charcoal pellet	63.6 ± 16.8	A. 11.05 B. 4.54 ± 0.37 C. 11.30 ± 0.95
thymidine	27.8 ± 8.6	A. N.D. B. N.D. C. 10.07 ± 0.66

Approximately 10^9 cells were pulsed with $[^{32}P]P_i$ (1 μCi/ml, 8 hours) and with $[^3H]$dThd (1 μCi/ml, 2 hours). Cells were transferred to 10^{-5} mol/l phosphate medium 24 hours prior to addition of $[^{32}P]P_i$. The acid-soluble fraction (row 1) was obtained as previously described (Materials and Methods). Aliquots were taken for determination of radioactivity and P_i. A third aliquot was treated with dextran-coated charcoal (2.5 mg Norit A + 0.25 mg dextran) and centrifuged (10 minutes, 1500g). The resultant supernatant (row 2) was similarly analyzed for radioactivity and P_i. The charcoal precipitate was then extracted with ammoniacal ethanol (8). A portion of this extract (row 3) was analyzed for radioactivity and P_i. A second portion was lyophilized and chromatographed as previously described. The fraction corresponding to a dTMP standard run in parallel was eluted with 0.01N HCl (row 4). Aliquots were again taken for counting and determination of P_i. Column 1 presents these fractions. Column 2 (recovery of $[^3H]$dThd) provides a marker for the distribution of nucleotides among soluble pool components. Radioactivity in the acid-soluble pool is taken as 100%. Column 3 (^{32}P specific activity) is obtained by dividing the observed ^{32}P dpm in the sample by the mass of phosphate detected. In column 3, A, B and C represent different experiments. N.D. means not done.

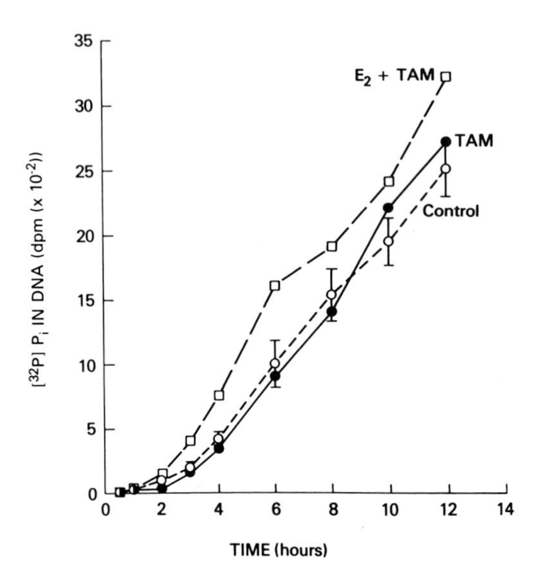

FIGURE 4. Hormonal regulation of specific activity of acid-soluble phosphate pools in MCF-7 cells.

Cells were transferred to 10^{-5} mol/l phosphate medium 8 hours before a second medium change and hormone addition at time 0. Samples were harvested at the times indicated and the acid-soluble fraction analyzed for phosphate content and ^{32}P. Values are normalized per unit protein in the sonicate and are the average of at least 2 determination ± 1 SD. Only the standard deviations of control samples are presented although standard deviations did not exceed 10% in any group. control (0———0); tamoxifen (---); estradiol + tam (0———0)

in specific activity of phosphate as a consequence of this decrease in pool size.

We conclude that intracellular phosphate pools in MCF-7 cells are affected by hormone treatment and that variability in the system makes independent determination of the specific activity of phosphate pools mandatory if extrapolation to DNA synthetic rates are to be made. Our data also suggest that the measurement of DNA synthesis should be based on a brief period of labeling time after equilibrium conditions have been established in DNA precursor pools.

Since ^{32}P is an effective precursor for a variety of cellular materials (22) it is critical to measure incorporation uniquely into DNA. The use of hydroxylapatite chromatography to isolate DNA is well known (19,20).

Table 3 shows the recovery profile from MCF-7 cells pulsed with either [^{3}H] dThd, [5-^{3}H] Urd, or [^{3}H] Leu. Column 1 notes the step in the fractionation process. The remainder of the table presents the percent of radioactivity recovered following acid-precipitation on millipore filters. Total acid-precipitable counts are taken as 100%. At this time only the top 4 rows of the table are of interest. HAP chromatography, RNAase + pronase incubation, or HAP chromatography subsequent to RNAase + pronase incubation all yield excellent recovery of DNA ([^{3}H] dThd incorporation, column 2). However considerable amounts of uridine and leucine label remained following RNAase + pronase treatment alone and uridine was a substantial contaminant when samples were directly applied to HAP columns without prior enzymatic hydrolysis. HAP chromatography and enzymatic hydrolysis together, however, reduce uridine and leucine label to background levels, yielding high recovery of DNA free from contamination with protein or RNA. This procedure was followed in all subsequent experiments.

TABLE 3. Recovery of radioactivity during purification of DNA and dTMP.

	Fraction	[^{3}H]dThd	[^{3}H]Urd	[^{3}H] Leu
1.	Total acid-precipitable	100 ± 18	100 ± 9	100 ± 22
2.	RNAase + pronase incubation	82 ± 17	15 ± 4	42 ± 11
3.	Hydroxylapatite chromotography	96 ± 16	11 ± 6	2 ± 0.7
4.	RNAase + pronase + hydroxylapatite chromotography	94 ± 11	1 ± 0.5	1 ± 0.2
5.	acid-hydrolyzed precipitate	61 ± 1	10 ± 4	12 ± 2
6.	TLC eluate (dThd, dTMP)	42 ± 6	0.5 ± 0.1	0.5 ± 0.1

Approximately 10^6 cells were pulsed with either [^{3}H]dThd (1 μCi/ml, 1 hour), [^{3}H]Urd (1 μCi/ml, 1 hour), or [^{3}H]Leu (1 μCi/ml, 1 hour). At steps 1-4 of purfication, acid-precipitable material was collected by the millipore filter technique for determination of radioactivity. Aliquots were taken directly for counting at steps 5 (differential acid hydrolysis of total acid-precipitable material) and 6 (eluate from thin-layer chromotography of fraction 5). All values are reported as the mean of 3 independent determinations divided by the value obtained for total acid-precipitable material (row 1) ± 1 SD and were normalized per unit protein in the sonicate.

To confirm that all ^{32}P eluted from HAP columns represents only label in DNA cells labeled with [^{32}P] P_i and [^{3}H] dThd were harvested and various cell fractions isolated (Table 4). The eluate from HAP chromatography + enzymatic hydrolysis was divided in half for acid-precipitation on millipore filters and acid-precipitation and hydrolysis to constituent nucleotides. A [^{14}C] dTMP standard was added to the acid-precipitate in order to estimate recovery during further fractionation. Following lyophilization of solubilized material samples were chromatographed. The fractions corresponding to appropriate markers (dTMP, dCMP) were identified, and counted directly. Values are presented as % of recovery (sonicate = 100%). Column 2 ([^{3}H] dThd) provides an index of the efficiency of DNA recovery through the hydrolysis step. Virtually all acid-precipitable thymidine label was recovered in the HAP eluate and following acid-hydrolysis. The pattern of recovery of ^{32}P label is presented in column 3. Approximately 2% of total ^{32}P is recovered in DNA. On further chromatography approximately 25% of ^{32}P in the HAP eluate can be identified in dTMP, (the anticipated value if the HAP eluate is highly purified DNA, and if thymidine represents approximately 25% of the total nucleotide bases in DNA).

The above results suggest that label eluted from hydroxylapatite following enzymatic hydrolysis is a valid index of net DNA synthesis. The conversion from dpm to mass units of incorporation is contingent upon determination of the specific activity of [^{32}P] P_i in acid-soluble pools.

Given the availability of 2 independent methods for determination of mass of DNA synthesized ([^{32}P] P_i incorporation and ethidium bromide fluorometry) it is possible to investigate the possible contribution of unscheduled DNA synthesis or repair to DNA synthesis in MCF-7 cells. Measurements based on ^{32}P incorporation could be an overestimate of net DNA synthesis under conditions in which repair is a significant part of incorporated label. A comparison between the mass of DNA synthesised (based on incorporation of ^{32}P of known specific activity into DNA over a 72 hour period) and measurements of mass (based on ethidium bromide fluorometry) at the start and at various points during that 72 hour period is presented in Table 5. Even under circumstances in which prolonged administration of ^{32}P may induce cell and DNA damage, the amount of repair is in fact not measureable against a background of replicative DNA synthesis. Possibly, under circumstances in which replicative DNA synthesis is blocked (i.e. drug administration), repair synthesis may be more important.

TABLE 4. Recovery of $[^{32}P]P_i$ from hydroxylapatite and thin-layer chromatography.

Fraction	$[^3H]$dThd %	$[^3H]$dThd dpm	$[^{32}P]P_i$ %	$[^{32}P]P_i$ dpm
soncicate	100 ± 10	213745	100 ± 13	1.71×10^7
acid-precipitable	81 ± 11	171587	28 ± 1	4.80×10^6
acid-soluble	19 ± 1		71 ± 2	
charcoal-treated supernatant	1 ± 0.2		33 ± 5	
charcoal pellet extract	11 ± 2.5		12 ± 2	
HAP eluate (DNA)	80 ± 8	170996	2 ± 0.1	
				167220 ± 8763
acid-hydrolysis	83 ± 4		2 ± 0.1	
lyophilization			2 ± 0.1	
chromatography			0.5 ± 0.05	
				35510 ± 3473
recovery standard $[^{14}C]$dTMP				0.81
dpm corrected for recovery				43840 ± 4339
% of dpm recovered from HAP				0.26

MCF-7 cells were transferred to 10^{-5} mol/l phosphate medium 24 hours prior to labeling with $[^3H]$dThd (1 μCi/ml, 2 hours) and $[^{32}P]P_i$ (1 μCi/ml, 8 hours). Dpm in the sonicate were taken as 100%. Cell sonicates were processed as previously described (Materials and Methods). Known amounts of $[^{14}C]$dTMP (0.1 μCi) were added to the acid-precipitable pellet from unlabeled MCF-7 cells just prior to acid-hydrolysis as an index of recovery for chromatography. All values are the mean of 3 determinations ± 1 SD and were normalized per unit protein in the sonicate.

TABLE 5. Contribution of repair DNA synthesis to measurements of net DNA based on ^{32}P incorporation.

Time (hours)	nmol DNA (^{32}P)	nmol DNA (EtBr)	Spec. Act. (dpm/nmol ^{32}P)	Ratio (^{32}P/EtBr)
0	0	31.33 ± 4.4	0	0
24	18.2 ± 3.6	53.6 ± 7.8	10.7 ± 1.26	0.816
48	29.6 ± 3.7	60.7 ±12.6	15.9 ± 3.7	1.006
72	34.0 ± 6.2	65.4 ± 7.0	11.4 ± 2.4	0.997

Cells were replicately plated in IMEM + 2.5% charcoaled calf serum + 10^{-7} mol/l insulin. When cells were subconfluent medium was replaced with IMEM containing 10^{-5} mol/l phosphate. At time 0 medium was again replaced with IMEM + 10^{-5} mol/l phosphate, 5 x 10^{-9} mol/l 17β-estradiol, and approximately 1 μCi/ml ^{32}P P_i. One group of samples was immediately harvested (time 0) and other groups were harvested at successive 24 hour intervals. DNA synthesis was quantitated by ethidium bromide fluorometry (column 3) and by conversion of ^{32}P incorporation to mass units based on the specific activity of ^{32}P in the acid soluble pool (column 4). Isolation of DNA was performed as previously described in Materials and Methods. The ratio of DNA synthesized (^{32}P) to the difference in mass between 0 time and the indicated time points (EtBr) is shown in column 5.

Given the validity of methodology for isolation of DNA and determination of specific activity of acid-soluble phosphate pools it is possible to develop an experimental protocol to measure DNA synthesis. In any given experiment each experimental group is divided into three; one subset is labeled for 6 hours with [^{32}P] P_i, a second is similarly labeled for 8 hours, and a third is labeled for 8 hours and in addition is pulsed for the 2 hours prior to harvest with [^{3}H] dThd. This protocol is necessitated by the extended period required for equilibration of [^{32}P] P_i with intracellular phosphate pools and by the fact that dThd administration (required for other experiments not discussed here) affects the pattern of net DNA synthesis thus is an independent variable in experimental design. A simple experiment in which the major variable is estrogen administration is shows in Table 6.

TABLE 6. Presentation of raw data derived from a typical experiment.

Group	Protein (mg)	DNA (ug)	Acid soluble (nmol P_i)	Acid soluble (dpm ^{32}P)	^{32}P (dpm)
Control					
6 hours	0.523	23.1	30.6	1.61	6685
8 hours	0.463	21.7	25.5	1.26	10115
8 hours + dThd	0.508	22.2	28.0	1.51	12670
Estradiol					
6 hours	0.869	27.9	45.0	2.75	8885
8 hours	0.876	28.6	43.6	2.63	16780
8 hours + dThd	0.793	29.0	41.1	2.47	18056

Experimental groups (column 1) are depicted as control and estradiol (5×10^{-9} mol/l 32 hours). Within each group 6 hour and 8 hour subgroups refer to the time of labeling with [^{32}P]P_i. In the 8 hour plus dThd subgroup cells are labeled for 8 hours with [^{32}P]P_i and for the last 2 hours with [^3H]dThd. Columns 2 and 3 refer to the protein and DNA content of the sonicate. Columns 4 and 5 refer to the mass and radioactivity of phosphate in the acid soluble fraction and are determined as previously described. Column 6 applies to radioactivity detected in DNA (eluted from hydroxylapatite columns). Standard deviations are not presented but were in the range of 10-15% of presented values.

These data transformed to mass (pmol) of phosphate incorporated are illustrated in Table 7.

TABLE 7. Reduction of raw data to mass units of incorporation.

Group	P_i (nmol/mg)	^{32}P (dpm/pmol P_i)	$[^{32}P]P_i$ (pmol/mg protein)	$[^{32}P]P_i$ (pmol/mg DNA)
Control				
6 hours	58.5	52.6	127.1	2822
8 hours	55.1	49.4	204.8	4492
8 hours + dThd	55.1	53.9	235.1	5361
Estradiol				
6 hours	51.8	61.1	145.4	5536
8 hours	49.7	60.3	278.2	8513
8 hours + dThd	51.8	60.0	300.9	8245

Column 1 refers to experimental groups as previously described (Table 6). Columns 2 and 3 present the mass and specific activity of phosphate in the acid soluble pool. Column 4 refers to the mass of precursor incorporated into DNA during the labeling period indicated in column 1. Mass units are derived by dividing the observed dpm for each precursor by its known or presumed specific activity ($[^{32}P]P_i$ = value presented in column 5).

The final analysis involves 3 sets of values; the effect of thymidine on the system (dThd/control), the effect of the experimental manipulation (estrogen/control) and the effect of thymidine coupled with the effect of estrogen (estrogen + dThd/control) are shown in Table 8.

TABLE 8. Data reduction to incorporation during a 2 hour period.

	Group	P_i/mg protein	P_i/mg DNA
1.	Control	77.7	1670
	Control + dThd	108	2539
	Estradiol	132.8	2977
	Estradiol + dThd	155.5	2709

With such an experimental design it is possible to investigate the hormonal regulation of net DNA synthesis. The effect of varying concentrations of estradiol on the incorporation of labeled precursor ([^{32}P] P_i) into DNA is shown in Figure 5. After 32 hours treatment, there was a dose related increase in the incorporation of phosphate into DNA of MCF-7 cells which was evident between 10^{-7} and 10^{-11} mol/l and maximal at 10^{-9} mol/l.

A time course in which experimental values are normalized against control cells at the same time point is shown in Figure 6. The effect of thymidine alone is limited to early time points (18 to 30 hours). Estrogen treatment alone induces a slight peak at approximately 18 hours and major effects at later stages of the incubation as values for net DNA synthesis rose progressively up to 5 fold over control levels. The coupled

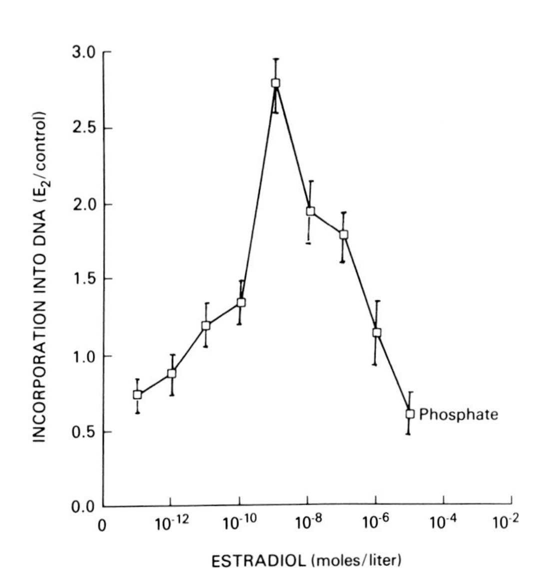

FIGURE 5. Effect of varying concentrations of estradiol on DNA systhesis in MCF-7 cells.

Cells were treated as previously described. 24 or 26 hours after time 0 [^{32}P] P_i (1 μCi/ml) was added to wells. Radioactivity in acid-soluble and DNA fractions was determined along with protein, DNA, and acid-soluble phosphate contents. Values for incorporation of phosphate (pmol/2 hours) were calculated and normalized per unit DNA. Results are presented as the ratio to control values.

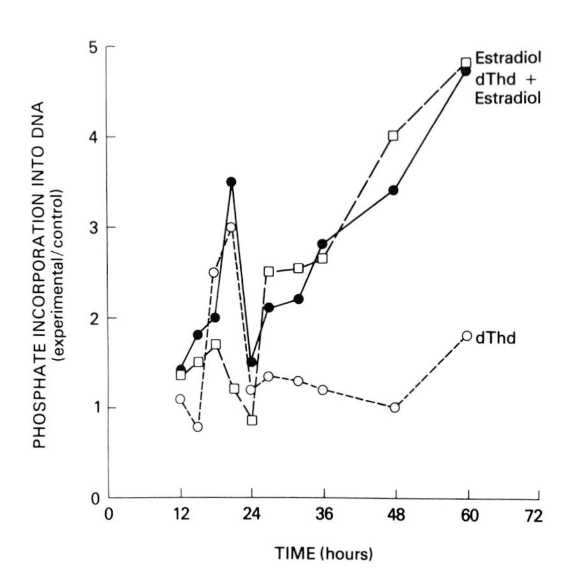

FIGURE 6. Effect of exogenous thymidine and estradiol on net DNA synthesis in MCF-7 cells.

Cells were treated with estradiol (5×10^{-9} mol/l) and normalized against control (untreated) cells. Standard deviations within any experimental group did not exceed 15%.

effect of estradiol and thymidine essentially lies in an increase in the magnitude of the early wave of DNA synthesis over either estrogen treatment or thymidine administration alone. The first wave of synthesis in MCF-7 cells is thus primarily associated with availability of exogenous thymidine and the second with availability of estrogenic hormones.

A similar experiment in which tamoxifen administration is the major variable is shown in Figure 7. Thymidine administration alone (10^{-7} mol/l, 5 µCi/ml) increased net DNA synthesis during early stages of the incubation. Tamoxifen profoundly inhibited incorporation of phosphate into DNA. Tamoxifen and thymidine (10^{-7} mol/l) together consistently permitted recovery of MCF-7 cells from tamoxifen arrest and in the experiment shown here produced an increase above control levels during the earlier wave of DNA synthesis. The ability of dThd to reverse tamoxifen inhibition of phosphate incorporation during later stages of the incubation required higher concentrations of thymidine.

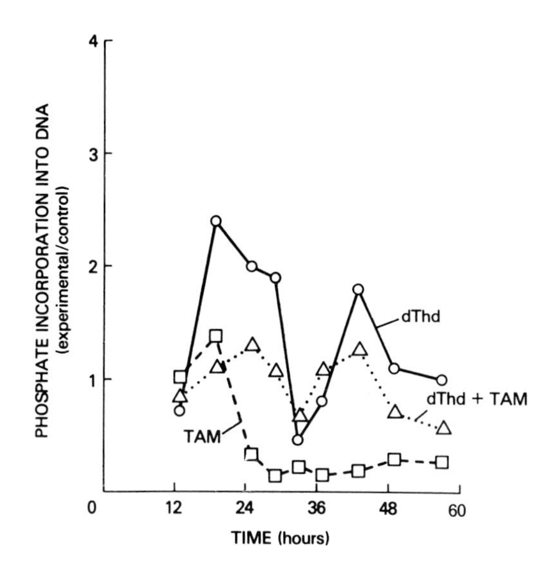

FIGURE 7. Effect of exogenous thymidine and tamoxifen on net DNA synthesis in MCF-7 cells.

Cells were treated with tamoxifen (2×10^{-6} mol/l) and data reduced as previously described. Standard deviations did not exceed 15%.

Thus, estrogens and anti-estrogens exert powerful regulatory effects on DNA synthesis in MCF-7 cells. The growth fraction is dramatically increased by estrogen administration. Cells frequently show two major peaks of DNA synthesis; an early wave which is associated with the availability of exogenous thymidine (18 to 30 hours) and a second wave of synthesis unaffected in magnitude by the administration of thymidine. The effects of estrogen alone are manifested in an amplification of these waves of DNA synthesis rather than in a change in the time frame within which cells enter DNA synthesis. Estrogen action relative to controls is especially evident during later phases of the serum-free incubation. Tamoxifen administration decreases net DNA synthesis within 18 hours after treatment and this effect can be reversed by simultaneous exposure to estradiol. Thymidine partially reverses the inhibition produced by tamoxifen during the early wave of DNA synthesis.

DISCUSSION

MCF-7 cells resemble other vertebrate cell lines in a number of features relating to the utilization of phosphate. The distribution of intracellular phosphate in other tissue culture systems is strikingly similar to data presented here for MCF-7 cells. Intracellular acid-soluble phosphate levels between 20 and 80 nmol/10^6 cells have been reported in other cell lines over a wide range of ambient phosphate concentrations (22). Approximately one-third to one-half of this is reportedly P_i and the remainder (organic phosphate) is primarily made up of ribonucleoside phosphates with a small deoxyribonucleoside phosphate component (1-2%) as well as sugar phosphates (3%). Equilibration among various soluble pool components is reported to be extremely rapid (24). Cunningham and Pardee (25) in a number of experiments in mouse 3T3 fibroblasts observed that approximately 50% of organic acid-soluble ^{32}P label could be localized in nucleotide components within 15 minutes of addition of label, that ^{32}P uptake was energy dependent, and that addition of tracer amounts of ^{32}P failed to influence pool size. It can also be noted that estimates of pool size of nucleotide components in other cell lines which are based on the assumption of a constant specific activity of ^{32}P in all such pools (26) give excellent agreement with independent estimates based on enzymatic assay (27,28).

The time requirement of equilibration of label in intra-cellular phosphate pools in other tissue culture systems ranges from 4 to 48 hours depending on growth rate and cell type (22). Contact-inhibited cells such as chick or mouse 3T3 fibroblasts display a slower equilibration (27) wheras rapidly growing cells will generally reach a steady state within 12 hours (29). The rate of uptake of phosphate into acid-soluble pools has been variously estimated as from 2.4 to 120 nmol/hour (22).

It is apparent that the usual size of the soluble phosphate pool in MCF-7 cells (30 nmol/mg protein), the rate of uptake (4-8 nmol/hour, Table 3), the time required for equilibration of label (6-8 hours, Figure 2) and the distribution of label among acid-soluble pool components (Table 2) are consistent with data reported for other vertebrate cell types in tissue culture.

The use of [^{32}] P_i to measure rates of DNA synthesis has been reported in a variety of tissue culture systems. It is generally accepted that the incorporation of 2 molecules of phosphage into DNA represents the synthesis of 1 base pair of nucleotides; hence 2 nanomoles of incorporated phosphage = 1

II. dThd/Control	1.39	1.52
E_2/Control	1.71	1.78
E_2 + dThd/Control	2.00	1.61

Column 1 refers to experimental groups. Column 2 refers to the differences in the mass of precursor incorporated into DNA between 6 and 8 hours of labeling. In section II data are still further reduced to a set of ratios relative to appropriate controls. The proper control for [^{32}P]P_i is found in row 1 of section 1 (control).

The systematic approach to the investigation of DNA synthesis described here, however, can be utilized in an extended study of the control of DNA synthesis in mammalian cells in tissue culture. The adaptation of a variety of techniques previously described in the literature to analysis of very small amounts of material permits such questions as hormonal regulation of DNA synthesis to be examined conveniently, repetitively and accurately. With respect to utilization of phosphate in DNA synthesis the MCF-7 line appears representative of other tissue culture systems and may provide an excellent and useful model system due to its known responsiveness to a variety of hormones, i.e. insulin (30), glucocorticoids (31), and estradiol (2).

The matter which is the primary concern of this paper is the specific regulation of DNA synthesis in MCF-7 cells by estradiol. It is clear that estrogen administration results in an increase in the growth fraction of MCF-7 cells. It is not clear however whether this is due to 1. a decrease in cell death, 2. an increased growth rate (decrease in length of the cell cycle), or 3. an increased recruitment of nondividing cells arrested in G_1 (G_0) into the actively dividing population. There is evidence that estrogen in vivo acts as a trophic hormone under physiological conditions by recruitment of cells from G_0 into G_1 as, for example, in the induction of uterine growth (32,33). However it has also been reported that estrogen administration in MCF-7 cells does in fact decrease cell cycle time and increased growth was attributed in the case to a specific decrease in the length of the G_1 phase of the cell cycle (34). In addition estradiol has been observed to increase the mitotic index of MCF-7 cells by 51% (serum-free conditions, 2×10^{-9} mol/l estradiol, 72 hours) and by 72 to 240% (1% charcoal-treated calf serum, 2×10^{-9} mol/l estradiol, 24 to 48 hours). The mitotic index was observed to vary between 1.28% and 15.34% depending on the circumstances of growth and incubation time (35).

The fact that MCF-7 cells represent a transformed cell line may be of considerable significance in interpretation of data relating to cell cycle kinetics. Transformed cells are known exhibit dormancy, being recruited from the actively dividing population into the non-dividing population (36). However the depth of their arrest is considered to be slight and they are comparatively easily returned to the cell cycle (37,38). It is well established that the growth requirements (i.e. serum, Ca^{++}) of transformed cells are characteristically less than those of non-transformed cell lines and this has been attributed in part to a reduction in the stringency of requirements for passage through G_1 at the so called restriction point (39).

The pattern of net DNA synthesis in MCF-7 cells and the effects of exogenous thymidine on that pattern yield some interesting suggestions on the possible role of salvage and de novo pathways of pyrimidine production in MCF-7 cells and the mechanisms by which estrogen may increase the rate of DNA synthesis. The early wave of DNA synthesis in MCF-7 cells is modulated by the availability of exogenous dThd. Thymidine administration increases DNA synthesis at early time points in controls, estrogen-treated cells and tamoxifen treated cells. Estrogen-treated cells have a greater ability to utilize thymidine as a trigger for net DNA synthesis than do controls. Tamoxifen-treated cells which are profoundly inhibited are rescued by exogenous thymidine during the early stage of the incubation. Thymidine kinase activity is increased by estrogens and decreased by anti-estrogens in MCF-7 cells (40). The increase in net DNA synthesis in estrogen-treated populations relative to controls at early incubation stages could therefore be due to increased thymidine kinase activity. The ability of thymidine to reverse tamoxifen inhibition implies that, although thymidine kinase activity is decreased, there is no functional impairment of salvage activity.

During later stages of the incubation, effects of thymidine on net DNA synthesis were not as apparent, indicating an increased dependency in MCF-7 cells on de novo pyrimidine production. Such a dependency may reflect 1. a depletion of available salvage sources of thymidine in the earlier wave of DNA synthesis or 2. an estrogen-dependent induction of critical de novo or intermediary enzymes.

In any event, these studies conclusively show that estradiol and tamoxifen have profound and opposite effects on true replicative DNA synthesis in human breast cancer cells in culture.

REFERENCES

1. Lippman, M.E., and Bolan, G. (1975) Nature 256, 592-593.
2. Lippman, M.E., Bolan, G, and Huff, K. (1976) Cancer Res. 36, 4595-4601.
3. Brooks, S.C., Locke, E.R., and Soule, H.D. (1973) J. Biol. Chem. 248, 6251-6253.
4. Horwitz, K.B., Costlow, M.E., and McGuire, W.L. (1975) Steroids 26, 785-795.
5. Rasmussen, R.E., and Painter, R.B. (1966) J. Cell Bio. 29, 11-19.
6. Soule, H.D., Vasquez, J., Long, A., Albert, S., and Brennan, M.J. (1973) J. Natl. Cancer Inst. 51, 1409-1416.
7. Richter, A., Sanford, K.K., and Evans, V.J. (1972) J. Natl. Cancer Inst. 49, 1705-1712.
8. Strobl, J.S., and Lippman, M.E. Cancer Res. in press.
9. Harper, M.J.K. and Walpole, A.L. (1967) J. Reprod. Fertil. 13, 101-119.
10. Lowry, O.H., Rosebrough, N.J., Farr, A.L., and Randall, N.J. (1951) J. Biol. Chem. 193, 265-275.
11. Lippman, M., Bolan, G., and Huff, K. (1976) Cancer Treat. Rep. 60, 1421-1429.
12. Fiske, C.H., and Subbarow, Y. (1925) J. Biol. Chem. 66, 375-405.
13. Tsuboi, K.K., and Price, T.D. (1959) Arch. of Biochem. and Biophys. 81, 223-237.
14. Randerath, K., and Randerath, E. (1964) J. Chromatog. 16, 111-125.
15. Chmielewicz, Z.F., and Acara, M. (1964) Anal. Biochem. 9, 94-99.
16. Ogur, M., and Rosen, G. (1950) Arch. Biochem. 25, 262-276.
17. Boer, G.J. (1975) Anal. Biochem. 65, 225-231.
18. Beers, P.C., and Wittliff, J.L. (1975) Anal. Biochem. 63, 433-441.
19. Markov, G.G., and Ivanov, I.G. (1971) Anal. Biochem. 59, 555-563.
20. Meinke, W., Goldstein, D.A., and Hall, M.A. (1974) Anal. Biochem. 58, 82-88.
21. Yamana, K., and Sibitani, A. (1960) Biochem. Biophys. Acta 41, 304-309.
22. Hauschka, D.V. (1973) in Methods in Cell Biol. (D.M. Prescott, ed.), Vol. 7, Academic Press, New York and London.
23. Colby, C., and Edlin, G. (1970) Biochemistry 9, 917-920.
24. Plagemann, P.G.W. (1972) J. Cell Biol. 52, 131-146.
25. Cunningham, D.D., and Pardee, A.B. (1969) Proc. Nat. Acad. Sci. U.S. 69,702-709.
26. Weber, M.J. (1971) J. Biol. Chem. 246, 1828-1833.
27. Lindberg, U., and Skoog, L. (1970) Anal. Biochem. 34, 152-160.
28. Skoog, K.L., Nordenskjold, B., and Bjursell, K.G. (1973) European J. Biochem. 33, 428-432.

29. Jeanteur, P., Amaldi, F., and Attardi, G. (1968) J. Mol. Biol. 33, 757-775.
30. Osborne, C.K., Bolan, G., Monaco, M.E., and Lippman, M.E. (1976) Proc. Natl. Acad. Sci. U.S. 73, 4536-4540.
31. Lippman, M.E., Bolan, G., and Huff, K. (1976) Cancer Res. 36, 4602-4609.
32. Hamilton, T.H. (1968) Science 161, 649-661.
33. Epifanova, O.I. (1971) in The Cell Cycle and Cancer (R. Baserga, ed.), Marcel Dekker, New York.
34. Weichselbaum, R.R., Hellman, S., Piro, A.J., Nove, J.J., and Little, J.B. (1978) Cancer Res. 38, 2339-2342.
35. Strobl, J.S. (1979) Equilibrium Binding Analysis of Estrogen Receptors and Pharmacokinetic Studies of the Dissociation of Estrogens and the Effects of Antiestrogens in Estrogen-responsive Human Breast Cancer Cells (MCF-7) in Tissue Culture. A Dissertation submitted to the Faculty of the Graduate School of Arts and Sciences of George Washington University in partial satisfaction of the requirements for the degree of Doctor of Philosophy.
36. Clarkson, B. (1973) in Antineoplastic and Immunosuppressive Agents for the Handbook of Experimental Pharmacology (D. Johns and A.C. Sartorelli, eds.), Springer-Verlag, New York.
37. Costlow, M., and Baserga, R. (1973) J. Cell Physiol. 82, 411-420.
38. Todaro, G. (1972) Nature New Biology 240, 57-60.
39. Pardee, A.B. (1974) Proc. Nat. Acad. Sci. U.S. (1974) 71, 1286-1290.
40. Bronzert, D.A., Monaco, M.E., Pinkus, L., Aitken, S.C., and Lippman, M.E. Submitted to Bioc. J.

Control Mechanisms in Animal Cells,
edited by L. Jimenez de Asua et al.
Raven Press, New York © 1980.

Transition Probability and the Regulation of the Cell Cycle

Robert Shields

Unilever Research, Sharnbrook, Bedfordshire MK44 1LQ, England

There is general agreement that the cell cycle times of cells both in vivo and in culture are highly variable although the origin of this variability has remained unclear. The variability is not due to trivial reasons such as genetic heterogeneity in the cell population as this variability is expressed within cell clones and even between sister cells (8).

Any idea as to the origin of this variability must take into account the following facts: (i) most of this variation occurs in the G1 phase of the cycle, (ii) this variation increases as cell proliferation rates decrease, (iii) sister cell cycle times are frequently correlated although the cycle times of mother and daughter cells are not, (iv) it is the length and variability of the G1 which is the major determinant of the length and variability of the cell cycle and so determines proliferation rates. The purpose of this chapter is to compare the principle deterministic models of the cell cycle with the transition probability (TP) model of the cell cycle to see if a choice between the models can be made on kinetic grounds.

Deterministic Explanations of the Distribution of Cell Cycle Times

Until recently it was generally accepted that the cell cycle time and its variability in a cell population were "deterministic." That is the cell cycle time of an individual cell is uniquely determined by its precise biochemical composition. Variation could arise because individual cells in the population differ in properties which determine their transit through the cell cycle. In a genetically homogeneous population such differences could occur because of unequal partition of cell constituents at division or random fluctuation in cell cycle important molecules present at low concentration. The precise form of the cell cycle distribution will depend on how these differences interact.

Cell cycle distributions can be modeled fairly well by the normal distribution of the logarithm of the cell cycle time (7). Such a distribution results from the multiplicative interaction of randomly distributed variables (3). A realistic distribution of generation times will also result if individual cells traverse the cycle at normally distributed rates. Since the time taken to complete the cycle will be inversely proportional to the rate of cell cycle traverse the normal distribution of rates given rise to reciprocal normal distribution of generation times (4). Published data on cell cycle distributions has been fitted to both logarithmic normal and reciprocal normal distributions (11).

Experimentally Determined Cell Cycle Distributions

The data shown in FIG. 1 shows the distribution of cell cycle times for exponentially growing Balbc 3T3 cells determined by time-lapse cine photography. Panel 1a shows that the distribution of cell cycle times is skewed towards longer time intervals, a feature that has been noted many times before (6). The other panels show the same data plotted as cummulative distributions on probit paper whose ordinate turns a cummulative normal distribution into a straight line. The results show that neither the normal gausian or logarithmic normal distribution fit the data, however the fit to a normal distribution of rates is excellent, the superior fit of this distribution has been remarked on before (4).

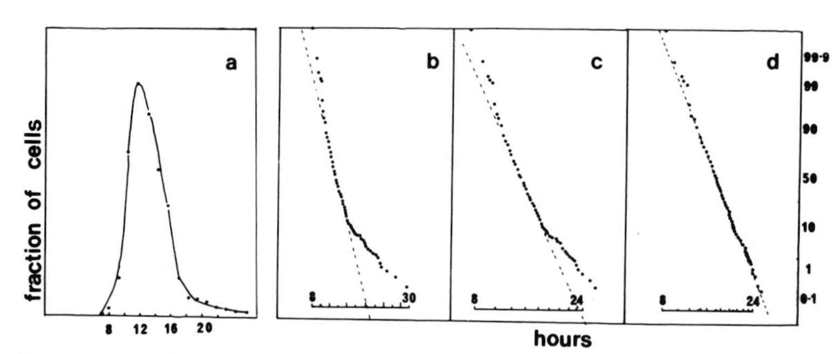

FIG. 1. The distribution of cell cycle times of Balbc 3T3 cells grown in 10% calf serum. The intermitotic times were determined by time lapse cine microscopy, (a) the frequency distribution of cell cycle times, (b) cummulative distribution of proportion of cells with cycle time longer than time shown on ordinate (plotted as probate), (c) same data with logarithm of cycle time on ordinate, (d) distribution of generation rates.

Any successful model of the cell cycle must fit the data at more than one proliferation rate. FIG. 2 shows results from an experiment where SV3T3 cells were grown in different concentrations of serum. Panel 2a shows that as the serum concentration decreases and cell proliferation rates drop the mean cell cycle

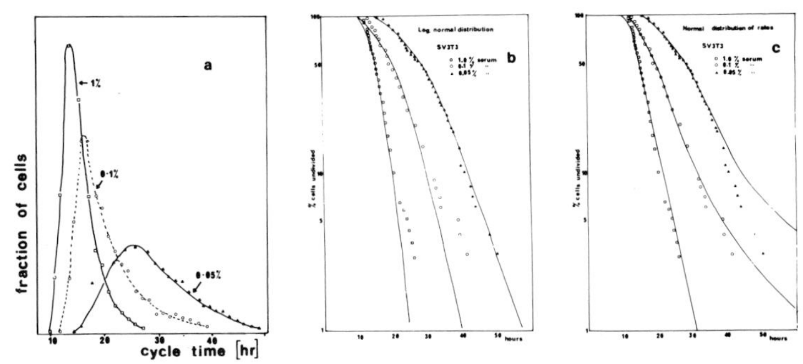

FIG. 2. Distribution of cell cycle times of SV3T3 cells in different levels of serum, (a) frequency distribution of generation times, (b) & (c) the same data plotted as α curves. α_t is the proportion of cells with a cell cycle time greater or equal to the time shown in the ordinate, (b) the solid lines are the best fit for the logarithmic distribution of generation times, (c) the solid lines show the best fit for the normal distribution of generation rates.

gets longer and the variability increases. However there is only a small change in the minimum cell cycle time. Panels b and c show the same data plotted as a cummulative distribution (in this case the ordinate is logarithmic rather than probate). The drawn lines show the best fit distributions for (b) logarithmic normal distributions of generation times, (c) normal distributions of generation rates. As can be seen neither distribution fits the data well although the fit is rather better for the logarithmic normal distribution. From these experiments and that shown in FIG. 1 it appears that neither of thes models are entirely satisfactory. Most importantly these models ignore the fact that most of the cell cycle variability is confined to G1 whereas these models assume that the variability is spread throughout the cycle.

Probablistic Explanation of the Distribution of Cell Cycle Times

The transition probability (TP) model of the cell cycle as originally proposed claimed that the initiation of each cell cycle depends on a single critical event or transition in G1 which occurs at random. Cells in a population were regarded as identical in that they had equal probabilities per unit time of undergoing this transition. The distribution of generation times resulted not because of phenotypic differences between the cells but because of the probablistic nature of the critical event (10). The cell awaiting to undergo the transition was said to be in an indeterminant state (the A state), after the transition the cell entered the deterministic B phase of the cycle. The model further proposed that although the B phase was variable in the population (an assumption necessary to give good fit to the distribution of cycle times) it was the same in sister cells. Thus sister cells differed only in the time they spent in the probablistic A state, and so the distribution of differences of sister cell cycle times (the β curve) would be exponential (5), (9). The exponential form of the β curve has been the most powerful argument in favour of a random event in the cell cycle.

The original formulation of the TP model had the strength that it placed the cycle variability in G1 and provided a quantitative explanation of the behaviour of sister-cells. However it did not explain why B phase should be variable, what form this variability takes and why B phase should be the same in sister cells. Also it did not account satisfactorily for the overall distribution of cell cycle times.

Recently a modified TP hypothesis has been proposed which accounts for B phase variability by postulating a second random transition in the cell cycle (2). This new model also predicts the form of the distribution of cell cycle times and quantitatively explains the sister-sister correlation. The derivation of the model is to be published elsewhere, in this paper I shall examine how well its predictions fit experimental data.

The Cell Cycle contains at least one Random Transition

One postulate of the TP hypothesis is that the probability of a cell undergoing the transition is independent of its age. As a consequence, if B phase is identical in sister cells, identical β curves will be obtained for sister cells grouped according to the youngest cell of the cell pair (9). The data in FIG. 3 shows the relationship between the cycle time of sister cells of Balbc 3T3 cells. The sister-sister correlation in this experiment is 0.604. The data was then divided into four equally sized groups according to the age of the youngest cell of the cell pair (the groups are shown by different solid and open symbols in FIG. 3). Separate β curves were plotted for each of these groupings and the result is shown in FIG. 4b. All the β curves are exponential and parallel with no sign of curvature. Similar results have been obtained with S.Albus (9) and SV3T3 cells (my own unpublished work). It must be stressed that this analysis will only provide meaningful results if each of the sub-groups contains a sufficient number of cells. In the experiment shown each group contained 56 cell pairs, attempts to do this analysis on as few as 15 cell pairs are statistically meaningless (e.g. 6).

I have shown previously that only cell cycle models containing an exponentially distributed phase can give truly exponential β curves. Attempts to simulate exponential looking β curves using other models of the cell cycle invariably give rise to lines with initial curvature (6); (8). The most physiologically plausible explanation for the exponential cell cycle phase is that the cell contains a random transition.

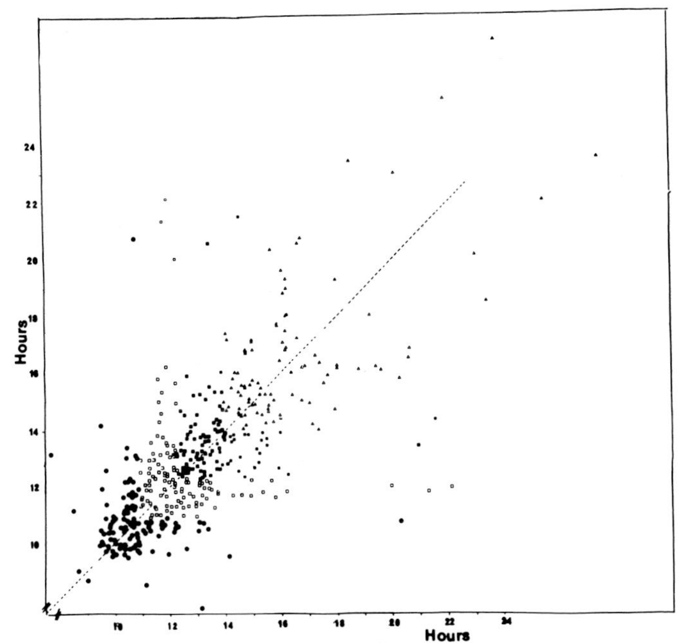

FIG. 3. The relationship between the cycle times of sister cells in the experiment shown in FIG. 1. r_{ss} was 0.604. The different symbols show the different groups of sister cells used in the construction of the β curves shown in FIG. 4(b).

Independent Confirmation of a Random Cell Cycle Transition

The TP hypothesis assumes that the overall variance in the cell cycle V_T is the sum of the variance of A and B states, ie. $V_T = V_A + V_B$. It may be shown analytically (2) that if the B phase is the same in sister cells then the sister-sister correlation coefficient r_{ss} is given by

$$r_{ss} = \frac{V_B}{V_T} = \frac{V_B}{V_A + V_B}$$

Since r_{ss} and V_T (= 7.285 $(\text{hr})^2$) are obtainable from experiment both V_A and V_B may be calculated. Now V_A is postulated to result from the exponential distribution of A state durations. If the rate constant for leaving the A state is K_A then $V_A = 1 / (K_A)^2$, ie. K_A may be calculated from V_A. When this is done the value of K_A (0.59) obtained is similar to the determined from the slope of the β curve (0.65). This is shown in Fig. 4A where the drawn β curve was calculated solely from r_{ss} and the overall population variance. The fact that the K_A obtained from calculation and from the β curve agree within 10% is a powerful validation of the model.

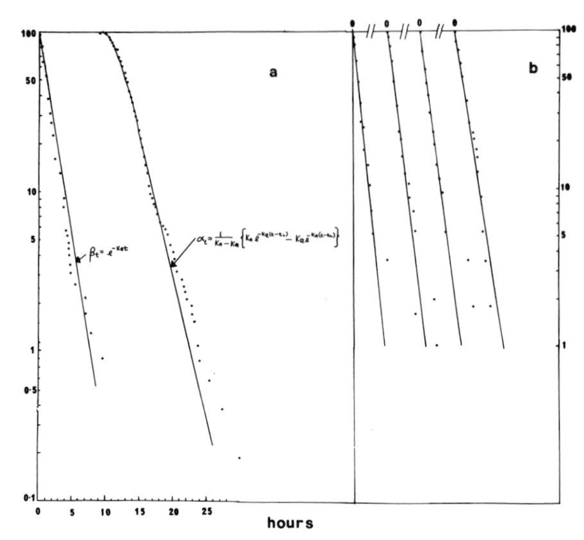

FIG. 4. Semi-logarithmic plots of the percentage of cells (α_t) undivided against age (t) for Balbc 3T3 cells (same experiment as FIG. 1), together with the corresponding plots of the percentage of sibling pairs (β_t) with a difference in intermitotic times $\geqslant t$. There are 514 observations of the cycle time for the α curve and 224 intersibling times in the β curve. (a) The points are the actual data for β (left) and α (right), the lines are calculated as explained in the text. (b) β plots obtained for groups of sister cells ranked according to Ti, the age at division of the earliest dividing cell of the cell pair. The groups correspond to these shown in FIG. 3 and are (left to right) $7.7\text{hr} < Ti \leqslant 10.8\text{hr}$, $10.8 < Ti < 12.2\text{hr}$, $12.2\text{hr} < Ti < 13.75$, $13.75\text{hr} \leqslant Ti$. There are 56 cell pairs in each group

Evidence that the Cell Cycle contains Two Transitions

The evidence outlined in the previous section shows that the cell cycle contains at least one exponentially distributed phase which is postulated to be the result of a random transition in the cell cycle. The evidence for the presence of a second transition comes partly from an examination of the distribution of B phases. The B phase distribution may be calculated from the overall distribution of cell generation times and the slope of the β curve (2). When the resulting distribution of B phases is examined it too appears to contain an exponentially distributed component. If this second exponential phase is due to an indeterminant phase from which the cell exists with a rate constant K_Q per unit time then the variance of B phase is $1/(K_Q)^2$. The value of K_Q may be determined either from the examination of the B phase distribution or from a knowledge of the sister-sister correlation coefficient and overall cell cycle variance as outlined in the previous section. It may be shown (2) that the form of the α curve for the two transition model is

$$\alpha_t = \frac{1}{K_A - K_Q} \left[K_A^{-K_Q(t-t_L)} - K_Q^{-K_A(t-t_L)} \right]$$

(α_t is the proportion of cells whose cell cycle is greater or equal to time t).

When the values of K_A and K_Q calculated from V_A and V_B are substituted in this equation along with an experimentally determined value of t_L the log α curve shown in FIG. 4a is obtained. As can be seen the fit to the experimental data is good.

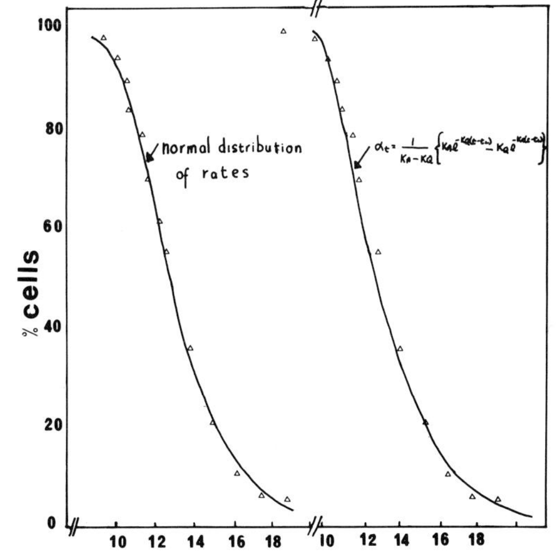

FIG. 5. Linear plots of the percentage of cells (α_t) undivided against age (experiment shown in FIG. 1). The symbols are actual data, the lines are (left) the normal distribution of rates giving the best fit to the data. The line was calculated from FIG. 1d. The right hand line is the distribution calculated for the two transition model of the cell cycle using K_A = 0.59 and K_Q = 0.48. The value of r_{ss} = 0.604. The values of t_L = 9.38hr was determined empirically to give the best fit to the data.

This goodness of fit is emphasised in FIG. 5 where the α curve is plotted with a linear ordinate and compared with the normal distribution of rate curve (obtained from FIG. 1d). The fit of both distributions to the α curve is good, however the two transition model is preferred on a number of grounds. Firstly the transition probability model explains the especially variable nature of G1 as being due to the probablistic nature of the A state (which is contained in G1). The rate normal distribution does not ascribe the variability to any cell cycle phase. Secondly the transition probability model accounts quantitatively for (i) the exponential distribution of differences of sister cells, (ii) the sister-sister correlation.

It should be noted that the α and β curves are not parallel although they may become quite close to being parallel depending on the values of K_A and K_Q. Also the slope of the α curve will be dependent on the age of cell grouping being considered. If the cell population is divided into sub-populations based on cell age the α curves will be different for each sub-group. Thus the finding that α curves are different for different sub-populations is not evidence for kinetic inhomogeneity in the population (6).

CONCLUSION

From the data presented in this paper it is concluded that the two transition model of the cell cycle can satisfactorily account for the distribution of cell cycle times in the cell population. Furthermore the model quantitatively explains the behaviour of sister cells. On kinetic grounds alone this model is to be preferred to deterministic models of the cell cycle. A possible biochemical explanation of the cell cycle kinetics will be presented elsewhere (1,2).

Acknowledgments: This work was carried out at the Imperial Cancer Research Fund during the tenure of a series of short term fellowships. I would like to thank R. Brooks and A.B. Pardee for communicating their results before publication.

References

1. Brooks, R.F. (1980). Soc. Exptl. Biol. Seminar Series (In Press).
2. Brooks, R.F., Bennett, D.C. and Smith, J.A. (1980). Cell in press.
3. Koch, A.L. (1966). J. Theoret. Biol. 12, 276-290.
4. Kubitschek, H.E. (1971). Cell and Tissue Kinetics, 4, 113-122.
5. Minor, P.D. & Smith, J.A. (1974). Nature, 248, 241-243.
6. Pardee, A.B., Shilo, B. & Koch, A.L. (1979). Cold Spring Harbor Symposium (in Press).
7. Schmid, P. (1967). Exp. Cell Res., 45, 471-486.
8. Shields, R. (1977). Nature, 267, 704-707.
9. Shields, R. (1978). Nature, 273, 755-758.
10. Smith, J.A. & Martin, L. (1973). Proc. Natn. Acad. Sci. USA, 70, 1263-1267.
11. Wheals, A.E. (1977). Nature, 267, 647

Control Mechanisms in Animal Cells,
edited by L. Jimenez de Asua et al.
Raven Press, New York © 1980.

Regulation of Early Cell Cycle Events by Serum Components

W. J. Pledger and Walker Wharton

Department of Pharmacology and Cancer Research Center, University of North Carolina School of Medicine, Chapel Hill, North Carolina 27514, U.S.A.

Cultured fibroblasts, such as BALB/c-3T3 cells, require serum for active proliferation. Growth arrest produced by serum deprivation can be achieved by either reducing the serum concentration in the growth media or by growing cells to a high density. Cells rendered quiescent by either means require 12 hours to begin DNA synthesis after the addition of media containing fresh serum. Critical cell cycle regulatory events under the specific control of serum components occur in this 12 hour Go/G_1 transition to S phase.

Pardee (9) described a "restriction control point" in the Go/G_1 phase of the BHK cell cycle that is sensitive to both nutrients and whole calf serum. Baserga (3) has proposed a scheme of biochemical events during Go/G_1 that cascade into the commitment of DNA synthesis. The elucidation of the regulatory events controlled by serum that allow replication has been difficult because of a lack of synchrony and the delay in commitment as well as because serum, a very complex melange, has been tactically used as if it were a single component. Relatively few proteins in serum are needed to propagate several cell lines, although those proteins may vary for different cells (13,19). Serum has been found to have separate components needed to either maintain cell viability (10) or to stimulate replication (7). Transport of nutrients or other essential factors may be facilitated by carrier proteins or hormones also found in serum.

Balk (2), Kohler and Lipton (8) and Ross et al. (14) demonstrated that plasma, the fluid portion of blood, supported viability but not replication in several cell lines. Their observations lead to the discovery that a growth factor released from platelets during clot formation was the mitogen in serum that induced the replication of fibroblasts. This factor, termed platelet-derived growth factor (PDGF), is a basic (pI 9.7) polypeptide, MW 13,000, purified first by Antoniades et al. (1) and later by Heldin et al. (6). Crude PDGF can be obtained by preparing heat-treated (100°C for 10 minutes) extracts of human platelets (subsequently referred to as platelet extract).

COMPETENCE AND PROGRESSION: CONTROL BY SERUM COMPONENTS

Several concentrations of platelet extract were added to quiescent BALB/c-3T3 cells in media containing varying concentrations of platelet-poor plasma (PPP). PPP was prepared by heating plasma to 56° for 30 minutes after all formed bodies had been removed. The percentage of cells that entered S during a 36 hour period was a function of both the amount of platelet extract and the PPP concentration (11). Even though the higher concentration of platelet extract alone induced some cells to enter S, the maximum percentage of cells that synthesized DNA was obtained when optimal PPP was present in the growth media.

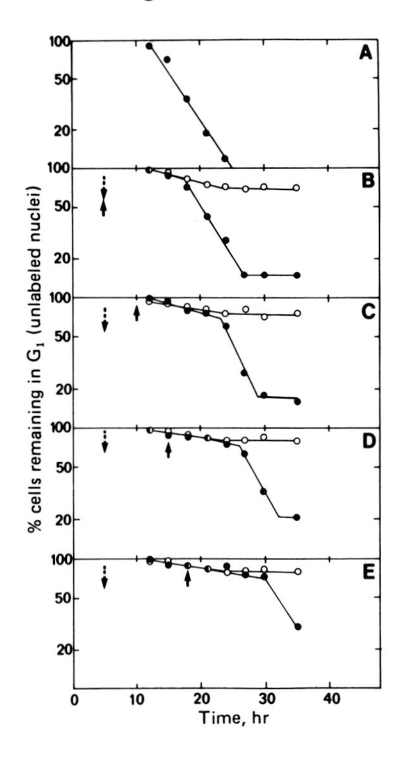

FIG. 1. (A) Cultures were treated with 50 μg of platelet extract at 37° in 0.2ml of medium containing 5% platelet-poor plasma and [^{3}H]dThd. At the indicated times, cultures were fixed and processed for autoradiography. (B-E) Cultures were treated with 50 μg of platelet extract in 0.2 ml of medium for 5 hr (↓) at 37°, washed, and returned to 0.2 ml of medium containing [^{3}H]dThd but lacking platelet-poor plasma. At the times indicated by (↑) the medium was supplemented with platelet-poor plasma (●,5%; O,0.25%). The cultures were fixed and processed for autoradiography at time intervals (11).

A transient exposure of quiescent cells to platelet extract has been shown to allow cells to respond to PPP and enter S phase. If cells were treated with platelet extract, then thoroughly washed and transferred to media alone, there was no stimulation of DNA synthesis. However, if PPP was added to the media after treatment with platelet extract, the cells entered S phase. The total number of cells that initiated DNA synthesis and the rate of entry into S phase was dependent on the PPP concentration (11). Platelet extract was thus said to make cells "competent". Competent cells were able to respond to factors in PPP and undergo "progression" of Go/G_1 and enter S phase. Cells that were not competent could not respond to the "progression factors" in PPP and therefore could not traverse the cell cycle.

Quiescent cells exposed to platelet extract in the presence of PPP entered S after 12-14 hours (Fig. 1a). Other cultures were exposed to platelet extract in the absence of PPP, then transferred to media alone. PPP was added at various times after the removal of platelet extract. In each case the rapid entry of cells into S phase occurred 12 hours after the addition of PPP (Fig. 1b-e). Platelet extract induced a stable state in which the cells were competent to replicate DNA and which was not dependent on the continued presence of the growth factor, although progression through G_1 did not occur until PPP was present.

By the use of a transient exposure of competent cells to PPP for various times, two distinct growth arrest points have been observed (12). These plasma dependent arrest points are illustrated in Figure 2.

Therefore three distinct growth arrest points have been located in the Go/G_1 transition to S phase. One point, competence formation, is controlled by PDGF; the traverse of two points, V and W, are in part controlled by plasma derived components.

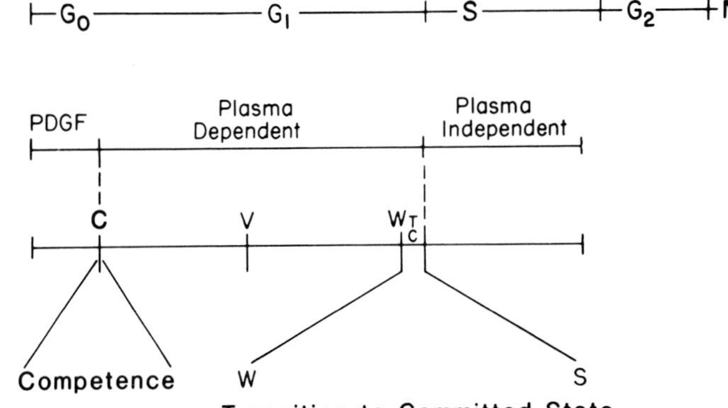

FIG. 2. Model of several sequential events in Go/G_1 that precede DNA synthesis.

When competent cells were exposed to media containing PPP

obtained from either normal rats or rats that had been hypophysectomized, the PPP from hypophysectomized rats was much less active in the support of progression through Go/G_1, than was that obtained from normal rats (16). However, in either type of PPP, cells initiated DNA synthesis 12 hours after the addition of plasma. The addition of purified somatomedin C to PPP from hypophysectomized rats increased the number of cells that entered S, but did not alter the 12 hour lag period for Go/G_1 progression. Recently it has been shown that competent cells placed in PPP from hypophysectomized animals became arrested at the V point; e.g. addition of somatomedin C to such cells allowed entry into S after only 6 hours (18). It should be pointed out that PPP components other than somatomedin C were necessary for the progression of competent cells to V, that is no progression of Go/G_1 was found when competent cells were transferred to media containing only somatomedin C (16). Stiles et al. (17) have shown that limitation of essential amino acids also brought about growth arrest at the V point. Recently in collaboration with Stiles et al. (16), we proposed a dual control model for cell growth in which substances such as PDGF which induce competence formation, act on specific cells to initiate cell replication. Progression factors work to control growth once cells have been brought into readiness for replication, and, in general, would display a wider range of specificity in order to maintain the balanced growth in the total organism.

COMPETENCE AND PROGRESSION SPECIFIC EVENTS

Quiescent cells must be rendered competent before undergoing progression of the cell cycle. Therefore, we suggest that PDGF must induce one or more unique reactions not induced by other serum components. Such putative reactions initiate cell replication and could include the synthesis of a specific protein or RNA, alterations in enzyme activities, nutrient uptake, protein phosphorylation or some other cellular event. PPP may also induce specific biochemical events. These events may respond in competent cells. An example of a unique action of PDGF is the induction of the preferential synthesis of proteins not induced by PPP (manuscript in preparation). On the other hand somatomedin C binding to 3T3 cells was increased only after sequential addition of PPP to PDGF-treated cells (4). This is apparently an example of the action of plasma factors on only competent cells. Regulation of uridine kinase activity provides an example of an alternate system of control by serum components.

SERUM COMPONENTS AND REGULATION OF THYMIDINE AND URIDINE KINASE

Uridine kinase and thymidine kinase activity in extracts of 3T3 cells were determined after quiescent cells were stimulated to undergo replication. Uridine kinase activity increased rapidly from 3 to 6 hr. following the stimulation of quiescent cells with fresh serum, and remained at the increased level for at least 12 hours. The increase in activity of uridine kinase was dependent on the concentration of serum (Fig. 3).

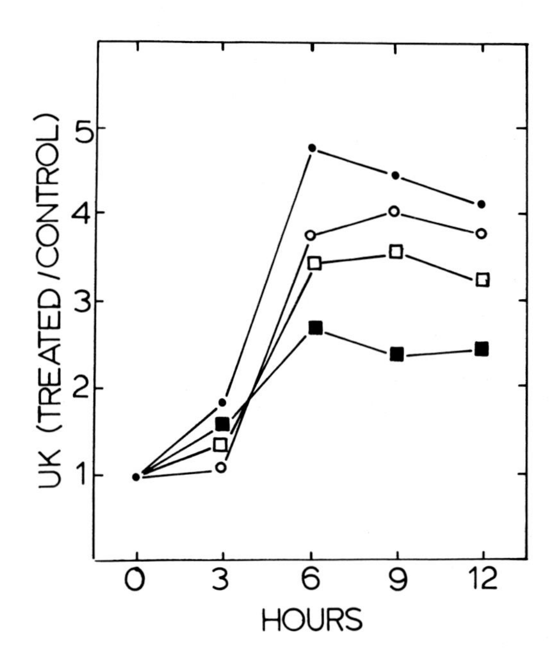

FIG. 3. At 0 hrs. media containing either 2.5 (■), 5 (□), 7.5 (0) or 10% (●) calf serum was added to density-inhibited 3T3 cells grown in 35 mm dishes. At the indicated times cellular extracts were made and uridine kinase activity was measured according to the method of Estes and Huang (5).

When quiescent cells were treated with either PPP or partially purified PDGF (see Figure Legends), the uridine kinase activity increased during the 3-6 hour time period. However, unlike in serum-treated cultures, the activity began to decrease after 6 hours (Fig. 4). When cells were treated with both PPP and PDGF, the uridine kinase activity increased from 3 to 6 hours and remained constant at least over the next 6 hours, similar to the effects following stimulation with serum. Therefore, although both PDGF and PPP induced uridine kinase activity, both were needed to sustain the activity. These factors interact to maintain high enzyme activity which may be reflective of requirements in cells traversing the G_1 phase. We also do not know whether the high levels are due to a continuing synthesis of the enzyme or to a stabilization of existing kinase activity.

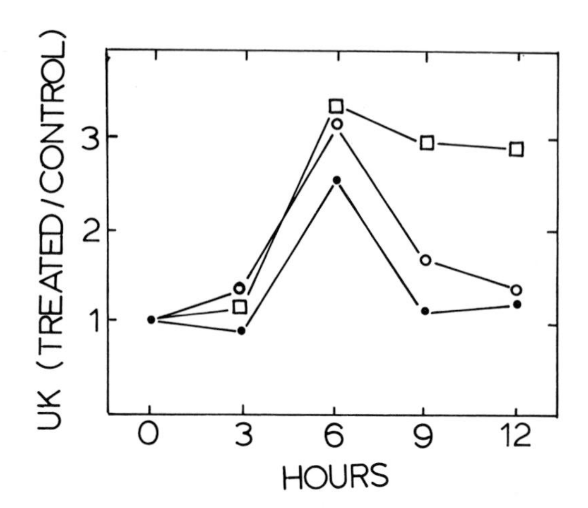

FIG. 4. Uridine kinase activity was measured in cellular extracts prepared at various times after the addition of media containing 5% PPP (O), 10 μg PDGF preparation (●) or 5% PPP + 10 μg PDGF preparation (□) to quiescent 3T3 cells. PDGF was partially purified by passing platelet extract over CM Sephedex as described by Antionaides et al. (1).

In an effort to separate the plasma components that may have induced uridine kinase from those that were needed to sustain the uridine kinase activity, the small molecular weight components were removed from the plasma. PPP was dialyzed against 1M acetic acid and passed over a G100 Sephadex column. The large molecular weight proteins were collected in the void volume, concentrated, and dialyzed against culture medium. We refer to this plasma as stripped PPP. When quiescent 3T3 cells were exposed to stripped PPP there was no induction of uridine kinase (Fig. 5). However, if the cells were also treated with PDGF to induce uridine kinase activity, the stripped PPP allowed the activity to remain at the induced level over the period from 6 to 12 hours. These data show that PPP has components that induce uridine kinase activity and separate components that will sustain the increased activity induced in the presence of PDGF. It has already been shown that PPP contains separate components needed for progression of either late or early G_1.

These data indicate that the uridine kinase activity is induced by factors present in both the preparation of PDGF and in PPP. However, the presence of continued elevated activity is cell cycle dependent. Once cells are made competent, factors present in stripped plasma allow continuous expression of uridine kinase activity.

Unlike uridine kinase activity, thymidine kinase activity is not induced directly by PDGF or PPP. The induction of this

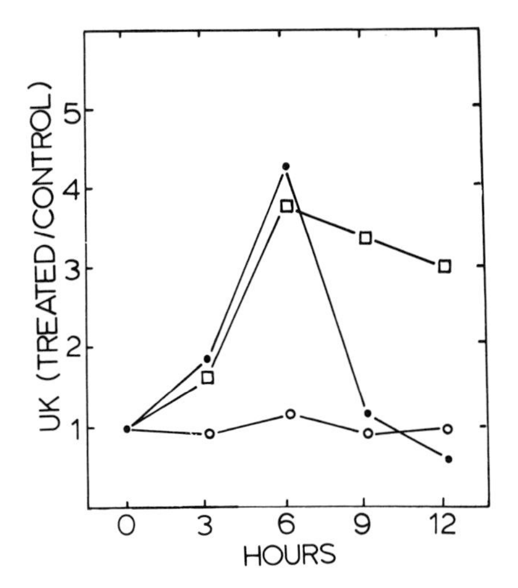

FIG. 5. 15% stripped PPP (O), 10 μg PDGF preparation (●) or 15% stripped PPP + 10 μg PDGF preparation (■) was added to quiescent 3T3 cells at 0 hrs. At the indicated times cellular extracts were made and uridine kinase activity was measured.

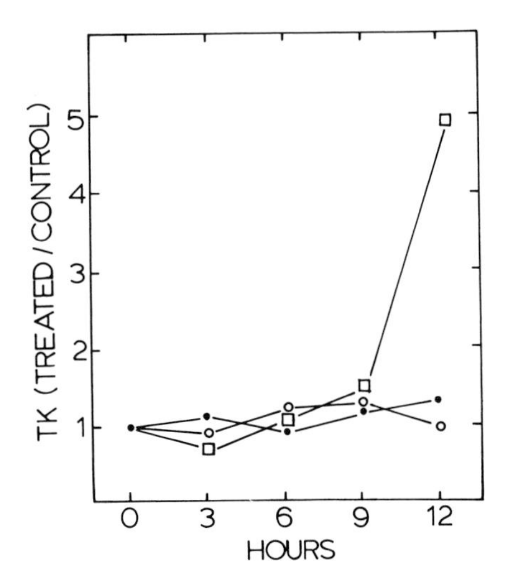

FIG. 6. Thymidine kinase activity, assayed according to the method of Estes et al. (5), was measured in extracts of 3T3 cells taken at the indicated times after the addition of media containing 10 μg PDGF preparation (●), 5% PPP (O) or 10 μg platelet extract + 5% PPP (□).

activity in late G_1 requires both PDGF and PPP, (Fig. 6), suggesting that the induction of thymidine kinase activity is cell cycle-dependent as suggested by others (reviewed in 3).

The use of PDGF and factors in plasma will probably yield more regulatory arrest points (restriction points) in the cell cycle. Also the use of such growth factors will allow the investigation of the biochemical processes that control the traverse of these putative points. Since transformation with SV40 alters the requirement for various serum facotrs (15,16), the elucidation of growth control by these factors may reveal how transformation alters cell regulation. (Supported by NIH grants CA16086, CA24193).

REFERENCES

1. Antoniades,H.N., Scher,C.D. and Stiles,C.D. (1979) P.N.A.S. USA 76, 1809-1813.
2. Balk,S.D. (1971) P.N.A.S. USA 68, 271-275.
3. Baserga,R. (1976) in: Multiplication and Division in Mammalian Cells (Marcel Dekker, New York), pp. 53-77.
4. Clemmons,D., Van Wyk,J.J. and Pledger,W.J. (1979) J. Endoc. abstract.
5. Estes,J., Huang,E.S. (1977) J. Virol. 24, 110-118.
6. Heldin,C.-H., Westermark,B., and Wasteson,A. (1979) P.N.A.S. USA 76, 3722-3726.
7. Holley,R.W. and Kiernan,J.A. (1971) Growth Control in Cell Cultures, eds. Wolstenholme,G.E.W. and Knight,J. (Churchill and Livingstone, London) Ciba Foundation Symposium, pp. 3-10.
8. Kohler,N., and Lipton,A. (1974) Exp. Cell Res. 87, 297-301.
9. Pardee,A.B. (1974) P.N.A.S. USA 71, 1286-1290.
10. Paul,D., Lipton,A., Klinger,I. (1976) P.N.A.S.USA 68, 634-648.
11. Pledger,W.J., Stiles,C.D., Antoniades,H.N. and Scher,C.D. (1977) P.N.A.S. USA 74, 4481-4485.
12. Pledger,W.J., ibid. (1978) 75, 2839-2843.
13. Rizzino,A., Stats,G. (1978) P.N.A.S. USA 75, 1844-1848.
14. Ross,R., Glomset,B., Kariya,B. and Harker,L. (1978) P.N.A.S. USA 71, 1207-2110.
15. Scher,C.D., Pledger,W.J., Martin,P., Antoniades,H., and Stiles, C.D. (1978) J. Cell Physiol. 97, 371-380.
16. Stiles,C.D., Capone,G.T., Scher,C.D., Antoniades,H.N., Van Wyke,J.J. and Pledger,W.J. (1979) P.N.A.S. USA 76, 1279-1283.
17. Stiles,C.D., Isberg,R.R., Pledger,W.J., Antoniades,H.N. and Scher,C.D. (1979) J. Cell Physiol. 99, 395-406.
18. Stiles,C.D., Pledger,W.J., Van Wyk,J.J., Antonidaes,H.N., and Scher,C.D. Hormones and Cell Culture. Cold Spring Harbor Conferences on Cell Proliferation, Vol. 6. (1979).
19. Taub,M., Chuman,L., Saier,N.H., and Sato,G. (1979) P.N.A.S. USA 76, 3338-3342.

Control Mechanisms in Animal Cells,
edited by L. Jimenez de Asua et al.
Raven Press, New York © 1980.

An Ordered Sequence of Temporal Steps Regulates the Rate of Initiation of DNA Synthesis in Cultured Mouse Cells

Luis Jimenez de Asua

Friedrich Miescher-Institute, CH-4002 Basel, Switzerland

The understanding of how the proliferation of animal cells is controlled requires the elucidation of the sequence of signals and molecular events leading up to the initiation of DNA synthesis and cell division (2, 22,31,45). In higher organisms, which consist of an ordered collection of cells arranged in tissues and specific organs, the rate of cell proliferation is regulated by mechanisms of diverse complexity (31,35). In normal cells, these regulatory mechanisms are adjusted through the interaction of cells with growth factors, hormones, other macromolecular compounds, ions and nutrients, and also by changes in the architectural assortment of cells into tissues that give organs their precise structure (1,6,10,11,15,18,20). In contrast, malignant cells or cells transformed in vitro by oncogenic viruses or chemicals have lost their basic mechanisms for regulating their rate of proliferation (5,6,10,11). They become almost independent of changes in their extracellular environment as well as of the architechtural assortment and arrangement in different tissues of the organism (5,6,10,11). Thus, one approach to the understanding of malignant transformation is to elucidate the basic regulatory mechanisms that operate in normal cells to change the rate of initiation of DNA synthesis and cell division.

To Ignacio and Isabel

Within the last two decades, two important advances have been made which are fundamental for the unravelling of the complex cellular regulatory mechanisms. First, the isolation of clonal lines of cultured mammalian cells provides an *in vitro* system in which cell proliferation can be studied under constant environmental conditions (1,18-20). Second, and more recently, several growth factors and other macromolecular compounds have been identified and purified enabling us to search for their mode of action on cells and their interaction with well characterized hormones in regulating the events that set in motion DNA replication and cell division (15,18).

There is some evidence that in cultured mouse Swiss 3T3 cells, growth factors, such as prostaglandin $F_{2\alpha}$ ($PGF_{2\alpha}$) (25), fibroblastic growth factor (FGF) (14) and epidermal growth factor (EGF) (7), and conventional hormones, like insulin and hydrocortisone, regulate the rate of initiation of DNA synthesis through separate sequence of cellular events (31,53). The main sequence of events which leads to the initiation of DNA synthesis is controlled by growth factors, while hormones like insulin or hydrocortisone, which do not stimulate the initiation of DNA synthesis, seem to act via secondary regulatory sequences. These, in turn, adjust the stimulatory effect of growth factors by altering the rate of cellular entry into the S phase (31). Also, it has been shown that growth factors and hormones interact with cells during the lag phase in an ordered series of temporal steps (31).

In what follows I shall discuss some of our results and ideas about the mode of interaction of growth factors and hormones. Some of these ideas come from previous experimental work, while others constitute a useful framework for future experimental attack. I will present the following topics: 1. Growth factors stimulate the initiation of DNA synthesis by delivering two signals. 2. Hormones modify the action of growth factors through a cooperative phenomenon. 3. The time of exposure to the growth factors required for cells to become competent for DNA synthesis. 4. The rate of initiation of DNA synthesis is regulated by an ordered sequence of temporal steps. 5. Variations in the model of interaction between growth factors and hormones. 6. Growth factors may act through different sequences of events. 7. The rate-limiting step. Biochemical and molecular mechanisms.

MATERIALS AND METHODS

Cell Cultures

Swiss mouse 3T3 cells (58) were propagated in Dulbecco's modified Eagle's medium containing 100 μg/ml streptomycin, 100 units/ml penicillin and supplemented with 10% fetal calf serum (FCS) as previously described (29).

Assay for the Initiation of DNA Synthesis

Cells were plated at 1.5×10^5 per 35 mm dish in 2 ml of Dulbecco's modified Eagle's medium supplemented with 6% FCS and with low molecular weight nutrients as before (29). Three days after seeding, the cells were given fresh medium supplemented as indicated (29) and then allowed to become confluent and quiescent in 3-4 days. Such 3T3 cultures gave low rates of cellular entry into S (29). For determination of the labeling index cells were exposed to 1 μM (3 μCi/ml) (methyl-^{3}H)thymidine from the time of additions until the times indicated in each experiment. Duplicate cultures were then processed for autoradiography (29). To measure the incorporation of radioactive precursor into DNA cells were exposed to 3 μM (3 μCi/ml) (methyl-^{3}H)thymidine from 0 until 28 hr after additions. Triplicate cultures were processed for scintillation counting (25).

Determination of Rate Constant k for Entry into S-Phase

The percentage of resting cells in G_1 that remained unlabeled (y) in a given time (t) were calculated from the labeling index (29). The results were plotted as $\log_{10}y$ against time (t) in hours. Straight lines given by $\log_{10}y = a - bt$ fitted these data well (29). First order rate constants (k) were then calculated by geometrical methods from the gradients of the curves (b), since $k = \log_e 10 \cdot b$. In cases where additions were made at 15 hr, the rate was calculated for times greater than 24 hr. Prior to this, the rate constant was gradually increasing. The maximum and minimum values of k were within 10-20% and the lag phase was estimated to within 1 hr. The value for nonstimulated cells was measured in quiescent cultures which were continuously labeled over a period of 7 days.

Materials

PGE_1 and $PGF_{2\alpha}$ were the generous gift of Dr. J. Pike, Upjohn, FGF and EGF were obtained from Collaborative Research. Crystalline insulin and hydrocortisone were purchased from Sigma. (Methyl-^{3}H)thymidine was purchased from the Radiochemical Centre, Amersham.

RESULTS AND DISCUSSION

1. Growth Factors Stimulate the Initiation of DNA Synthesis by Delivering Two Signals

Addition of serum or growth factors, such as $PGF_{2\alpha}$, FGF or EGF, to confluent quiescent 3T3 cells, arrested in the G_0/G_1 phase or A state of the cell cycle stimulates an abrupt increase in the rate of initiation of DNA synthesis after a constant lag phase of 13-15 hours. The initiation of DNA synthesis follows first order kinetics and can be quantified by a rate constant k (3,4,29,30,31,54,56,57).

The stimulatory effect of these growth factors has been interpreted as evidence for the existence of two signals, signal 1 and signal 2 (29,31,53). Signal 1 sets in motion the events necessary for the progression through the lag phase. The length of the lag phase is independent of the growth factor concentration ($PGF_{2\alpha}$ or EGF (Table 1 and Fig. 1)).

TABLE 1. Effect of two concentrations of $PGF_{2\alpha}$ added at different times on the rate constant (k)

Additions	Time of addition (hr)	Rate constant (x 10^2 / hr)
None	-	0.04
$PGF_{2\alpha}$ (60 ng/ml)	0	1.10
$PGF_{2\alpha}$ (300 ng/ml)	0	1.80
$PGF_{2\alpha}$ (60 ng/ml) + $PGF_{2\alpha}$ (240 ng/ml)	0 8	1.70
$PGF_{2\alpha}$ (60 ng/ml) + $PGF_{2\alpha}$ (240 ng/ml)	0 15	1.60
FCS (10%)	0	25.90

Rate constant k was calculated as described in Materials and Methods. The duration of the lag phase was 15 hr. (Reprinted from Jimenez de Asua *et al.* (29)).

Signal 2 regulates the rate of initiation of DNA synthesis (i.e. the value of k), and increasing concentrations of $PGF_{2\alpha}$ or EGF give increasing values of rate constant k. The effect of signal 1 and signal 2 can be separated. Signal 2 does not necessarily need to be triggered with signal 1. Addition of subsaturating concentrations of EGF (or $PGF_{2\alpha}$) initiates a constant lag phase of 14-15 hours and causes a small increase in the value of k for cellular entry into S. The maximum response can be achieved by adding a saturating concentration 5-8 hours or at the end of the lag phase (Table 1, Fig. 1).

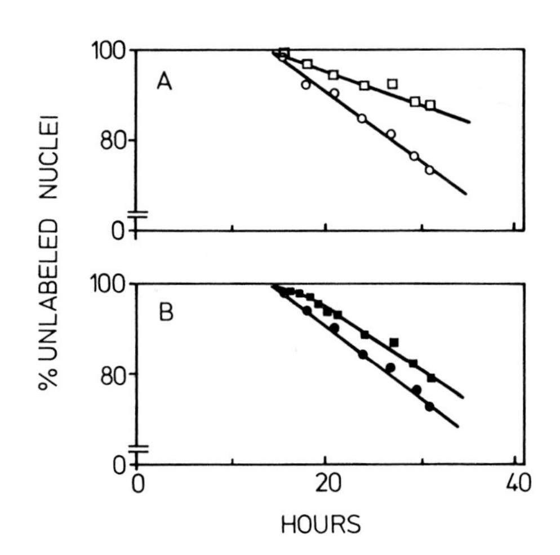

FIG. 1 Fraction of cells that remain unlabeled after different additions of EGF. A. EGF (4 ng/ml) (□) or EGF (20 ng/ml) (O) was added at the start. B. EGF (4 ng/ml) was added at the start and then EGF (16 ng/ml) after 8 hr (●) or after 15 hr (■). Final values of k ($x10^{-2}$/hr) in A were 0.88 (□) and 1.67 (o), in B were 1.69 (●) and 1.48 (■). Duration of the lag phase was 15 hr. Data from Otto, A. et al. (43).

However, addition at the end of the lag phase results in a gradient increase in k during 4 to 6 hr until a similar final value of k is reached. Similar results were obtained with FGF in Swiss 3T3 cells (31, 48). The nature of the signal and signal 2 triggered by each specific growth factor is not yet known.

2. Hormones Modify the Action of Growth Factors Through a Cooperative Phenomenon

A basic distinction between the mode of action of growth factors and hormones can be made. Growth factors, like EGF, FGF or $PGF_{2\alpha}$ (31,53) trigger the events necessary to set in motion the lag phase and increase the rate of cellular entry into S. In contrast, hormones like insulin or hydrocortisone, at physiological concentrations, can act only by changing the value of k already triggered by the growth factor(s) (Table 2). The effect of these hormones can be positive or negative. Insulin always has a synergistic effect on increasing the value of k (Table 2). The amplification effect of insulin is unlikely to be identical to signal 2, since the maximum value of k attained with saturating levels of any growth factor alone is always increased with insulin. Also, insulin at physiological levels fails to deliver signal 1, since under our culture conditions it does not initiate DNA synthesis (Table 2). Only at pharmacological concentrations does insulin alone stimulate DNA synthesis in Swiss 3T3 cells (31,43,48).

Another interesting aspect of the interaction of insulin with growth factors is the positive cooperative effect on the dose response curve of the mitogens for the initiation of DNA synthesis (Fig. 2). When $PGF_{2\alpha}$ was added to resting cells and the incorporation of (methyl-^{3}H)-thymidine into DNA was measured as a function of different concentrations of $PGF_{2\alpha}$, a sigmoidal curve was observed. In the presence of insulin the curve was hyperbolic, and the minimum concentration required for the maximum effect of $PGF_{2\alpha}$ was reduced 10-fold (Fig. 2). An identical relationship was observed with the fraction of cells synthesizing DNA measured by autoradiography (53).

The nature of the sigmoidal relationship between the incorporation of (methyl-^{3}H)-thymidine and the concentration of $PGF_{2\alpha}$ suggests that the stimulation of DNA synthesis in the cell population by a growth factor may be a cooperative phenomenon by analogy with enzyme kinetic analysis. Re-plotting the data according to the Hill equation gives a coefficient of interaction for $PGF_{2\alpha}$ alone of 1.8 and for $PGF_{2\alpha}$ and insulin of 1.00 (Fig. 2). The reduction of the value in the Hill coefficient produced by insulin suggests that the polypeptide hormone may have a positive cooperative effect

TABLE 2. Effect of Insulin and Hydrocortisone on the Initiation of DNA Synthesis Stimulated by EGF, FGF or $PGF_{2\alpha}$

Additions	Lag phase (hr)	Rate constant ($\times 10^2$ / hr)
a None	--	0.04
Insulin	--	0.15
Hydrocortisone	--	0.10
Insulin + hydrocortisone	--	0.21
EGF	14	1.60
EGF + insulin	14	3.87
EGF + insulin + hydrocortisone	14	1.80
b FGF	14	1.50
FGF + insulin	14	4.70
FGF + hydrocortisone	14	3.00
FGF + insulin + hydrocortisone	14	8.01
c $PGF_{2\alpha}$	15	1.60
$PGF_{2\alpha}$ + insulin	15	6.00
$PGF_{2\alpha}$ + hydrocortisone	15	0.61
$PGF_{2\alpha}$ + insulin + hydrocortisone	15	3.00
Serum	15	23.90

The duration of the lag phase for insulin, hydrocortisone, or insulin plus hydrocortisone, was greater than 34 hr. Additions were made at the following concentrations: Insulin (50 ng/ml), hydrocortisone (30 ng/ml), EGF (20 ng/ml), FGF (50 ng/ml), $PGF_{2\alpha}$ (300 ng/ml) and FCS 10%. Data presented in a is from Otto et al. (43) and b and c reprinted from Jimenez de Asua et al. (31).

upon the stimulation of DNA synthesis in the cell population by $PGF_{2\alpha}$. Similar results were observed with insulin and increasing concentrations of EGF or FGF. Although the application of this analysis and the Hill equation may be useful here, it is not yet possible to ascertain whether the interaction of $PGF_{2\alpha}$ and insulin occurs at the level of the growth factor-binding receptor or at some subsequent metabolic step.

A similar analysis of the kinetics of the initiation of DNA synthesis stimulated by serum has been made by applying the Michaelis-Menten equation for enzyme kinetics (13).

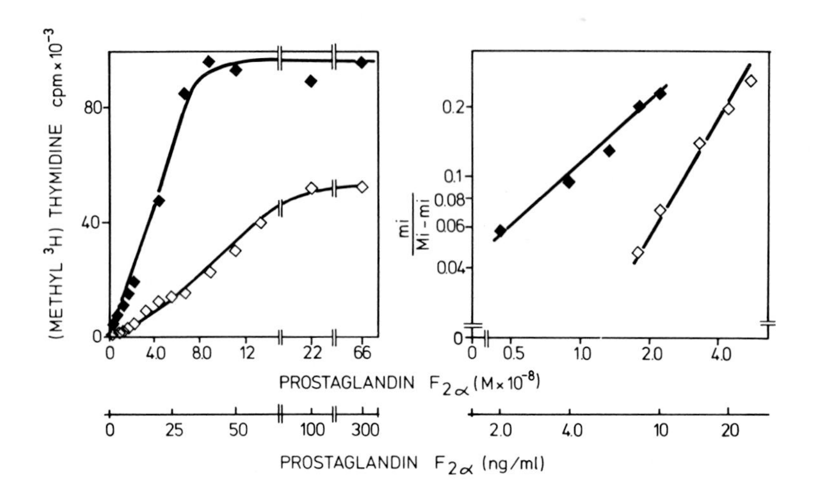

FIG. 2. Effect of different concentrations of $PGF_{2\alpha}$ alone or plus insulin on the stimulation of DNA synthesis. (A) Increasing concentrations of $PGF_{2\alpha}$ (◇) or $PGF_{2\alpha}$ plus insulin (50 ng/ml) (◆) were added to resting cells. The cpm incorporated into DNA per dish corresponded to 2.6×10^5 cells. (B) Results were expressed according to the Hill equation. The $\log_{10}$ of $PGF_{2\alpha}$ concentration (ng/ml or $M \cdot 10^{-8}$) was plotted against $\log_{10}$ mi/Mi-mi. Mi is the maximum incorporation of (methyl-^{3}H)thymidine at saturating $PGF_{2\alpha}$ concentration, mi is cpm incorporated at each subsaturating concentration of $PGF_{2\alpha}$.

Another hormone, hydrocortisone, has either a positive or negative effect on the value of k depending on the growth factor which triggers the initiation of DNA synthesis (Table 2). Physiological concentrations of hydrocortisone added with $PGF_{2\alpha}$ alone, $PGF_{2\alpha}$ plus insulin or EGF plus insulin inhibit the initiation of DNA synthesis by decreasing the value of k (Table 2). In contrast, hydrocortisone, added with FGF alone or FGF plus insulin (Table 2) increases the rate of cellular entry into S. In the latter case, the glucocorticoid also shows a positive cooperative effect on the dose-response curve of FGF for stimulating DNA synthesis (48). Furthermore, the synergistic effect of hydrocortisone also seems to be exerted by a different mechanism from the signal 2 delivered by the FGF alone or with insulin.

3. The Time of Exposure to the Growth Factor Required for Cells to Become Competent for DNA Synthesis

How long does a growth factor need to be present to make resting cells competent to initiate DNA synthesis? Exposure of quiescent 3T3 cells to $PGF_{2\alpha}$ from 0 to 8 hours, and determining the labeling index after 28 hours, reveals that only 10% of the cells are in the S phase (Fig. 3). As shown in the figure, 6-8 hours of exposure to $PGF_{2\alpha}$ is the minimum time required to make cells competent to initiate DNA synthesis. Addition of

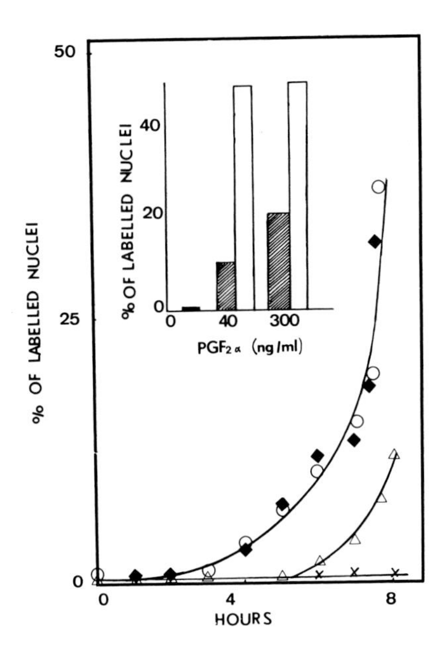

FIG. 3. Time of exposure to $PGF_{2\alpha}$ required for cells to become competent to initiate DNA synthesis. $PGF_{2\alpha}$ (300 ng/ml) (△) was removed after different times; insulin (50 ng/ml) was added at the time of removal (◆) or at 8 hr, irrespective of the time of removal (O). Insulin (50 ng/ml) alone was removed after different times (×). $PGF_{2\alpha}$ or insulin was removed from the culture medium by aspiration and cells were washed 2 x with Dulbecco's medium. Conditioned medium retrieved from parallel cultures was added to the treated cells. Insert: Continuous exposure to $PGF_{2\alpha}$ (hatched bars) or $PGF_{2\alpha}$ plus insulin (50 ng/ml) (empty bars), to insulin alone (black bar). Radioactively labeled cultures were processed for autoradiography at 28 hr.

insulin at the time of the removal of $PGF_{2\alpha}$ showed that the presence of the hormone shortened the time for competence by 2 to 3 hours and gave a maximum labeling index of 32%. Similar results were obtained when insulin was added at 8 hours regardless of the time of removal of $PGF_{2\alpha}$. In contrast, exposure and removal of insulin alone did not have any stimulatory effect on DNA synthesis (Fig. 3). Continuous exposure to two concentrations of $PGF_{2\alpha}$, alone or with insulin resulted in higher values of the labeling index (Fig. 3, insert). Similar observations showed that EGF with human fibro blasts required 8 hours to make cells competent for initiation of DNA synthesis (55). In contrast, in Balb c/3T3 Pledger et al. (47) have shown that the minimum time of exposure to a partially purified preparation of platelet factor to induce competence is one hour.

4. The rate of initiation of DNA synthesis is regulated by an ordered sequence of temporal steps

Several lines of evidence strongly suggest that growth factors, hormones and hormone-like substances regulate the rate of cellular entry into S by interacting during the lag phase in an ordered sequence of temporal steps. As a working hypothesis, one can envisage three basic types of interactions during the lag phase. (1) At the start, (2) at any time or (3) at specific times as shown in Fig. 4. These interactions may bring changes in the kinetics of initiation of DNA synthesis in three different ways: a) by changing the duration of the lag phase, b) by changing the rate of cellular entry into S (i.e. the value of k), c) by changing both of these parameters (31,53).

The first type of interaction, whereby the kinetics can only be influenced by the additions made at the start of the lag phase, is seen in the stimulatory effect of the growth factors EGF, FGF or $PGF_{2\alpha}$ in delivering signal 1 (Table 1, Fig. 1). Hormones like insulin or hydrocortisone can display their synergistic or inhibitory effect when added after the lag phase has been triggered by anyone of these growth factors. $PGF_{2\alpha}$, EGF or FGF added at different times stimulate a constant lag phase from the time of their addition, even if insulin or insulin and hydrocortisone had been added for different periods of time prior to the addition of any of the growth factors (Fig. 5A, 6A)(31,48). The interaction at the start was essentially what Pardee initially proposed as the "restric-

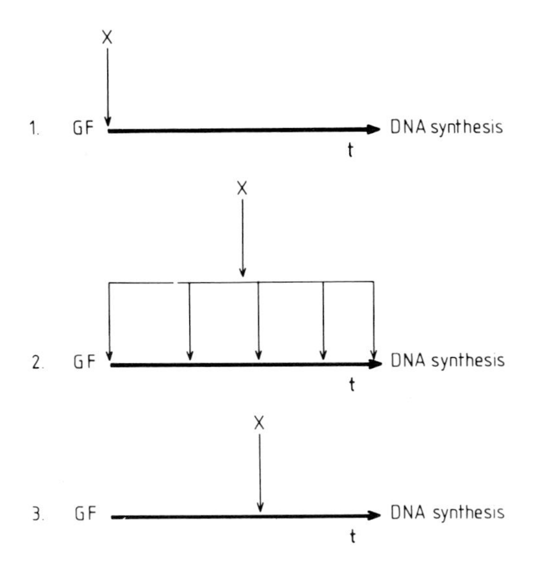

FIG. 4. Possibilities of interactions during the lag phase (GF) represents a growth factor. (X) represents a further addition of growth factor, hormone or other compound. In 1, GF and X only interact at the start; in 2, the interaction occurs at any time; and in 3, only at specific times. Anyone of the postulated interactions results in changes in the kinetics of the initiation of DNA synthesis.

tion point" (44-46). However, according to our observations, the restriction point occurs within two hours of the end of the lag phase. The interaction at the start and hence the delivering of the signal 1, however, is related to the competence phenomenon (47)(see Section 3). The restriction point appears to be the rate-limiting step at the end of the lag phase and related to the signal 2, which regulates the rate of entry into S (31,44-46).

The second type of interaction, at any time of the lag phase, is exemplified by the synergistic effect of insulin observed in combination with $PGF_{2\alpha}$, EGF or FGF (28-31). Insulin, at physiological concentration, added up to 8 hours after stimulation of cells by $PGF_{2\alpha}$ (Fig. 5B) or by EGF (Fig. 6B) produced almost the same value of k as if insulin was added with $PGF_{2\alpha}$ or with EGF at the start. Also addition of insulin at the end of the lag phase, triggered either by $PGF_{2\alpha}$, EGF or FGF

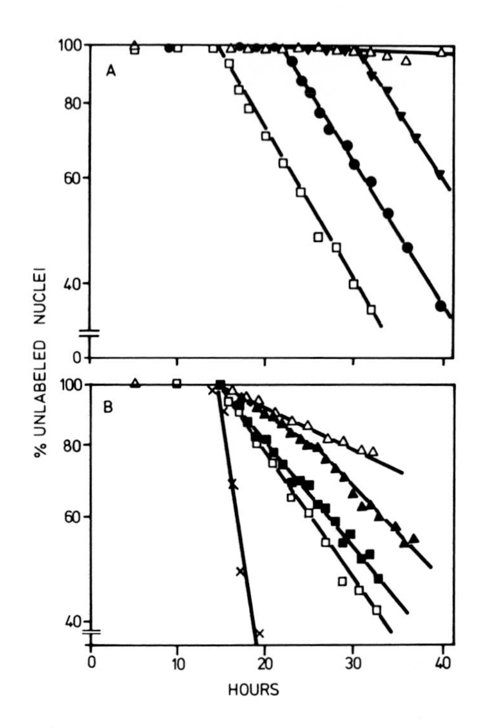

FIG. 5. Fraction of cells that remain unlabeled after nonsynchronous additions of $PGF_{2\alpha}$ and insulin. (A) Insulin (50 ng/ml) was added at zero time (△), then $PGF_{2\alpha}$ (300 ng/ml) was added at the start (□), after 8 hr (●), or after 15 hr (▼). (B) $PGF_{2\alpha}$ (300 ng/ml) was added at zero time (△), then insulin (50 ng/ml) was added at the start (□), after 9 hr (■) or after 15 hr (▲). FCS (10%) was added alone (x). Final values of k ($x10^2$/hr) in A were 0.15 (△), 6.0 (□), 5.6 (●) and 5.3 (▼); in B were 1.6 (△), 4.7 (□), 3.9 (■), 3.3 (▲) and 24.0 (x). Duration of the lag phase in A and B was 15 hr. (Reprinted from Jimenez de Asua et al. (29 and 30)).

(31,48) produced a gradual increase in the value of k reaching to its maximum effect after 8 to 9 hours. However, an early increase can be detected as early as two hours after insulin addition. This strongly suggests that the sequence of events controlled by any of these growth factors and the sequence of events controlled by insulin can rapidly interact to change the rate of initiation of DNA synthesis (31).

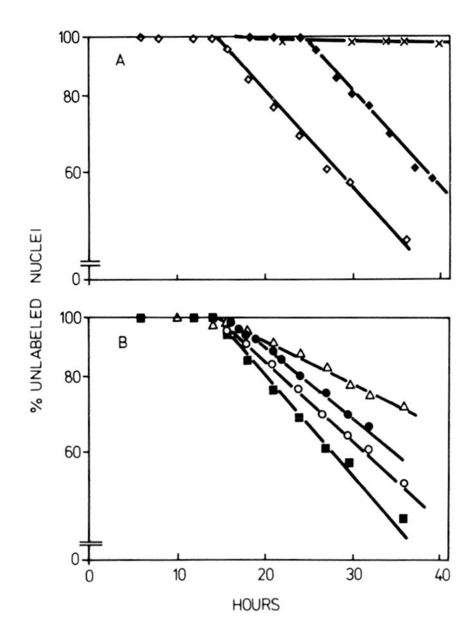

FIG. 6. Fractions of cells that remain unlabeled after nonsynchronous additions of EGF and insulin. (A) Insulin (50 ng/ml) was added at zero time (x), then EGF (20 ng/ml) was added at the start (◇) or after 10 hr (◆). (B) EGF (20 ng/ml) was added at zero time (△), then insulin (50 ng/ml) was added at the start (■), after 8 hr (o) or after 15 hr (●). Final values of k ($x10^2$/hr) in A were 0.12 (x), 3.80 (◇) and 3.78 (◆); in B were 1.5 (△), 3.87 (■), 3.10 (o) and 3.0 (●). Duration of the lag phase in A and B was 14 hr. (Otto et al., in preparation (43)).

Another example of the interaction at any time of the lag phase is the synergistic effect of hydrocortisone on increasing the value of k triggered by FGF (Table 3). Physiological concentrations of the glucocorticoid have a synergistic effect when added at any time after FGF, and also produces delayed increase in the value of k when added at the end of the lag phase (31,32,48). Furthermore, hydrocortisone also further increases the value of k when added with FGF and insulin at the start. Addition of hydrocortisone 8 hours after FGF plus insulin results in almost identical rate constant as if all three compounds had been added at the start. Likewise, insulin is able to exert its synergistic effect when added 8 hours after FGF plus hydro-

TABLE 3. Effect of nonsynchronous additions of hydrocortisone and insulin to FGF on the initiation of DNA synthesis

Additions	Time of addition (hr)	Rate constant (x 10^2/hr)
None	-	0.06
FGF	0	1.50
FGF + insulin	0	4.70
FGF	0	
+ insulin	8	3.90
FGF	0	
+ insulin	15	3.40
FGF + hydrocortisone	0	2.90
FGF	0	
+ hydrocortisone	8	2.90
FGF	0	
+ hydrocortisone	15	2.30
FGF + hydrocortisone + insulin	0	7.50
FGF	0	
+ hydrocortisone + insulin	8	7.60
FGF + hydrocortisone	0	
+ insulin	8	9.00
FGF + hydrocortisone	0	
+ insulin	15	5.10
FGF + insulin	0	
+ hydrocortisone	8	10.00
FCS	0	25.00

The rate constant (k) was calculated as described in Materials and Methods. In all cases, the duration of the lag phase was 13-14 hr. Additions were made at the following concentrations: FGF (50 ng/ml), insulin (50 ng/ml), hydrocortisone (100 ng/ml), FCS (10%). The same effects were observed with a lower concentration of hydrocortisone (20 ng/ml). Data reprinted from Jimenez de Asua et al. (31).

cortisone (Table 3). Insulin added after 15 hours causes a gradual increase in the rate of entry into S, resulting in a slightly lower final value of k, which was not obtained until 6 hours after insulin addition (Table 3). Also, insulin and hydrocortisone added together 8 hours after FGF enhances the value of k as if they had been present from the start (Table 3). These results indicate that insulin and hydrocortisone act through separate but convergent pathways (31).

The third type of interaction, i.e. specific times of the lag phase, can be either inhibitory or stimulatory (31,42). Hydrocortisone, at physiological concentration, inhibits the rate of initiation of DNA synthesis stimulated by $PGF_{2\alpha}$ or $PGF_{2\alpha}$ plus insulin only if added within the first 3 hours of the lag phase (Fig. 7) (27,29,31,32). At later times, 5, 15 or 27 hours, the glucocorticoid has no effect. Hydrocortisone also inhibited the effect of EGF alone or with insulin during the first 9 hours of the lag phase (43).

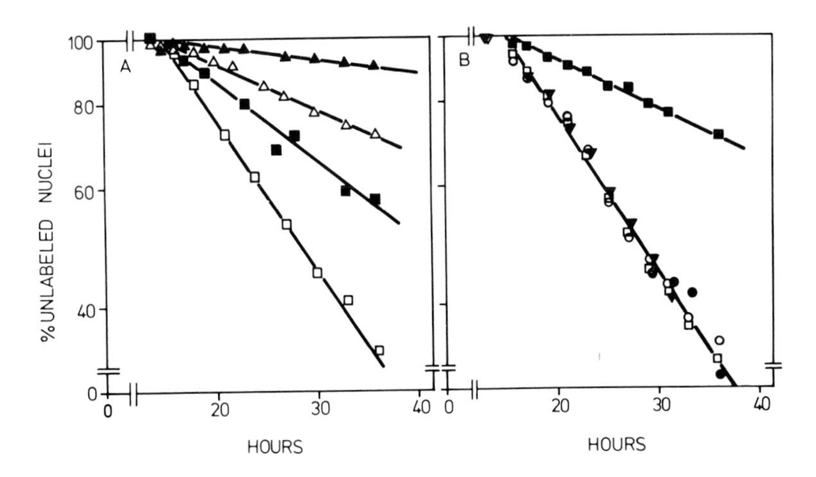

FIG. 7. Effect of hydrocortisone on the fraction of unlabeled cells remaining after the addition of $PGF_{2\alpha}$ or $PGF_{2\alpha}$ plus insulin. Concentrations used in A and B were for $PGF_{2\alpha}$ 300 ng/ml, for insulin 50 ng/ml and for hydrocortisone 20 ng/ml. A) Effect of synchronous addition of $PGF_{2\alpha}$, insulin and hydrocortisone. $PGF_{2\alpha}$(△), $PGF_{2\alpha}$ + hydrocortisone (▲), $PGF_{2\alpha}$ + insulin (□), $PGF_{2\alpha}$ + insulin + hydrocortisone (■). B) Effect of nonsynchronous addition of hydrocortisone to cells stimulated by $PGF_{2\alpha}$ and insulin. $PGF_{2\alpha}$ + insulin at time 0 (□), with addition of hydrocortisone at time 0 (■), after 5 hr (o), after 15 hr (▼) or after 27 hr (●). Final values of k (x 10^{-2} hr) were in A: 1.60 (△), 0.54 (▲), 5.3 (□) and 2.7 (■); in B: 5.3 (□), 1.7 (■), 5.3 (o), 5.4 (▼) and 5.4 (●). Duration of the lag phase in A and B was 15 hr. (Reprinted from Jimenez de Asua et al. (29)).

Another example of an inhibitory effect at specific times is the inhibition of DNA synthesis by prostaglandin E_1 (PGE_1). PGE_1, which markedly increases the intracellular levels of cyclic AMP, decreases the value of k only when added within the first 8 hours, whether the lag phase has been set by $PGF_{2\alpha}$ alone or with insulin (29) or by FGF alone or in combination with the hormones (48). However, when added with FGF at the start, PGE_1 can lengthen the lag phase (48). Later additions of PGE_1 (29,31) or dibutyryl cyclic AMP (60) do not inhibit and may even stimulate the rate of cellular entry into S (31,60).

An example of a stimulatory effect at specific times during the lag phase is the increase in the value of k produced by cytoskeleton-disrupting drugs. Colchicine or colcemid added within the first 8 hours after the addition of $PGF_{2\alpha}$, EGF, or FGF markedly enhances the rate of entry into the S phase. Additions at later times, 8 or 15 hours, do not have any stimulatory effect (42, see Otto, Ulrich and Jimenez de Asua, this book).

According to the times and type of interactions of $PGF_{2\alpha}$ or FGF with insulin, hydrocortisone and/or PGE_1 during the lag phase two different models can be drawn (Fig. 8) (31). The evidence presented here strongly suggests that a major sequence of events leading up to the initiation of DNA synthesis is controlled by growth factors. Hormones and other compounds, which do not stimulate DNA synthesis, can act only through minor or secondary pathways to amplify or reduce events occurring in the major sequence. During the lag phase there may be multiple regulatory steps at which growth factors, hormones and nutrients interact with cells to produce changes in the rate of initiation of DNA synthesis. This model has been proposed as a working hypothesis for studying the effect of growth factors and hormones on Swiss 3T3 cells.

5. Variations in the model of interaction between growth factors and hormones

The model for the type and times of interaction of growth factors and hormones (Fig. 8) is as yet too simple to account for changes in the physiological states of Swiss 3T3 cell variants as opposed to their parental cell lines. I have found that in a variant of Swiss 3T3 cells (3T3-L) the mode of interaction of EGF and insulin in initiating DNA synthesis is different from that in the parental Swiss 3T3 cell line. In the

A

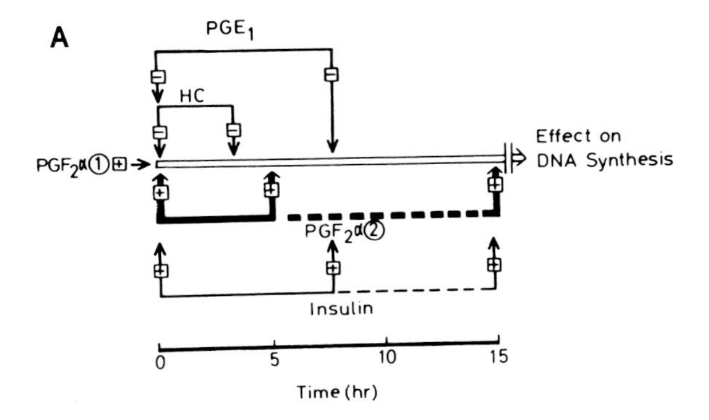

B

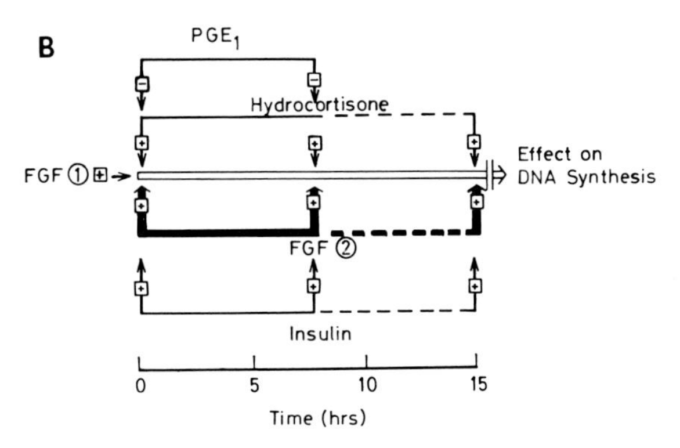

FIG 8. Times when growth factors and hormonal agents are required for alterations in the kinetics of the initiation of DNA synthesis in Swiss 3T3 cells. A. Interactions between $PGF_{2\alpha}$, insulin, hydrocortisone and PGE_1. B. Interactions between FGF, insulin, hydrocortisone and PGE_1. In A and B the solid lines represent the times of addition of the growth factors controlling agents which give immediate positive ([+]) or negative ([−]) alterations in the value of the rate constant k at the end of the lag phase. Broken lines represent the time of the additions which, at the end of the lag phase, provides nonlinear increases in the first order kinetics for entry into S. These became linear after a few hours. The numbers represent the hypothetical signals delivered by $PGF_{2\alpha}$ or FGF and/or hormones. Reprinted from Jimenez de Asúa et al. (31).

parental cell line EGF and insulin have a mode of interaction which resembles that between $PGF_{2\alpha}$ or FGF and insulin (31). In contrast, in the variant (3T3 L) neither EGF nor insulin alone can produce an increase in the labeling index within 28 hours but when both are added together, an increase is observed (Table 4).

TABLE 4. Effect of EGF and insulin on DNA synthesis in a parental and variant Swiss 3T3 cell

Additions	% Labeled nuclei	
	Parental	Variant
None	0.2	0.2
EGF	15.0	1.0
Insulin	0.5	0.5
EGF + Insulin	39.0	35.0
FCS	98.0	95.0

Additions were made as follows: EGF (20 ng/ml), insulin (50 ng/ml) and FCS (10%). Cultures were radioactively labeled with (methyl-^{3}H)thymidine from 0 until 28 hours after additions and then processed for autoradiography as in Materials and Methods. Lag phase was 14 hours in both parental and variant 3T3 cells.

However, in the variant $PGF_{2\alpha}$ or FGF have the same way of interaction as in the parental cell line.

Furthermore, in another clonal cell line, Balb c/ 3T3 (A31), O'Farrell et al. (40) have shown that $PGF_{2\alpha}$ alone or with insulin has no stimulatory effect, while FGF and hydrocortisone have synergistic effects on the stimulation of DNA synthesis.

Finally, the application of this model to cells transformed by oncogenic viruses is not possible, since these cells are locked in continuous proliferation and are almost independent of changes in the external environment (5,10,45). Even under serum-free, defined medium conditions, transformed cells either replicate or die. Two different mechanisms may explain their loss of growth control. One is the constitutive initiation of DNA synthesis by the direct interaction of virally coded protein(s) such as T antigen with the cellular genome, thereby bypassing those events controlled by external growth factors and hormones (16). The second is the continuous synthesis and release of growth factors by transformed cells, which probably act mainly at their own surface membrane, delivering the positive signal(s) to ensure continuous proliferation (see articles by Bürk and Todaro in this book).

6. Growth factors may act through different sequences of events

Do growth factors stimulate DNA synthesis through a common sequence of events or by independent pathways? Results presented above demonstrate that $PGF_{2\alpha}$ as well as EGF can initiate DNA synthesis by setting in motion the lag phase and increasing the rate of entry into the S phase, i.e. by delivering signal 1 and signal 2. Basically, two possibilities can be envisaged: 1. that $PGF_{2\alpha}$ and EGF act through the same intracellular pathway, meaning that signal 1 and signal 2 are identical for both growth factors, or 2. that they operate through different sequences of events (31). The first possibility implies that there can be no additive or synergistic effect of one growth factor on the other at saturating concentrations, while with the second possibility of separate mechanisms a synergistic effect between growth factors could be expected. Indeed, I have observed that $PGF_{2\alpha}$ and EGF, which alone give 18% and 15% labeled nuclei, respectively, together result in 50% labeled nuclei after 28 hours. This supports the idea that each growth factor stimulates a different sequence of events to initiate DNA synthesis. However, these pathways must interact at the level of the rate-limiting step that governs the rate of cellular entry into the S phase. This does not exclude that different sequences may have events in common at other times during the lag phase.

7. The rate-limiting step - Biochemical and molecular events

Many lines of evidence indicate that the initiation of DNA synthesis is controlled by a rate-limiting step towards the end of the lag phase (3,4,29-32,53-56). Our experiments show that addition of saturating concentrations of $PGF_{2\alpha}$, EGF or FGF 15 hours after a subsaturating concentration of the same growth factor results in an increase in the value of k as early as 2-4 hours. Also, addition of insulin 15 hours after $PGF_{2\alpha}$, EGF or FGF produces a rapid increase in the rate of cellular entry into the S phase as does the addition of hydrocortisone alone or with insulin 15 hours after FGF. It can be postulated that the rate-limiting step is responsible for the loss of synchrony upon entry into the S phase. Upon stimulation by $PGF_{2\alpha}$ the entire cell population becomes insensitive to the inhibitory effect of hydrocortisone within a 2 hour period. However, only a small percentage of the cell population initiates DNA synthesis within a 2 hour period after the end of the lag phase. Similarly, the synergistic effect of colchicine or colcemid can only be seen when added within the first 8 hours of the lag phase triggered by $PGF_{2\alpha}$, FGF or EGF. This means that while cells progress relatively synchronously through the lag phase, they initiate DNA synthesis nonsynchronously.

Which biochemical and molecular events stimulated by growth factors and hormones regulate the initiation of DNA synthesis? When we turn to the biochemical changes at the level of the surface membrane following stimulation by growth factors and hormones in resting cells, the correlation of each single event with the initiation of DNA synthesis becomes difficult (12,31). It has been shown that after addition of $PGF_{2\alpha}$, insulin, hydrocortisone to resting cells there is no correlation between a number of early events which do not require protein (24,26,50) synthesis and the eventual stimulation of DNA synthesis (31,34). Such events include rapid increases in the activity of uridine and phosphate uptake, early glucose transport, influx of ^{86}Rb (a measure of the activity of (Na^+, K^+) membrane ATPase) and decrease in the level of cyclic AMP (31,33,36,50,51). For instance, insulin triggers these events as well as the increase in glycolysis in Swiss 3T3 cells, but fails to initiate DNA synthesis (9). The only correlation so far found is between the late, protein synthesis-dependent increase in glucose transport and the initiation of DNA synthesis triggered by $PGF_{2\alpha}$ (30,31). Also, it has been shown that there is a correlation between the synergistic effect of $PGF_{2\alpha}$ and insulin on the stimulation of the late increase in glucose transport and the synergistic effect on the rate of entry into S phase (30,31). Ouabain which inhibits ^{86}Rb influx in 3T3 cells stimulated by serum (31-34) or $PGF_{2\alpha}$ and insulin also decreases the rate of initiation of DNA synthesis if added in the earliest part of the lag phase (50,51). Certainly, much more work has to be done to disclose whether these early biochemical events are an obligatory prerequisite for DNA synthesis or whether they are permissive events dissociable from separate signalling systems which normally regulate the onset of DNA synthesis in mammalian cells (12,31). However, since they represent a rapid response to the effect of growth factors and hormones, they constitute useful markers for monitoring the primary interaction of mitogens with the plasma membrane (50).

Several lines of evidence strongly suggest that the synthesis of new proteins is required for stimulated cells to enter into the S phase (2,4,21-23,31,49). Furthermore, it has been shown that the stimulation of DNA synthesis in quiescent cells correlates with a rapid increase in the rate of phosphorylation of the 40S ribosomal protein S6 (59) followed by a 3-fold increase in the rate of protein synthesis (52). It is still unclear whether the requirement for the initiation of DNA synthesis is a coordinate increase in the synthesis of all cellular proteins, a preferential synthesis of specific proteins or both (49).

Using high resolution two-dimensional gel electrophoresis (41), it has been shown that following stimulation of resting 3T3 cells by $PGF_{2\alpha}$ and insulin there are preferential changes in the metabolism of two non-histone nuclear polypeptides (31,39). Polypeptide 1 (P_1, 30,000 M_r) which is present in resting cells, is mainly synthesized between 2.5 and 7.5 hours of the lag phase, declining rapidly thereafter (Table 5). Polypeptide 2 (P_2, 36,000 M_r) only appears near the end of the lag phase, its increased synthesis correlates with the increase in the rate of cellular entry into the S phase (Table 5). Hydrocortisone, which inhibits the effect of $PGF_{2\alpha}$ and insulin by decreasing the value of k, markedly reduces the synthesis of polypeptide 2, while polypeptide 1 is not affected (Table 5).

TABLE 5. Effect of $PGF_{2\alpha}$, insulin and hydrocortisone on the appearance of two nuclear non-histone polypeptides[a]

Addition	Time of labeling (hr)	Appearance of P_1	Appearance of P_2
A. None	5	+	-
B. $PGF_{2\alpha}$ + insulin	0- 5	+++	-
C. $PGF_{2\alpha}$ + insulin	25-30	+	+++
D. $PGF_{2\alpha}$ + insulin + hydrocortisone	25-30	+	+

[a]Cells were labeled for period of 5 hr with (^{35}S)methionine and the nuclear proteins were separated by two-dimensional gel techniques as previously described (31).

Interestingly, the increased appearance of both polypeptides exhibits some correlation with the times at which signal 1 and signal 2 operate to set in motion the lag phase and increase the rate of cellular entry into the S phase. These observations (31) are consistent with recent reports by Riddle et al. (49) of the appearance of different cytoplasmic and nuclear proteins in resting cells stimulated by serum.

Finally, two recent observations may provide a new insight into this field. One is the discovery of tetra-P-adenosine, which seems to stimulate DNA replication in permeabilized BHK cells (17). The other is that DNA polymerase α seems to be the enzyme responsible for initiating DNA synthesis in mammalian cells (8). It remains to be investigated whether this nucleotide and this polymerase are specifically induced upon stimulation by growth factors.

In the future, the isolation of those compounds which may be responsible for controlling DNA synthesis in stimulated cells and their introduction into quiescent cells by micro-injection or by cell fusion may prove to be useful in elucidating the mode of growth factors and hormones in regulating DNA synthesis.

ACKNOWLEDGEMENTS

I am deeply indebted to Dr. Angela Otto for encouragement and stimulating discussions. I am also grateful to Dr. Veronica Richmond and Dr. George Thomas for criticisms and revision of the manuscript, to Dr. Minnie O'Farrell, Department of Biology, University of Essex, Dr. Julia Lever, Department of Biochemistry and Molecular Biology, University of Texas, USA, and Dr. P. S. Rudland, Ludwig Institute for Cancer Research, London, for their invaluable contribution to my scientific research, to Ms. A.M. Kubler for skilful technical assistance. L. J. de A. was a Special Fellow of the Leukemia Society of America, Inc., while part of this research was carried out at the Imperial Cancer Research Fund Laboratories during 1975-1977.

REFERENCES

1. Armelin, H.A., Nishikawa, K., and Sato, G.H. (1974): In: Control of Cell Proliferation in Animal Cells, edited by B. Clarkson and R. Baserga, pp. 1-97. Cold Spring Harbor Laboratories.
2. Baserga, R. (1976): Multiplication and Division in Animal Cells. Dekker, New York.
3. Brooks, R.F. (1975): J. Cell Physiol., 86:369-378.
4. Brooks, R.F. (1977): Cell, 12:311-317.
5. Burger, M.M. (1971): In: Growth Control of Cell Cultures, Ciba Foundation Symposium, edited by G. E.W. Wolstenholme, and J. Knight, pp. 45-69. Churchill, Livingston, London.
6. Bürk, R.R. (1973): Proc. Natl. Acad. Sci.USA, 70: 369-372.
7. Cohen, S., Taylor, J.M., and Savage, C.R. (1974): Rec. Progr. Horm. Res., 30:533-574.
8. Castellot, J.J., Miller, M.R., Lehtomaki, D.M., and Pardee, A.B. (1979): J. Biol. Chem., 253:6904-6908.
9. Diamond, I., Legg, A., Schneider, J.R., and Rozengurt, E. (1978): J. Biol. Chem., 253:866-867.
10. Dulbecco, R. (1970): Nature, 227:802-806.
11. Dulbecco, R. (1969): Science, 166:962-968.
12. Dulbecco, R. (1975): Proc. R. Soc., London, B 189: 1-14.
13. Ellem, K.A., and Gierty,J.E(1978): J. Cell Physiol., 92:381-400.
14. Gospodarowicz, D. (1974): Nature, 249:123-127.
15. Gospodarowicz, D., and Moran, J. (1976): Annu.Rev. Biochem., 45:531-558.
16. Graessman, A., Graessman, M., and Mueller, C. (1976): Proc. Natl. Acad. Sci. USA, 73:366-370.
17. Grummt, F. (1978): Proc. Natl. Acad. Sci.USA, 75: 371-375.
18. Holley, R.W. (1975): Nature, 249:487-490.
19. Holley, R.W., and Kiernan, J. (1974): Proc. Natl. Acad.Sci.USA, 71:2908-2911.
20. Holley, R.W., and Kiernan, J. (1974): Proc. Natl. Acad. Sci.USA, 71:2942-2945.
21. Kidwell, W.R., and Mueller, G.C. (1969): Biochem. Biophys. Res. Comm., 36:756-763.
22. Jazwinski, S.M., Wang, J.L., and Edelman, G.M. (1976): Proc. Natl. Acad. Sci.USA, 73:2231-2235.
23. Jazwinski, S.M., and Edelman, G.M. (1976): Proc. Natl. Acad. Sci. USA, 73:3933-3936.

24. Jimenez de Asua, L., and Rozengurt, E. (1974): Nature, 251:624-626.
25. Jimenez de Asua, L., Clingan,D., and Rudland, P.S. (1975): Proc. Natl. Acad. Sci.USA, 72:2724-2728.
26. Jimenez de Asua, L., Rozengurt, E., and Dulbecco,R. (1977): Nature, 265:450-452.
27. Jimenez de Asua, L., Carr, B., Clingan, D., and Rudland, P.S. (1977): Nature, 265:450-452.
28. Jimenez de Asua, L., O'Farrell, M.K., Clingan, D., and Rudland, P.S. (1977): Biochem. Soc. Trans., 5:937-939.
29. Jimenez de Asua, L., O'Farrell, M.K., Clingan, D., and Rudland, P.S. (1977): Proc. Natl.Acad. Sci.USA, 74:3845-3849.
30. Jimenez de Asua, L., O'Farrell, M.K., Bennett, D., Clingan, D., and Rudland, P.S. (1977): Nature, 265: 151-153.
31. Jimenez de Asua, L., Richmond, K.M.V., Otto, A., Kubler, A.M., O'Farrell, M.K., and Rudland, P.S. (1979): In: Hormones and Cell Culture, edited by R. Ross and G.H. Sato, pp. 403-424. Cold Spring Harbor Laboratories.
32. Jimenez de Asua, L., Richmond, K.M.V., Kubler, A.M., Ulrich, M.O., O'Farrell, M.K., and Otto, A.M. (1980): Proc. First Int. Congress on Hormones and Cancer, (in press).
33. Lever, J.E., Clingan, D., and Jimenez de Asua, L. (1976): Biochem. Biophys. Res. Commun. 71:136-143.
34. Lever, J.E., Otto, A.M., Kubler, A.M., and Jimenez de Asua, L. (manuscript in preparation).
35. Monod, J., and Jacob, F. (1961): Cold Spring Harbor Symp. Quant. Biol.,26:389-401.
36. Moroney, J., Smith, A., Tomei, L.D., and Wenner,C.H. (1978): J. Cell Physiol., 95:287-294.
37. Mueller, G. (1969): Fed. Proc., 28:1780-1789.
38. Mueller, G. (1971): The Cell Cycle and Cancer, edited by R. Baserga, pp. 269-307. Dekker, New York.
39. O'Farrell, M.K. (1977): Biochem. Soc. Trans.,5: 940.
40. O'Farrell, M.K., Clingan, D., Rudland, O.S., and Jimenez de Asua, L. (1978): Exp. Cell Res., 118: 311-321.
41. O'Farrell, P.H. (1975): J. Biol. Chem., 250:4007-4021.
42. Otto, A.M., Zumbé, A., Gibson, L.J., Kubler, A.M., and Jimenez de Asua, L. (1979): Proc. Natl. Acad. Sci. USA (in press).

43. Otto, A.M., Kubler, A.M., Ulrich, M.O., and Jimenez de Asua, L. (in preparation).
44. Pardee, A.B. (1974): Proc. Natl. Acad. Sci. USA, 71:1286-1290 .
45. Pardee, A.B., Jimenez de Asua L., and Rozengurt,E. (1974): In: Control of Cell Proliferation in Animal Cells, edited by B. Clarkson and R. Baserga, pp. 547-561. Cold Spring Harbor Laboratories.
46. Pardee, A.B., Dubrow, L., Hamlin, J.L., and Kletzien, R.F. (1978): Ann. Rev. Biochem., 47:715-750.
47. Pledger, W.J., Stiles, C.D., Antoniades, H.R., and Scher, C.D. (1977): Proc. Natl. Acad. Sci. USA, 75:2839-2843.
48. Richmond, K.M.V., Kubler, A.M., Martin, F., and Jimenez de Asua, L. (1980): J. Cell. Physiol. (in press).
49. Riddle, V.G.H., Dubrow, R., and Pardee, A.B. (1979): Proc. Natl. Acad. Sci. USA, 76:1298-1302.
50. Rozengurt, E. (1976): J. Cell. Physiol., 89:627-632.
51. Rozengurt, E., and Heppel, L. (1975): Proc. Natl. Acad. Sci. USA, 72:4492-4494.
52. Rudland, P.S. (1974): Proc. Natl. Acad. Sci. USA, 71:750-754.
53. Rudland, P.S., and Jimenez de Asua, L. (1979): Biochim. Biophys. Acta, 560:91-133.
54. Rudland, P.S., Seifert, W., and Gospodarowicz, D. (1974): Proc. Natl. Acad. Sci. USA, 71:2600-2604.
55. Shechter, Y., Hernaez, L., and Cuatrecases, P. (1978): Proc. Natl. Acad. Sci. USA, 75:5788-5791.
56. Shields, R., and Smith, J.A. (1977): J. Cell. Physiol., 91:345-356.
57. Smith, J.A., and Martin, L. (1973): Proc. Natl. Acad. Sci. USA, 70:1263-1267.
58. Todaro, G.J., and Green, H. (1963): J. Cell. Biol., 17:299-313.
59. Thomas, G., Siegmann, and Gordon, J. (1979): Proc. Natl. Acad. Sci. USA, 76:3952-3956.
60. Willingham, M.C., Johnson, G.S., and Pastan, I.H. (1972): Biochim. Biophys. Res. Commun., 48:743-748.

Control Mechanisms in Animal Cells,
edited by L. Jimenez de Asua et al.
Raven Press, New York © 1980.

Interactions Between Two Growth Factors During the Prereplicative Period in Swiss 3T3 Cells

K. M. Veronica Richmond

Friedrich Miescher-Institute, CH-4002 Basel, Switzerland

The control of mammalian cell growth is a fundamental problem in cell biology which is as yet poorly understood. A large number of specific growth factors have now been described, active on a variety of cell lines of different origins (21). One interesting question is whether these growth factors stimulate growth via similar mechanisms, or whether each has its own characteristic mode of action. The Swiss mouse 3T3 cell line offers a useful approach to this problem, since it is responsive to several growth factors under controlled environmental conditions.

Confluent, quiescent Swiss 3T3 cells in culture are arrested in G_0/G_1 or the A-state of the cell cycle (4, 22,23). Addition of serum or growth factors, such as prostaglandin $F_{2\alpha}$ ($PGF_{2\alpha}$)(9), fibroblast growth factor (FGF)(6,8,20) or epidermal growth factor (EGF)(5), results in the stimulation of initiation of DNA synthesis after a lag phase of c. 13-15 hr, at a rate characteristic of the growth factor and the culture and nutritional conditions (2,7,9,14,15). The initiation of DNA synthesis in the population is consistent with first order kinetics, and can be quantified by a first order rate constant, k (10,11).

A number of agents are known which do not stimulate growth alone, but which modify k when added together with a growth factor. Several of these have specific "windows" in the lag phase, during which they must be added to produce their effect (12). Insulin enhances k at whatever time it is added (12,13), while cytoskeleton-disrupting drugs enhance (16,17), and PGE_1 inhibits (12) k, only when added during the first 8 hr of the lag phase. Hydrocortisone enhances k at any time when added in combination with FGF (12,18), but in combination with $PGF_{2\alpha}$ it inhibits during the first 4 hr, and thereafter has no effect (11). These "windows" suggest that the lag phase consists of a sequence of events required for growth stimulation (12), and the difference in hydrocortisone action in combination with FGF or $PGF_{2\alpha}$ suggests a difference in the events stimulated by the two factors (18).

This paper presents an investigation of the interactions between $PGF_{2\alpha}$ and FGF in stimulating DNA synthesis. If the two factors act via identical mechanisms no enhancement in response would be expected from adding both together. If additive or synergistic effects were obtained, this would indicate that the two factors do act differently in some way.

MATERIALS AND METHODS

Swiss 3T3 cells were maintained and experimental cultures set up as described previously (10,12,15). Growth factors were added directly, without medium change, to cultures which had reached confluence and become quiescent after growth in 6% foetal calf serum (10,12,15). Incorporation of ^3H-thymidine was measured after incubation with 3 μM [methyl-^3H] thymidine (0.3 μCi/ml) from 0-25 hr after addition of growth factors. Cultures were harvested as previously described (10,12) and solubilized for scintillation counting in 0.5 M NaOH. For autoradiography, cells were exposed to 1μM [methyl-^3H] thymidine (3 μCi/ml) from the time of first addition of growth factor to the times indicated in each experiment. Pairs of cultures were fixed and processed for autoradiography as previously described (10,12).

RESULTS

FGF and $PGF_{2\alpha}$ are each capable of inducing DNA synthesis alone, although considerably less efficiently than serum (Fig. 1). A dose-response curve to $PGF_{2\alpha}$ in the presence or absence of saturating and sub-saturating concentrations of FGF is shown in Fig. 1. Addition of both growth factors at saturating concentrations gave 70% of the response to 15% serum. The two factors were here clearly synergistic, indicating that each stimulates at least one event which influ-

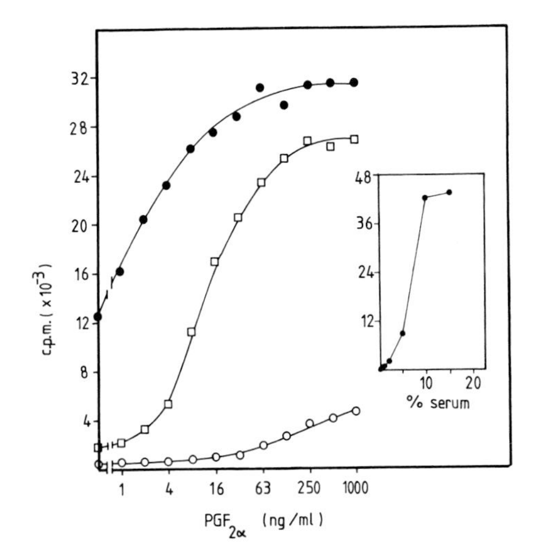

FIG. 1. Effect of FGF on the dose-response curve to $PGF_{2\alpha}$. $PGF_{2\alpha}$ was added to quiescent cultures at the concentrations shown, in the presence or absence of FGF. ^{3}H-Thymidine was added at time zero, and after 25 hr cultures were harvested and the cpm incorporated determined as described in Materials and Methods. (o) $PGF_{2\alpha}$ alone. (□) $PGF_{2\alpha}$ in the presence of 15 ng/ml FGF. (●) $PGF_{2\alpha}$ in the presence of 50 ng/ml FGF. Inset: Response (cpm x 10^{-3}) of the same cells to serum.

ences the initiation of DNA synthesis and which is not shared by the other. Synergy was also observed after investigation of the per cent labelled nuclei 25 hr after the addition of the growth factors (0 hr, Fig. 2). Here $PGF_{2\alpha}$ alone gave 4%, FGF alone 20%, and both together at 0 hr gave 35% labelled nuclei.

Investigation by sampling at a single time point gives no information about the kinetics of the interactions. Several simple predictions may be made as to the kinetics of interaction of $PGF_{2\alpha}$ and FGF, although these do not exclude the possibility of more complex interactions.

(1) Addition of a second growth factor together with or after the first may have a similar effect to addition of a k-modifying agent, such as insulin or hydrocortisone (12,13). The second addition does not change the lag phase, and may increase k either when added during a defined window in the lag phase

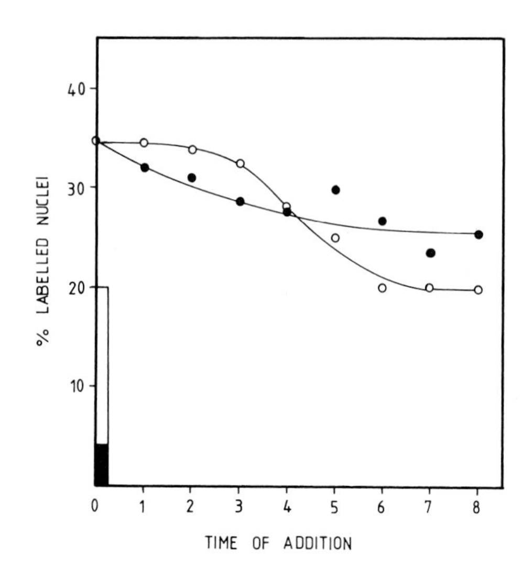

FIG. 2. Labelling index at 25 hr following addition of a second growth factor at varying times after the first. ^{3}H-Thymidine was present throughout, and cultures were fixed and processed for autoradiography as described in Materials and Methods. Dark bar: Response to $PGF_{2\alpha}$ alone. Light bar: Response to FGF alone. (o) Addition of $PGF_{2\alpha}$ (500 ng/ml) at varying times after FGF (50 ng/ml) at 0 hr. (●) Addition of FGF (50 ng/ml) at varying times after $PGF_{2\alpha}$(500 ng/ml) at 0 hr.

or when added at any time.

(2) Co-operation between two growth factors may allow faster completion of events required for initiation of DNA synthesis, and thus result in a shortening of the lag phase.

(3) Each growth factor may set its own independent lag phase, so that, for example, addition of a second growth factor 10 hr after the first would give a value of k characteristic of the first growth factor between 15 hr and 25 hr, and a change to a value of k characteristic of both together at about 25 hr.

Turning first to the question of whether there is a "window" during the lag phase for the interaction between $PGF_{2\alpha}$ and FGF: In the experiment shown in Fig. 2, the lag phase was set by the addition of FGF, and $PGF_{2\alpha}$

was added at various times thereafter, the labelling index was determined 25 hr after FGF addition. The complementary experiment, with $PGF_{2\alpha}$ added first and FGF at later times is also shown. In both cases the enhancement of the labelling index was less when the second factor was added later than the first, and when $PGF_{2\alpha}$ was added 6-8 hr after FGF the labelling index was the same as for FGF alone.

If this experiment is to be interpreted quantitatively, it must be shown that only one parameter is varying, i.e. either the lag phase or the value of k. It is thus necessary to follow the time course of cellular entry into S in a similar experiment. Fig. 3 shows the effect on the lag phase and the value of k

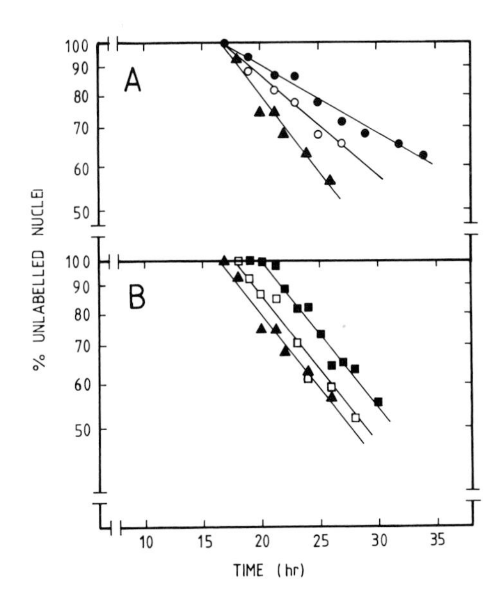

FIG. 3. Kinetics of the response to $PGF_{2\alpha}$ and FGF separately and together. FGF (50 ng/ml) and $PGF_{2\alpha}$ (500 ng/ml) were added to confluent quiescent Swiss 3T3 cells as described below. ^{3}H-Thymidine was added at 0 hr and pairs of dishes were fixed and processed for autoradiography at the times indicated. Additions were as follows: A) (●) $PGF_{2\alpha}$ alone ($k=2.9 \times 10^{-2}$/hr), (o) FGF alone ($k=4.1 \times 10^{-2}$/hr), (▲) $PGF_{2\alpha}$+ FGF together at 0 hr ($k=6.3 \times 10^{-2}$/hr). B) (▲) $PGF_{2\alpha}$+ FGF together at 0 hr, (□) $PGF_{2\alpha}$ at 0 hr + FGF at 5 hr ($k=6.3 \times 10^{-2}$/hr), (■) $PGF_{2\alpha}$ at 0 hr + FGF at 10 hr ($k=6.1 \times 10^{-2}$/hr). (Figure taken from ref. 19).

of adding $PGF_{2\alpha}$ and FGF together and of adding FGF at different times after $PGF_{2\alpha}$. When both factors were added simultaneously (Fig. 3A), the lag phase was the same as for each factor separately. The two factors did not co-operate to shorten the lag phase, only to increase the value of k. Addition of FGF 5 hr or 10 hr after $PGF_{2\alpha}$ gave the unexpected result of lengthening the lag phase; by 1.2 hr and 3.6 hr, respectively. The value of k was in each case the same as for simultaneous addition. A similar lengthening of the lag phase was observed when $PGF_{2\alpha}$ was added after FGF (19).

The interactions between FGF and $PGF_{2\alpha}$ thus do not follow any of the simple predictions outlined above. Cells which would have initiated DNA synthesis in response to $PGF_{2\alpha}$ alone between (in this case) $16^1/2$ and 20 hr were prevented from doing so by the addition of FGF at 10 hr. Once initiation of DNA synthesis began in the population, it proceeded at an exponential rate, i.e. with a single k value, similar to that obtained when both factors were added simultaneously.

DISCUSSION

The results presented here show that $PGF_{2\alpha}$ and FGF do not trigger identical sets of events. Addition of both factors together always gives a greater response than either alone. The response to both factors was reproducible over many experiments, giving an average k value of 6.5×10^{-2}/hr (data not shown), while the response to the individual growth factors was more variable, so that synergy was not found in every case (Fig. 3A). The lengthening of the lag phase by late addition of a second growth factor also argues for a difference between the growth factors, since this effect is not observed after a second addition of the same growth factor. Addition of saturating growth factor concentrations at various times after the lag phase is set by a sub-saturating concentration of the same growth factor increases k, but does not prolong the lag phase (12,13), and when the lag phase is set with a saturating concentration, further additions of the same growth factor have no effect (unpublished results).

The events triggered by the two growth factors are not completely independent for the following two reasons. Firstly, late addition of a second growth factor interferes with the response to the first, and postpones the initiation of DNA synthesis until after the end of a new lag phase. Secondly, the end of the new lag phase is not a constant time from the addition of the second growth factor. Addition of FGF at 5 hr, or 10 hr, after $PGF_{2\alpha}$ prolongs the lag phase by 1.2 hr, or 3.6 hr, respectively. There is thus some co-operation between the events triggered by the growth factors, such that cells at a later stage of the $PGF_{2\alpha}$ lag phase can complete the necessary events triggered by FGF in a shorter time.

What is the significance of these variations in the lag phase? It has been shown previously (10,12,13,18) that the ability to set a lag phase is a special property of a growth factor, and indeed, this property provides a definition of a growth factor for this system. Other agents which are able to increase k, such as insulin or (in combination with FGF) hydrocortisone, do not shorten the lag phase when added together with or later than the growth factor, and do not set the lag phase when added before the growth factor (12,13). When added at 15 hr, they do not postpone the initiation of DNA synthesis, but give a gradual increase in k, which reaches its final value about 6 hr later (11,12). The results reported here are further evidence for a special role for the growth factor throughout the lag phase, and imply that a growth factor stimulates some event(s) which have to be completed before the cell can initiate DNA synthesis.

It has been postulated that a rate-limiting step exists at the end of the lag phase, which accounts for the first-order entry into the S phase (1,12). This step is apparently sensitive to many environmental and metabolic factors, including the nature and concentration of growth factors (12), the presence or absence of insulin, hydrocortisone, or other agents mentioned above which modify k (12), the presence of vitamin B_{12} and hypoxanthine (11,15) and the level of protein synthesis in the cell (3). It may also be this step which is sensitive to processes triggered by the late addition of FGF, so that k does not change in two steps, in response to the two factors individually, but in a single step at the end of a new lag phase.

The biochemical basis of the differences in action of $PGF_{2\alpha}$ and FGF, and of their interaction remains to be elucidated and provides wide scope for future experiments.

ACKNOWLEDGEMENTS

I would like to thank Dr. John Pike (Upjohn Co.) for the generous gift of $PGF_{2\alpha}$; Miss F.N. Richmond for technical assistance; Drs. A. Otto and L. Jimenez de Asua for stimulating discussions, and Drs. J. Davis and P. Simons for critical reading of the manuscript.

REFERENCES

1. Brooks, R.F. (1975): J. Cell Physiol., 86:369-378.
2. Brooks, R.F. (1976): Nature, 260:248-250.
3. Brooks, R.F. (1977): Cell, 12:311-317.
4. Burns, F.J., and Tannock, I.F. (1970): Cell Tissue Kinet., 3:321-324.
5. Cohen, S., and Savage, C.R.,Jr. (1974): Recent Prog. Horm. Res., 30:551-574.
6. Gospodarowicz, D. (1974): Nature, 249:123-127.
7. Holley, R.W. (1975): Nature, 258:487-490.
8. Holley, R.W., and Kiernan, J.A. (1974): Proc. Natl. Acad. Sci.USA, 71:2908-2911.
9. Jimenez de Asua, L., Clingan, D., and Rudland,P.S. (1975): Proc. Natl. Acad. Sci. USA, 72:2724-2728.
10. Jimenez de Asua, L., O'Farrell, M.K., Clingan, D., and Rudland, P.S. (1977): Proc. Natl. Acad. Sci. USA, 74:3845-3849.
11. Jimenez de Asua, L., O'Farrell, M.K., Bennet, D., Clingan, D., and Rudland, P.S. (1977): Nature, 265:151-153.
12. Jimenez de Asua, L., Richmond, K.M.V., Otto, A.M., Kubler, A.-M., O'Farrell, M.K., and Rudland, P.S. (1979) In: Hormones and Cell Culture. Cold Spring Harbor Conferences on Cell Proliferation. Vol. 6, edited by R. Ross and G.H. Sato, pp. 403-423. Cold Spring Harbor Laboratory.
13. Jimenez de Asua, L., (this volume).
14. Mierzejewski, K., and Rozengurt, E. (1976): Biochem. Biophys. Res. Comm., 73:271-278.
15. O'Farrell, M.K., Clingan, D., Rudland, P.S., and Jimenez de Asua, L. (1979): Exp. Cell Res., 118: 311-321.
16. Otto, A.M., Zumbé, A., Gibson, L., Kubler, A.-M., and Jimenez de Asua (1979): Proc. Natl. Acad. Sci. USA, (in press).

17. Otto, A.M., Ulrich, M.-O., and Jimenez de Asua, L. (this volume).
18. Richmond, K.M.V., Kubler, A.-M., Martin, F., and Jimenez de Asua, L. (1980): J. Cell Physiol., (in press).
19. Richmond, K.M.V., and Jimenez de Asua, L. (manuscript in preparation).
20. Rudland, P.S., Seifert, W., and Gospodarowicz, D. (1974): Proc. Natl. Acad. Sci. USA, 71:2600-2604.
21. Rudland, P.S., and Jimenez de Asua, L. (1974): Biochim. Biophys. Acta, 560:91-133.
22. Shields, R., and Smith, J.A. (1977): J. Cell Physiol., 91:345-356.
23. Smith, J.A., and Martin, L. (1973): Proc. Natl. Acad. Sci. USA, 70:1263-1267.

Control Mechanisms in Animal Cells,
edited by L. Jimenez de Asua et al.
Raven Press, New York © 1980.

Does the Cytoskeleton Play a Role in the Initiation of DNA Synthesis Stimulated by Growth Factors?

*Angela M. Otto, Marie-Odile Ulrich, and Luis Jimenez de Asua

Friedrich Miescher-Institute, CH-4002 Basel, Switzerland

A basic feature of animal cells is their ability to replicate their genetic material (DNA) and to divide in response to changes in the extracellular environment (1,9,18). There is cumulative data to suggest that in vivo and in vitro the rate of proliferation of mammalian cells is controlled mainly by growth factors, hormones, ions, nutrients and other defined compounds. Of central importance for the understanding of growth control by growth factors is the elucidation of

1. How is a stimulatory signal delivered at the level of the surface membrane ?
2. Which events participate in transmitting the signals to intracellular targets involved in the initiation of DNA synthesis and cell division ?

Several lines of evidence have led to the idea that the cytoskeleton in association with proteins of the cell surface membrane (surface modulating activity) may play a regulatory role in the transmission of the mitogenic signals (6,14). The main approach to these questions is to examine the effect of cytoskeleton-disrupting drugs on the stimulation of initiation of DNA synthesis.

Mouse Swiss 3T3 cells offer a convenient experimental system to study growth regulation, since under

---

* Dedicated to the memory of my father.

defined conditions these cells become quiescent, i.e. arrested in the G_0/G_1 phase or A state of the cell cycle. Such synchronised cultures can be stimulated by serum or defined growth factors to initiate DNA synthesis, which occurs after a constant lag phase of 13-15 hr before the abrupt increase in the rate of initiation can be observed (3,11,19,22,23). The kinetics of initiation follows a first order process and can be quantified by a rate constant k. The lag phase can be dissected into a temporal sequence of events by further additions of growth factors and hormones or other compounds, which by themselves do not initiate DNA synthesis, but can produce positive or negative changes in the rate constant (see 11,19 and Jimenez de Asua, this volume).

In this paper, we present evidence that cytoskeleton-disrupting drugs can enhance the stimulatory effect of different growth factors on the initiation of DNA synthesis, suggesting that the intact cytoskeleton is not essential, but may exert a restrictive control on the rate at which 3T3 cells stimulated by growth factors can enter into the S phase.

MATERIALS AND METHODS

Swiss mouse 3T3 (25) cells were maintained as described before (10). To assay for the initiation of DNA synthesis, cells were plated at 1.5 x 10^5/35 mm dish in 2 ml Dulbecco's modified Eagle's medium (DEM) supplemented with low molecular components (11) and 6.5% fetal calf serum. Cultures were allowed to become quiescent 3-4 days after an intermediate medium change before additions were made directly to the culture medium (10). Cells were labelled for autoradiography by exposing cultures to 1 μM (3 μCi/ml) (methyl-^{3}H)thymidine from the time of addition of growth factors and hormones. For determination of the fraction of labelled nuclei, cultures were processed for autoradiography after 28 hr (10). For studying the kinetics of initiation of DNA synthesis cells were processed at times indicated in each experiment.

In experiments where colchicine or colcemid was present only part of the time, these drugs were removed from the culture medium by aspiration and cells were washed twice with DEM prewarmed at 37°C. Conditioned medium retrieved from parallel quiescent cultures was added to the experimental dishes with the growth factor and hormones at the same concentration as at the start.

For determination of the rate constant k, the percentage of resting cells that remain unlabelled (y) in a given time (t) was calculated from the labelling index. These results were plotted as the $\log_{10}y$ against t in hours. The data fitted well with the straight lines given by $\log_{10}y = a-bt$. First-order rate constants (k) were then calculated by geometrical methods from the slope of the lines (b), because $k = \log_e 10 \cdot b$ (10).

$PGF_{2\alpha}$ was a generous gift of John Pike, Upjohn Company, Kalamazoo, USA. FGF and EGF were obtained from Collaborative Research. Crystalline insulin, hydrocortisone, colchicine, colcemid and vinblastine were purchased from Sigma. (Methyl-^{3}H)thymidine was from the Radiochemical Centre, Amersham, England.

RESULTS

Cytoskeleton-Disrupting Drugs Enhance the Stimulatory Effect of Different Growth Factors

The effect of colchicine, colcemid or vinblastine added with different growth factors, alone or in combination with hormones, to confluent quiescent Swiss 3T3 cells is shown in Table 1. Addition of each drug results in an increase in the fraction of radioactively labelled nuclei stimulated by prostaglandin $F_{2\alpha}$ ($PGF_{2\alpha}$) alone or with insulin. With epidermal growth factor (EGF) alone or plus insulin, colchicine, colcemid or vinblastine produce a similar effect (Table 1). With either growth factor cytoskeleton-disrupting drugs increase the labelling index to the same percentage as the addition of insulin, and the combination of the hormones and a drug with the growth factor gives the maximum effect. Colchicine, colcemid or vinblastine also enhance the stimulation by fibroblast growth factor (FGF), however, the maximum effect is obtained without hormones. Addition of FGF with insulin and hydrocortisone also results in the maximum labelling index, which cannot be further increased by a cytoskeleton-disrupting drug (Table 1). Thus, in quiescent 3T3 cells three different cytoskeleton-disrupting drugs can enhance the stimulatory effect of growth factors, which differ either in their chemical nature or their interaction with hormones. Neither colchicine, colcemid nor

TABLE 1. Effect of cytoskeleton-disrupting drugs on the stimulation of DNA synthesis by growth factors[a]

Additions	% Labelled nuclei			
	---	Colchicine	Colcemid	Vinblastine
None	0.1	0.2	0.1	0.3
$PGF_{2\alpha}$	17	43	42	42
$PGF_{2\alpha}$ + insulin	47	78	75	78
EGF	15	38	39	40
EGF + insulin	40	70	74	75
FGF	25	77	76	78
FGF + insulin + hydrocortisone	77	78	77	77
Insulin	0.2	0.4	0.4	0.3
Insulin + hydrocortisone	0.3	0.5	0.4	0.4
Serum	97	98	97	99

[a]Concentrations used were for $PGF_{2\alpha}$ (400 ng/ml), EGF (20 ng/ml), FGF (50 ng/ml), insulin (50 ng/ml), fetal calf serum 15%, colchicine, colcemid and vinblastine (1 μM). (Reprinted from 15,17).

vinblastine stimulates the initiation of DNA synthesis in the absence of a growth factor (Table 1). Also, lumicolchicine, an inactive analog of colchicine, has no effect on the stimulation (17).

Addition of colchicine or colcemid changes the dose-response curves of a growth factor for the stimulatory effect. Fig. 1A shows that colchicine added with increasing concentrations of $PGF_{2\alpha}$ markedly decreases the amount of growth factor required for half-maximal stimulation. Insulin also shifts the dose-response curve to lower concentrations, and the addition of colchicine further enhances this effect with a concomitant increase in the labelling index at saturating concentrations of $PGF_{2\alpha}$. The change in the dose-response curves of EGF by insulin and/or colcemid are very similar to those of $PGF_{2\alpha}$. In contrast, the concentration curves of FGF with colchicine and/or insulin and hydrocortisone change only by the amount of FGF required for maximum stimulation (Fig. 1B).

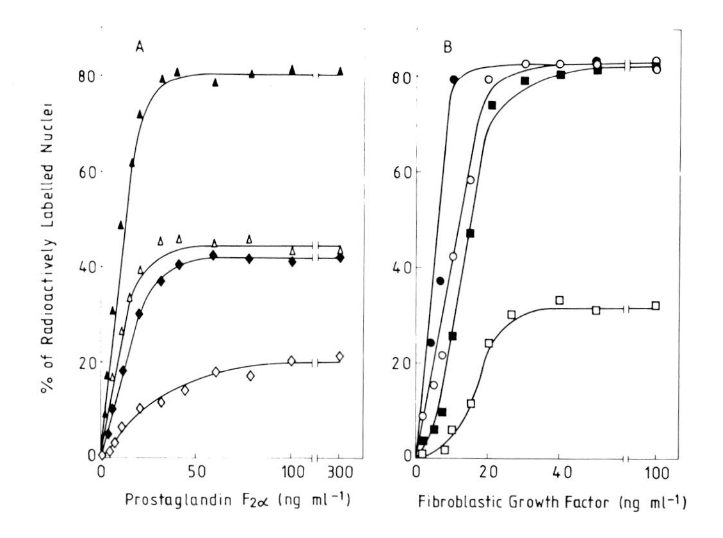

FIG. 1. Effect of colchicine on the labelling index stimulated by different concentrations of $PGF_{2\alpha}$ or FGF. Concentrations used were for insulin (50 ng/ml), hydrocortisone (100 ng/ml) and colchicine (2 μM). A. (◇) $PGF_{2\alpha}$, (◆) $PGF_{2\alpha}$+ insulin, (△) $PGF_{2\alpha}$+ colchicine, (▲) $PGF_{2\alpha}$+ insulin + colchicine. B. (□) FGF, (■) FGF + colchicine, (o) FGF + insulin + hydrocortisone, (●) FGF + insulin + hydrocortisone + colchicine. (Reprinted from 17).

How and When Do Microtubule-Disrupting Drugs Affect the Kinetics of DNA Synthesis ?

The stimulation of the initiation of DNA synthesis can be divided into two parameters: the length of the lag phase and the change in the rate of initiation(11). Which parameter is changed to produce the synergistic effect ? Fig. 2 shows the kinetics of DNA synthesis of cells stimulated by FGF. Synchronous addition of colchicine increases the rate constant, similar to the addition of insulin and hydrocortisone, without changing the length of the lag phase (Fig. 2A). The addition of colchicine with $PGF_{2\alpha}$ also does not change the lag phase, but exerts its effect by increasing the rate constant similar to insulin, and the addition of colchicine with $PGF_{2\alpha}$ plus insulin further increases the value of k (Fig. 3A). With EGF colchicine gives similar results as with $PGF_{2\alpha}$ (15). Thus, for all three growth factors colchicine exerts its synergistic

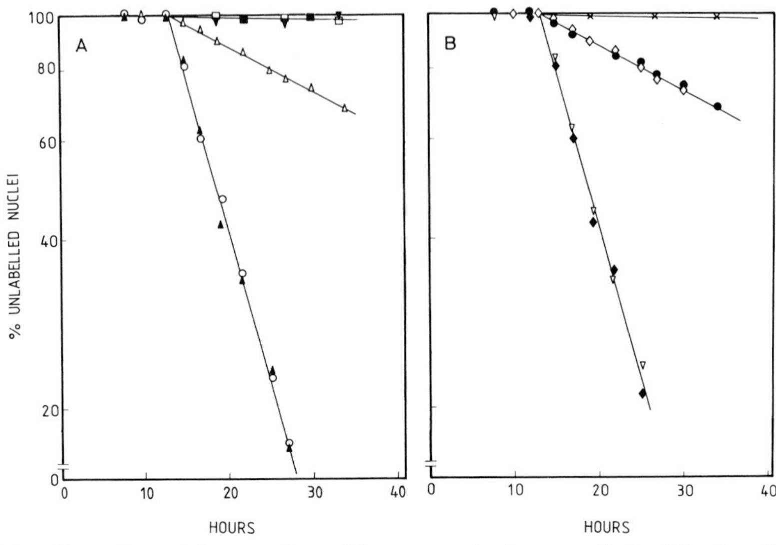

FIG. 2. Fraction of cells remaining unlabelled after addition of FGF, alone or with insulin and hydrocortisone, in the absence or presence of colchicine. Concentrations used were for FGF (50 ng/ml), insulin (50 ng/ml), hydrocortisone (100 ng/ml), colchicine (1 μM). A. (□) Hydrocortisone, (■) insulin, (▼) insulin + hydrocortisone, (△) FGF, (▲) FGF + colchicine, (o) FGF+ insulin + hydrocortisone. B. (◇) FGF + (colchicine at 8 hr), (●) FGF + (colchicine at 15 hr), (◆) FGF + (colchicine 0-5 hr), (▽) FGF + insulin + hydrocortisone + colchicine, (x) insulin + hydrocortisone + colchicine. The values of k (x 10^{-2}/hr) were (□,■,▼,x) 0.05, (△,◇,●) 1.84, (▲,o,◆,▽) 13.3. The duration of the lag phase is 13 hr. (Reprinted from 17).

effect by increasing the value of k (16).

What is the effect of colchicine added or removed at later times during the lag phase ? As shown in Fig. 2B and 3B, with neither growth factor did colchicine change the lag phase or enhance the rate constant when added at 8 or 15 hr. This means that events early in the lag phase are affected by colchicine to enhance the rate of initiation of DNA synthesis. Colchicine present only from 0-5 hr of the lag phase has the same synergistic effect as if it remains continuously in the culture. Even if cells are preincubated with colchicine for 1 hr and it is removed prior to addition of growth factor, the full synergistic effect is observed (not shown).

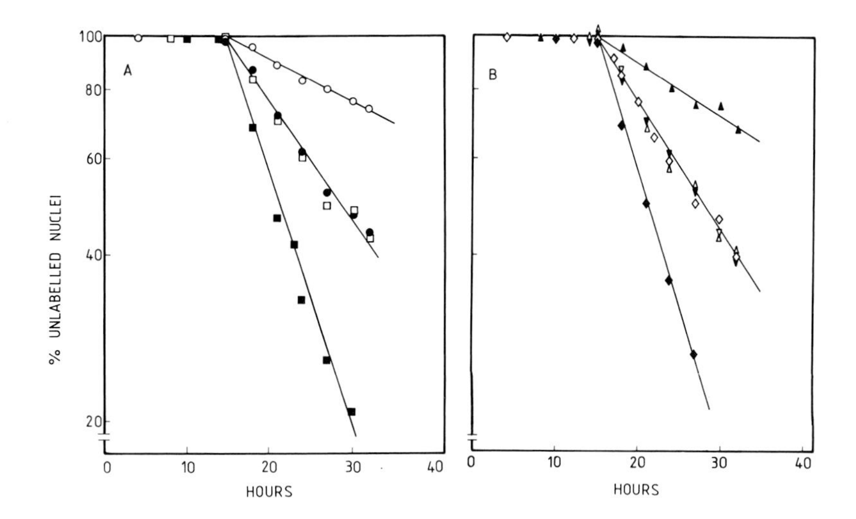

FIG. 3. Fraction of cells remaining unlabelled after addition of $PGF_{2\alpha}$, alone or with insulin and/or colchicine. Concentrations used were for $PGF_{2\alpha}$ (400 ng/ml), insulin (50 ng/ml) and colchicine (1 μM). A. Synchronous addition of colchicine: (o) $PGF_{2\alpha}$, (●) $PGF_{2\alpha}$ + insulin, (□) $PGF_{2\alpha}$+ colchicine, (■) $PGF_{2\alpha}$ + insulin + colchicine. B. Later additions and removal of colchicine: (▲) $PGF_{2\alpha}$ + (colchicine at 8 hr), (▼) $PGF_{2\alpha}$ + insulin + (colchicine at 8 hr), (△) $PGF_{2\alpha}$ + insulin + (colchicine at 15 hr), (◇) $PGF_{2\alpha}$ + (colchicine 0-5 hr), (◆) $PGF_{2\alpha}$ + insulin + (colchicine 0-5 hr). The values of k (x 10^{-2}/hr) were (o,▲) 1.8, (●,□,▼,▲,◇) 5.5, (■,◆) 11.2. The duration of the lag phase was 14 hr. (Reprinted from 17).

Colchicine as well as colcemid disrupt microtubules. However, there are differences in the capability of the microtubules to reassemble depending on the drug used. Indirect immunofluorescence shows that upon removal of colchicine, microtubules remain disrupted throughout most of the lag phase. (A. Zumbé, personal communication). In contrast, upon removal of colcemid, microtubules are reassembled within 1-2 hr (2,15).

The following experiments were carried out with EGF and colcemid to investigate the effect of another microtubule-disrupting drug on the kinetics of DNA synthesis. As shown in Table 2, the kinetics with EGF alone or plus insulin and/or colcemid follows the same basic pattern as described for $PGF_{2\alpha}$ with insulin and/or colchicine. Likewise, colcemid has no effect when added at 8 or 14 hr of the lag phase (Table 2).

TABLE 2. Effect of EGF alone or with insulin and/or colcemid on the initiation of DNA synthesis[a]

Additions	rate constant k ($\times 10^{-2}$/hr)
None	0.04
A. Synchronous addition of colcemid	
EGF	1.15
EGF + insulin	3.45
EGF + colcemid	3.44
EGF + insulin + colcemid	13.80
B. Later addition of colcemid	
EGF + (colcemid at 8 hr)	1.15
EGF + (colcemid at 14 hr)	1.14
EGF + insulin + (colcemid at 8 hr)	3.46
EGF + insulin + (colcemid at 14 hr)	3.45
C. Removal of colcemid	
EGF + (colcemid 0-5 hr)	2.75, 1.08[b]
EGF + insulin + (colcemid 0-5 hr)	7.16, 3.75[b]
EGF + insulin + (colcemid 0-8 hr)	8.96

[a]Concentrations used were for EGF (20 ng/ml), insulin (50 ng/ml) and colcemid (1 μM). The duration of the lag phase is 15 hr in all cases. [b]First value is during 15-19 hr, the second value for times after 19 hr. (Reprinted from 15).

However, if colcemid is present only from 0-5 hr of the lag phase, the rate constant is increased only between 15-19 hr, thereafter being the same as if colcemid had not been added (Table 2). Removing colcemid after 8 hr also results in a partial loss of the synergistic effect. Since microtubules are completely reassembled 1-2 hr after removal of colcemid (2,15), this result suggests further that the synergistic effect of the cytoskeleton-disrupting drugs is somehow related to their disruption of the microtubules.

The removal of colchicine or colcemid should allow the cells to complete the cell cycle, i.e. to go through cell division. Table 3, indeed, shows an increase in cell number of stimulated cultures. However, if colchicine or colcemid remains present in the cultures, the cell number in the monolayer decreases below the control count (no addition) due to detachment and cell death. The addition of colchicine or colcemid at 1-2 μM to quiescent cultures up to 48 hr has little effect on the morphology and no effect on the cell number in the monolayer.

TABLE 3. Effect of microtubule-disrupting drugs on cell number[a]

Addition	cell number per dish ($\times 10^5$)
A. None	5.0
Colchicine	4.9
$PGF_{2\alpha}$	6.4
+ colchicine	3.5
+ colchicine 0-5 hr	7.4
+ lumicolchicine	6.3
$PGF_{2\alpha}$ + insulin	9.8
+ colchicine	4.9
+ colchicine 0-5 hr	12.2
FGF	8.0
+ colchicine	2.6
+ colchicine 0-5 hr	9.5
10% Serum	18.0
B. None	7.0
Colcemid	7.1
EGF	9.2
+ colcemid	3.8
+ colcemid 0-5 hr	9.6
EGF + insulin	11.4
+ colcemid	4.2
+ colcemid 0-5 hr	11.9
10% Serum	25.0

[a]For determination of cell number, Swiss 3T3 cells were plated at 1.5×10^5 cells in 60 mm dishes in culture conditions as for determination of DNA synthesis (Materials and Methods). Additions were made about 4 days after plating when no mitotic figures were observed. Concentrations used were for $PGF_{2\alpha}$ (400 ng/ml), FGF (50 ng/ml), EGF (20 ng/ml), insulin (50 ng/ml), colchicine (1 μM), lumicolchicine (1 μM), and colcemid (1 μM). After 48 hr of addition, cells were removed from the dish with a trypsin solution (0.05%), resuspended in isotonic buffer and counted in a Coulter Counter. (Reprinted from 15,17).

DISCUSSION

The experiments presented here seem to indicate that depending on the time of action, cytoskeleton-disrupting drugs can affect the initiation of DNA synthesis by increasing the rate constant. The rate constant can only be increased when colchicine or colcemid is added at the beginning of the lag phase initiated by a growth factor. The synergistic effect is gradually lost with later additions, until after 8 hr the disruption of microtubules does not affect the kinetics of initiation of DNA synthesis at all (Fig. 4A). On the other hand, removal of colcemid before 8-10 hr, which allows for reassembly of the microtubules before 10-12 hr of the lag phase, also results in a loss of the synergistic effect (Fig. 4B). Thus, microtubules must be disrupted beyond 8 hr of the lag phase for the full synergistic effect. It can be postulated that at least

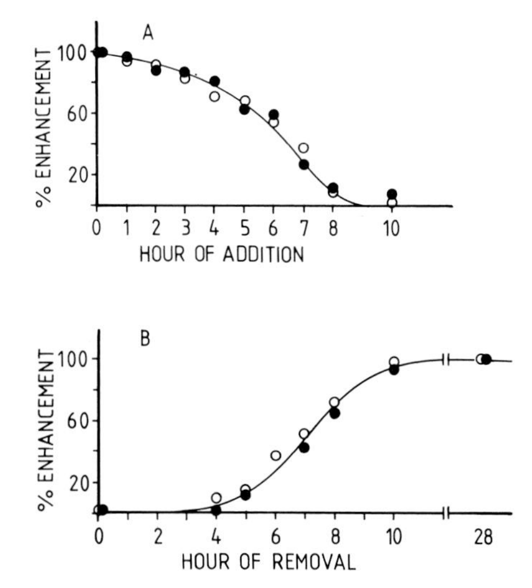

FIG. 4. Time dependency of colchicine and colcemid action during the lag phase on stimulation of DNA synthesis. A. Loss of synergistic effect with later additions of colchicine after $PGF_{2\alpha}$ (o) or $PGF_{2\alpha}$ plus insulin (●). B. Loss of synergistic effect with early removal of colcemid when added with EGF (o) or EGF plus insulin (●).

one event early in the lag phase is sensitive to the organization of the cytoskeleton. Disruption of the microtubules enhances the induction or action of this putative event and thereby increases the rate constant at the end of the lag phase. However, even though the putative event can no longer be enhanced after 7-8 hr, the microtubules apparently must remain disrupted beyond this time to ensure the complete accomplishment of this event, possibly by inhibiting its decay.

Other reports for the enhancing effect of colchicine and colcemid on DNA synthesis stimulated by growth promoting factors have appeared (7,24,27). However, inhibitory effects of these drugs on the DNA synthesis have also been reported (14,28). In view of the differential effect of these drugs on the enhancement depending on the growth factor (comparing e.g. $PGF_{2\alpha}$ and FGF) and that these drugs can act only at specific times during the lag phase, it can be proposed that in tissue culture experiments cytoskeleton-disrupting drugs may be simulating physiological substances acting in the animal, which could control the state of the cytoskeleton and thereby the responsiveness to growth factors. Indeed, in extracts of mammalian brain at least two peptides have been found, which interact with the colchicine-binding site on tubulin (13).

Which event(s) related to the initiation of DNA synthesis is sensitive to microtubule-disrupting drugs ? The following possibilities can only give hints.
1. The occupation and/or interaction of high affinity receptors, which are processes thought to be involved in making the cells competent for the initiation of DNA synthesis (21), may be increased. However, colchicine does not seem to affect the lateral movement of proteins in the surface membrane (20). 2. Also, internalization of the growth factor-receptor complex and/or its degradation (4,5) may be delayed. 3. At the cytoplasmic level, cytoskeleton-disrupting drugs may release the attachment of polyribosomes to the cytoskeleton (12) and could thus enhance new synthesis of total or specific proteins, which have been correlated with changes in DNA synthesis (11). 4. Disruption of microtubules has been shown to result in retraction of the mitochondria from the cell periphery to the region around the nucleus, which probably changes the distribution of ATP, divalent cations and other constituents of mitochondrial metabolism (8). 5. Furthermore, colchicine-treatment deciliates centrioles, and the deciliation and duplication of these structures seems to be

related to the stimulation of DNA synthesis (26).

Although the questions posed in the introduction cannot yet be answered, these results suggest that the polymerised state of the microtubules exerts a restrictive control on some event(s) leading to the initiation of DNA synthesis.

ACKNOWLEDGEMENTS

The authors are very grateful for stimulating discussions and critical reading of the manuscript to Drs. D. Monard, G. Thomas and C. Natoli.

REFERENCES

1. Baserga, R. (1976): Multiplication and Division in Animal Cell. Dekker, New York.
2. Brinkley, B.R., Fuller, G.M., and Highfield, D.P. (1976): In: Cell Motility, edited by R. Goldman, T. Pollard and J. Rosenbaum, pp. 435-456. Cold Spring Harbor Laboratory.
3. Brooks, R.F. (1976): Nature, 260:248-250.
4. Carpenter, G., and Cohen, S. (1976): J. Cell Biol., 71:159-171.
5. Das, M., and Fox, C.F. (1978): Proc. Natl. Acad. Sci. USA, 75:2644-2648.
6. Edelman, G.M., and Yahara, J. (1976): Proc. Natl. Acad. Sci. USA, 73:2047-2051.
7. Friedkin, M., Legg, A., and Rozengurt, E. (1979): Proc. Natl. Acad. Sci. USA, 76:3909-3912.
8. Heggeness, M.H., Simon, M., and Singer, S.J. (1978): Proc. Natl. Acad. Sci. USA, 75:3863-3866.
9. Holley, R.W. (1975): Nature, 258:487-490.
10. Jimenez de Asua, L., O'Farrell, M.K., Clingan, D., and Rudland, P.S. (1977): Proc. Natl. Acad. Sci. USA, 74:3845-3849.
11. Jimenez de Asua, L., Richmond, K.M.V., Otto, A.M., O'Farrell, M.K., and Rudland, P.S. (1979): In: Hormones and Cell Culture, edited by G.H. Sato and R. Ross, pp. 403-424. Cold Spring Harbor Laboratory.
12. Lenk, R., Ranson, L., Kaufmann, Y., and Penman, S. (1977): Cell, 10:67-78.
13. Lockwood, A.H. (1979): Proc. Natl. Acad. Sci. USA, 79:1184-1188.

14. McClain, D.A., D'Eustachio, P., and Edelman, G.M. (1977): Proc. Natl. Acad. Sci. USA 74:666-670.
15. Otto, A.M., Ulrich, M.O., Zumbé, A., and Jimenez de Asua, L. (1979): Manuscript submitted for publication.
16. Otto, A.M., Zumbé, A., Gibson, L., Kubler, A.M., and Jimenez de Asua, L. (1979): In: The Cytoskeleton: Membranes and Movement, Cold Spring Harbor Meeting on Cell Motility, p. 61. Cold Spring Harbor Laboratory.
17. Otto, A.M., Zumbé, A., Gibson, L., Kubler, A.M., and Jimenez de Asua, L. (1979): Proc. Natl. Acad. Sci. USA, in press.
18. Pardee, A.B., Dubrow, L., Hamlin, J.L., and Kletzien,R.F. (1978): Ann. Rev. Biochem., 47:715-750.
19. Rudland, P.S., and Jimenez de Asua, L. (1979): Biochim. Biophys. Acta, 560:91-133.
20. Schlessinger, J., Axelrod, D., Koppel, D.E., Webb, W.W., and Elson, E.L. (1977): Science, 195: 307-310.
21. Shechter, Y., Hernaez, L., and Cuatrecasas, P. (1978): Proc. Natl. Acad. Sci. USA, 75:5788-5791.
22. Shields, R., and Smith, J.A. (1977): J. Cell. Physiol., 91:345-356.
23. Smith, J.A., and Martin, L. (1973): Proc. Natl. Acad. Sci. USA, 70:1263-1267.
24. Teng, M.-K., Bartholomew, J.C., and Bissell, M.J. (1977): Nature, 268:739-741.
25. Todaro, G.J., and Green, H. (1963): J. Cell Biol. 17:299-313.
26. Tucker, R.W., Pardee, A.B., and Fujiwara, K. (1979): Cell, 17:527-535.
27. Vasiliev, J.M., Gelfand, I.M., and Guelstein, V.I. (1971): Proc. Natl. Acad. Sci. USA, 68:977-979.
28. Walker, P.R., Boyton, A.L., and Whitfield, J.F. (1977): J. Cell. Physiol., 93:89-98.

Control Mechanisms in Animal Cells,
edited by L. Jimenez de Asua et al.
Raven Press, New York © 1980.

Properties of Sarcoma Growth Factors (SGFs) Produced by Mouse Sarcoma Virus-Transformed Cells in Culture

George J. Todaro and Joseph E. De Larco

Laboratory of Viral Carcinogenesis, National Cancer Institute, National Institutes of Health, Bethesda, Maryland 20205, U.S.A.

In previous reports we have described that murine sarcoma virus-transformed cells have greatly diminished or absent epidermal growth factor (EGF) receptors on their cell surfaces (34,9). The effect is seen with transforming RNA viruses but not with DNA virus transformation nor with most chemical carcinogen induced transformation. Over the years we have accumulated cells transformed by a variety of agents, including DNA viruses such as simian virus 40 (SV40) and polyoma, RNA viruses such as murine and avian sarcoma viruses, chemical carcinogens, and radiation, as well as cells which have become transformed spontaneously during passage in cell culture. These have been obtained from Swiss 3T3, Balb/3T3, and other mouse and rat cell systems. In collaboration with Stanley Cohen, these transformed cells were tested for their ability to bind ^{125}I-labeled EGF (34). As shown in Table 1, it was seen that only cells transformed by the murine sarcoma viruses (MSV) changed in their ability to bind exogenous EGF. MSV-transformed mouse, rat or mink cells all show the same properties, specific loss of EGF receptors while normal levels of other surface receptors are maintained. In Table 2 our experience examining a variety of normal and transformed clones of mouse origin is summarized. It is seen that only the MSV-transformed clones, not the DNA virus-transformed clones, show this loss of EGF binding ability. Of 47 independently isolated, chemically transformed cells, 5 show a pattern like the MSV-transformed cells, i.e., almost complete loss of EGF receptors with normal levels of other receptors maintained. The chemically transformed cells without detectable EGF receptors have not yet been further characterized for what growth factors they may be producing. They represent, however, a distinct minority of chemically transformed cells that, with respect to this phenotype, behave like the MSV-transformed cells. The basis for this finding appears to be the production by the sarcoma virus-transformed cells of a family of growth factors called "sarcoma growth factors" (SGFs) (8). These factors are able to compete with ^{125}I-labeled EGF for cell surface receptors, are potent inducers of cell division, produce morphologic transformation of cells in monolayer and support anchorage-independent growth of normal cells in semi-solid medium, such as agar. The protein with these properties is produced by sarcoma virus-transformed cells and sufficient quantities are released into serum-free medium to allow its partial purification and characterization.

TABLE 1. Binding of mouse ^{125}I-EGF to normal and virus-transformed cells[a]

Cell Line	Transformed By:	Designation	Binding ^{125}I cpm/10^6
BALB/3T3	None	A31-CL7	6700
	SV40	SV-A31-CL1	6900
		SV-A31-CL6	5900
	M-MSV[b]	M-A31-CL71	310
	RSV[c]-B77	B77-A31-CL1	4300
	None	A31 (producing R-MuLV[d])	7400
Swiss 3T3	None	3T3-CL4	4700
	SV40	SV-3T3-CL6	4300
	Polyoma	Py-3T3-41	5200
	SV40+Polyoma	SV-Py-3T3-11	4600
	M-MSV	M-3T3-CL4	230

[a]The binding assays were performed on subconfluent cell culture monolayers in 60 mm petri dishes. Approximately 1 x 10^6 cells per dish were seeded 24 hr. before binding ^{125}I-EGF. The cells were washed twice with binding buffer (34). Bindings were initiated by adding 1.5 ml of binding buffer containing 3 ng of ^{125}I-EGF (approximately 110,000 cpm). After incubation for 40 min. at 37°C, unbound ^{125}I-EGF was removed and the cells washed four times with cold binding buffer. The amount of ^{125}I-EGF bound to the cells was measured by lysing the washed monolayers and transferring the lysate to counting vials. The radioactivity present was determined using a Beckman 300 series gamma counter at 67% efficiency. All data given are the averages of duplicates, and have been corrected for nonspecific binding. In a typical experiment 1 x 10^5 cpm of ^{125}I-EGF was added. The value for nonspecific binding ranged between 50 cpm and 150 cpm above machine background.

[b] Moloney murine sarcoma virus.

[c] Rous sarcoma virus.

[d] Rauscher murine leukemia virus.

The starting material for the purification of SGF was serum-free conditioned media obtained from a Moloney MSV-transformed 3T3 clone, 3Bll-IC (2). The cells were grown in roller bottles (Falcon #3027; 850 cm^2) containing Dulbecco's modification of Eagle's medium (DMEM) (14) with 10% calf serum (Colorado Serum Co.) The cells were washed once for 1 hour and again for 12 hours with 100 ml of serum-free Waymouth's medium (GIBCO,

TABLE 2. Cell transformation and EGF receptor availability[a]

Mouse cells	% Clones with decreased EGF binding	Clones with decreased EGF binding
Untransformed parental clones	0/35	0
SV40	0/8	0
Polyoma	0/4	0
MSV	12/12	100
Chemical	5/47	11

[a]Cell lines with levels of EGF binding of less than 20% of those found in the untransformed control cell cultures were considered to have shown decreased receptor availability. Logarithmically growing cells, 24 hours after seeding at 2-5 x 10^5 cells per plate on 20 cm^2 plastic petri dishes, were tested with ^{125}I-EGF at room temperature for 2 hr., as described (34).

MD705/1) (20) per roller bottle each time. These washes were discarded and two subsequent serum-free 48-hour collections were harvested and are referred to as "sarcoma-conditioned media". The viability of the cells maintained either in medium containing 10% serum or in serum-free medium for 5 days with 4 changes of medium was greater than 80% as determined by trypan blue exclusion. The "sarcoma-conditioned media" were clarified by centrifugation at 100,000 x g for 45 min. The supernate was concentrated twentyfold in a hollow fiber apparatus (Amicon;DC2), and dialyzed against 5 changes consisting of 5 volumes each of 1% acetic acid. This material was lyophilized and extracted with 1 M acetic acid. One percent of the volume of the starting conditioned media was used. The extract was clarified by centrifugation at 100,000 x g for 30 min. Thirty milliters of this 1 M acetic acid extract was chromatographed through a Bio-Gel P-60 column (5 x 90 cm) that had been equilibrated in and eluted with 1 M acetic acid. The column was run at 4°C at a flow rate of 15 ml/hr. and 15 ml fractions were collected. Aliquots were lyophilized for protein determinations (21), EGF competition, stimulation of thymidine incorporation, and soft agar growth activity.

A series of disulfide-containing peptides, found in serum-free conditioned media from a murine sarcoma virus-transformed 3T3 clone were shown to cause anchorage independent growth in untransformed cells (8). Figure 1 shows the results of testing an aliquot from the odd numbered fractions of a Bio-Gel P-60 column chromatography of concentrated supernatant media obtained as described above. The majority of the protein is found in the void volume, while there is a sharp peak of growth stimulating activity that has an apparent molecular weight from 9,000 to 10,000 daltons and elutes between the RNase and insulin markers.

The data presented in Figure 1 show growth stimulating activity as measured by the ability to induce the rat fibroblastic cells, clone 49F (9), to form progressively growing colonies in agar. The peak of growth stimulating activity, as measured by thymidine incorporated into monolayer cultures of mouse 3T3 cell clones or rat NRK cell clones, as well as EGF-competing activity, coelutes with the agar growth stimulating activity. The peak activity (pool II) in Figure 1 has always been found as a major activity from preparations of supernatants of MSV-transformed cells. Less pronounced growth stimulating activities with higher apparent molecular weights (16,000-24,000), as well as activity with an apparent molecular weight of 6,000, are also seen in some column preparations, but they are considerably more variable than the pool II material.

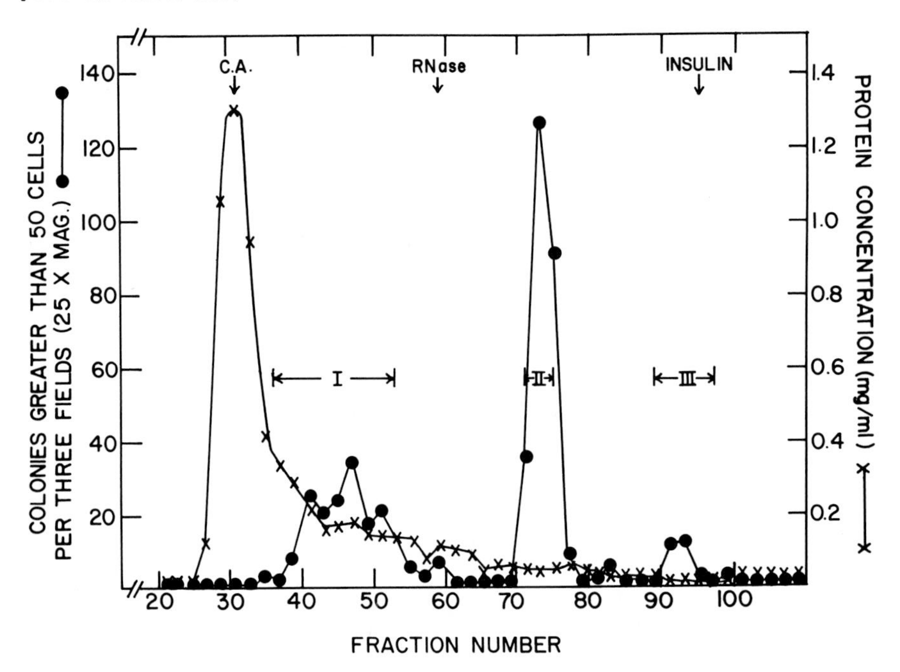

FIG. 1. Bio-Gel P-60 chromatography of the concentrate from the MSV-transformed cell line 3B11-IC (2) that was soluble in 1 M acetic acid. The soluble concentrate from 3 l of serum-free conditioned culture media in 30 ml of 1 M acetic acid was added to and chromatographed on a column of Bio-Gel P-60 (5 x 90 cm). Pool II was chromatographed on a column (0.9 x 15 cm) of DEAE-cellulose (Whatman: DE-52). Approximately 750 g of protein from pool II was lyophilized, dissolved in 10 ml of 0.005 M ammonium acetate (pH 7.5) and dialyzed against the same (3 changes of 50 volumes each). The dialyzed material was clarified by centrifuging for 90 min. at 100,000 x g. The clarified supernate was added to the DEAE-cellulose column that had been equilibrated

with the dialysis buffer. The column was washed with 50 ml of the equilibrating buffer, and a 200 ml linear salt gradient was developed, starting with 0.005 M ammonium acetate (pH 7.5) and finishing with 0.2 M ammonium acetate (pH 7.5). At the end of the salt gradient, the column was eluted with 50 ml of 1 M acetic acid. Fractions of 4 ml were collected in siliconized tubes. Aliquots from the even numbered fractions were lyophilized for protein determinations (1.5 ml) and for soft agar assays (0.5 ml). One hundred-fifty μliter samples were taken for conductance determinations. Soft agar assays were performed using the NRK fibroblastic clone 49F. Agar plates were prepared in 60-mm tissue culture dishes (Falcon, 3003) by first applying a 2 ml base layer of 0.5% agar (Difco, Agar Noble) in DMEM containing 10% calf serum. Over this basal layer, an additional 2 ml layer of 0.3% agar was added to the above medium containing the appropriate concentration of protein from the sarcoma conditioned medium, and 1×10^4 indicator cells. These dishes were incubated at $37^{\circ}C$ in a humidified atmosphere of 5% CO_2 in air and refed after 7 days with an additional 2 ml of 0.3% agar containing the appropriate supplements. Colonies were measured unfixed and unstained using an inverted microscope 14 days after seeding. Colonies with greater than 50 cells were scored as positive. The markers were: CA, carbonic anhydrase, 29,000; RNase 13,800; and insulin, 6,000. Proteins were determined according to the method of Lowry et al. (21).

The growth factor that is produced by the sarcoma virus-transformed cells has the property of making normal rat fibroblasts grow and form large colonies in soft agar. It also has a pronounced morphologic effect on normal fibroblasts, converting them to transformed cells that pile up and are virtually indistinguishable from those genetically transformed by sarcoma viruses (Figure 2). The soft agar growth promoting activity can be detected in supernatant fluids of serum-free medium from the sarcoma virus-transformed cells.

Previous results suggested that SGF exerted its biologic activity via the cell membrane EGF receptors. Cell clones derived from 3T3 cells that lack EGF receptors were generously provided by Dr. Harvey Herschman (UCLA). They had been selected by stimulating the 3T3 cells with EGF in the presence of colchicine (25). While 3T3 itself is responsive to the mitogenic action of EGF, the selected clone, NR6/6, lacks detectable EGF receptors, and is not responsive (25). 3T3 cells and the NR6/6 clone were tested in parallel for their response to calf serum and to a number of purified growth factors, including mouse EGF (7,26) and SGF purified as described above. The parental clone, 3T3/8, responded to all of the growth factors tested, as determined by ^{3}H-thymidine incorporation into DNA (Table 3). The NR6/6 clone, on the other hand, responded to all of the growth factors except EGF and SGF (DEAE-cellulose pool). DNA synthesis was stimulated in both clones when they were treated with calf

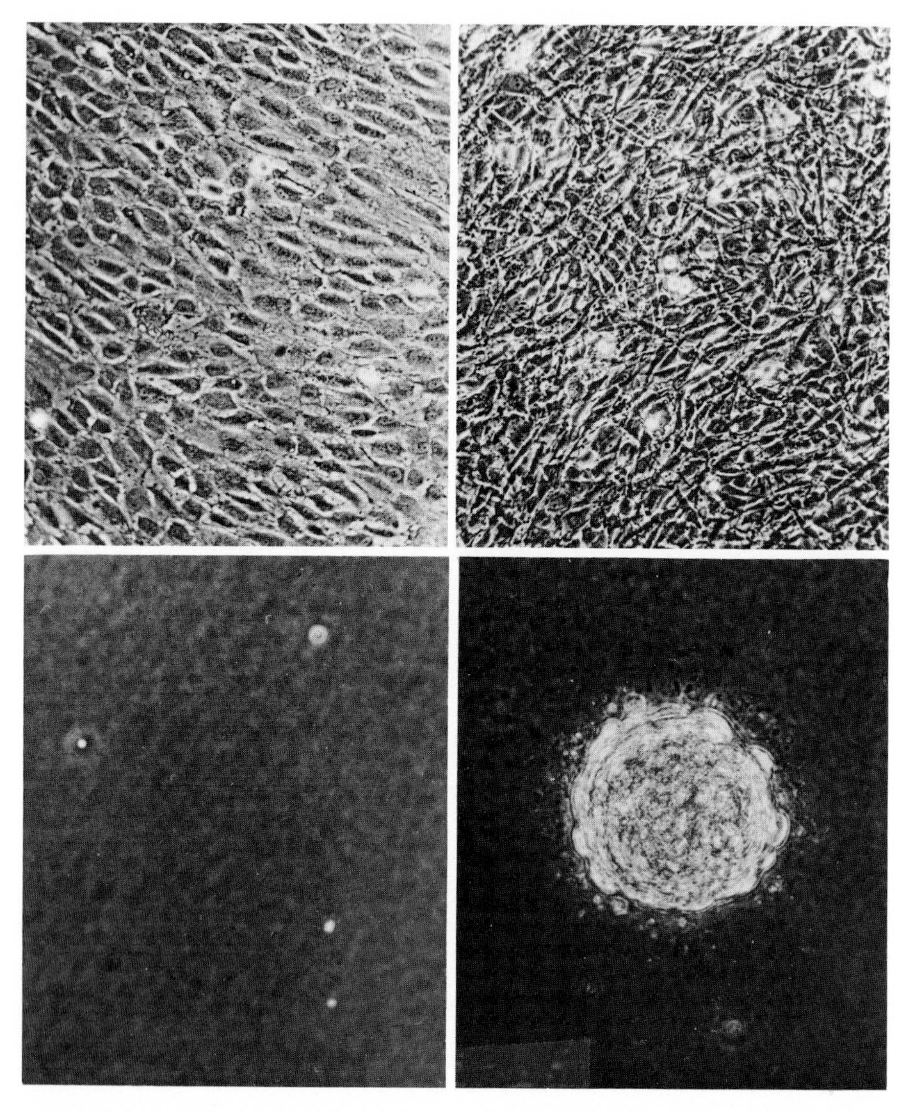

FIG. 2. (A) Untreated NRK cells. (B) NRK cells treated with an aliquot of SGF at 10 μg/ml and photographed 6 days later. The cells have grown to considerably higher cell density and display a morphology similar to that of virus-transformed cells. Magnification: A and B, 69X. (C) Untreated NRK cells plated in 0.3% soft agar. (D) NRK cells plated in 0.3% soft agar, treated with an aliquot of SGF at 10 μg/ml and photographed 2 weeks after treatment. The untreated cultures show primarily single cells with two or three cell colonies, but none of larger size. In the treated cultures, many colonies contained well over 500 cells. Magnification: C and D, 137X.

serum (65), FGF (16), MSA derived from rat liver cells (13,24), or the MSA-like activity released from a human fibrosarcoma line (10). The cells were also tested for their ability to grow in soft agar in the presence of the DEAE-cellulose column pool. In the soft agar assay, the 3T3/8 cells responded readily by developing a high percentage of large colonies (>50%) when treated with SGF, whereas the EGF receptor-negative cell, NR6/6, did not respond at all. The above results lead to the conclusion that available EGF receptors are required for SGF to exert its biologic effect in mouse 3T3 cells.

Table 3. Effect Of various growth factors on the induction of DNA synthesis in resting mouse 3T3 clones with and without EGF receptors[a]

Test Cells	^{3}H-Thymidine Incorporation ($X10^{-3}$) 3T3/8 ($EGF-R^+$)[b]	NR6/6 ($EGF-R^-$)	% of Control Ratio of ^{3}H-Thymidine Incorporation $EGF-R^-/EGF-R^+$ cells
Additions			
None	2.4	2.8	---
EGF (10 ng/ml)	33.5	3.1	1
SGF (1μg/ml)	63.7	2.9	1
Calf Serum (600 μg/ml)	55.3	63.7	115
FGF (10 ng/ml)	31.5	47.3	153
MSA (10 ng/ml)	18.2	14.3	73
Human MSA (200 ng/ml)	23.5	16.2	64

[a]The percent of the control thymidine incorporation was calculated by multiplying the amount of ^{3}H-thymidine incorporated into stimulated NR6/6 cells above the background cells (cells to which no additions were made) by 100 and dividing this product by the amount of ^{3}H-thymidine incorporated into the stimulated 3T3/8 cells above that found in unstimulated background cells.

[b]R^- denotes receptor-negative

R^+ denotes receptor-positive

An essential step in the purification and study of the properties of SGF produced by MSV-transformed mouse 3T3 cells utilized the human carcinoma cell line, A431, that has an exceptionally large number of EGF receptors (15,17). The A431 cells can be fixed with formaldehyde in tissue culture dishes and then used to specifically bind either mouse salivary gland EGF or SGF produced by virus-transformed cells. The unbound material from these preparations was removed, the cells washed several times, and the growth factors dissociated from their receptors with dilute acid. The materials released retain their biological activities and can rebind to their receptors. In these studies a fibroblastic clone, derived from a normal rat kidney (NRK) (9,12), that appears unusually sensitive to the growth stimulatory effect of SGF was used (8). Both EGF and SGF will induce cell division in monolayer cultures when allowed to bind to their receptors in responsive cells. SGF will also promote cell overgrowth and allow the stimulated monolayers to grow in multiple, criss-crossing layers as well as promote fibroblasts to form progressively growing colonies in agar (8).

In a previous report we showed that SGF selected by acid extraction and then passed over a Bio-Gel P-60 column was greatly enriched over the starting, unfractionated supernatant, but nevertheless, was less than 1% pure after this procedure; this will be referred to as "crude SGF". In Figure 3 we show isoelectric focusing columns run on ^{125}I-labeled EGF and iodinated preparations of crude SGF before and after cycling on fixed A431 cell monolayers. Isoelectric focusing of the ^{125}I-labeled EGF preparation prior to purification using fixed A431 cells (Fig. 3A) gave two sharp peaks. The smaller peak has a pI of 3.8 and the larger peak has a pI of 4.4. The labeled EGF that had been bound to and eluted from fixed cells gave a single sharp peak upon isoelectric focusing which had a pI of 4.4. (Fig. 3B); this value is consistent with that previously published for EGF (32). In contrast, isoelectric focusing of iodinated crude SGF gave a heterodisperse profile with the majority of the labeled material having acidic pI's (Fig. 3C). The radiolabeled peptide(s) from crude SGF which was bound to and eluted from the fixed A431 cells, however, gave a sharp peak upon isoelectric focusing which had a pI of between 6.8 and 7.0 (Fig. 3D).

Figure 4 shows the results of an experiment in which unlabeled, crude SGF was fractionated by isoelectric focusing, with carrier ^{125}I-SGF purified by cycling on the EGF receptor-rich human carcinoma cells. The fractions were then tested for their biological activity in the anchorage independent growth assay and for their ability to stimulate cell division in growth arrested rat fibroblastic (NRK) cell monolayers. The major activity is found in a narrow region with a pI of 6.8-7.0. Isoelectric focusing, then, can be used to further purify the growth factor from crude preparations with the retention of biological activity.

Table 4 shows the ability of different radiolabeled SGF preparations to bind to and be eluted from A431 cells. After two cycles of binding and eluting, the remaining counts bound with a much higher efficiency to A431 cells. In the case of the twice-cycled material, approximately 24% of the input counts bound specifically to the A431 cells; whereas only 0.12% of the input available for recovery by treating the cells with dilute acetic acid.

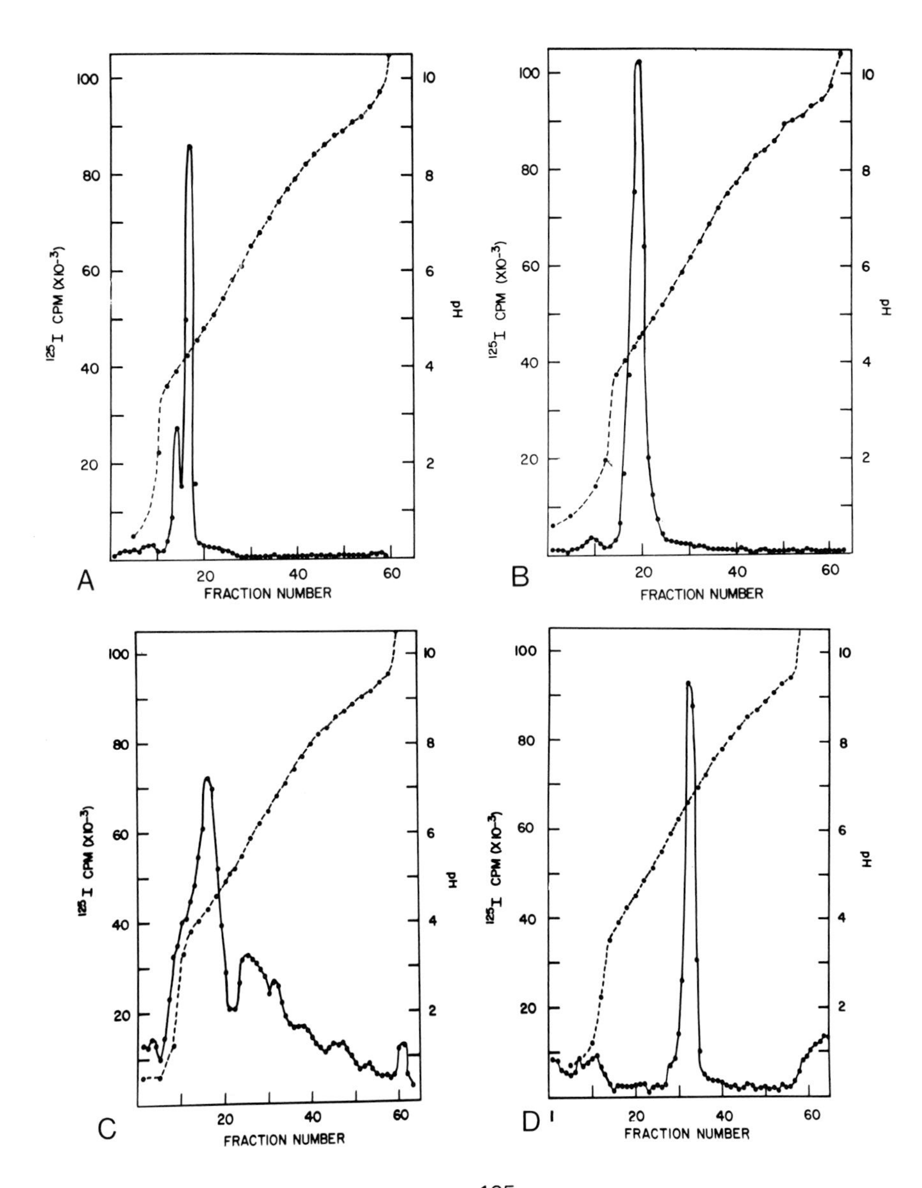

FIG. 3. Isolectric focusing of ^{125}I-labeled peptides. Isoelectric focusing was performed using a 110 ml LKB column. A 1% solution of H SO in 60% sucrose (w/v) was layered at the anode followed by a 40% to 0% (w/v) sucrose gradient containing 1% LKB ampholine ampholytes, pH 3.5-10. The lyophilized ^{125}I-labeled sample was resuspended in 1 ml of water mixed with 1 ml of the dense portion of the gradient and applied near the middle of the column. A 1% solution of NaOH was layered on top of the gradient

column. A 1% solution of NaOH was layered on top of the gradient in contact with the cathode. The column was run at 300 volts for 48 hr. at 4°C after which time 2 ml fractions were collected from the bottom of the column and the pH and content of ^{125}I-labeled material of each fraction was measured. A) ^{125}I-labeled EGF. B) ^{125}I-labeled EGF eluted from A431 cells fixed for 10 min. in 5% formalin in phosphate buffered saline (PBS) and exposed to ^{125}I-labeled EGF for 60 min. at room temperature in binding buffer. The unbound EGF was removed, the cells washed 4 times with binding buffer and the bound material eluted with three washes of 0.1% acetic acid in water. C) ^{125}I-labeled crude SGF. D) ^{125}I-labeled SGF eluted from formalin-fixed A431 cells as described in (B).

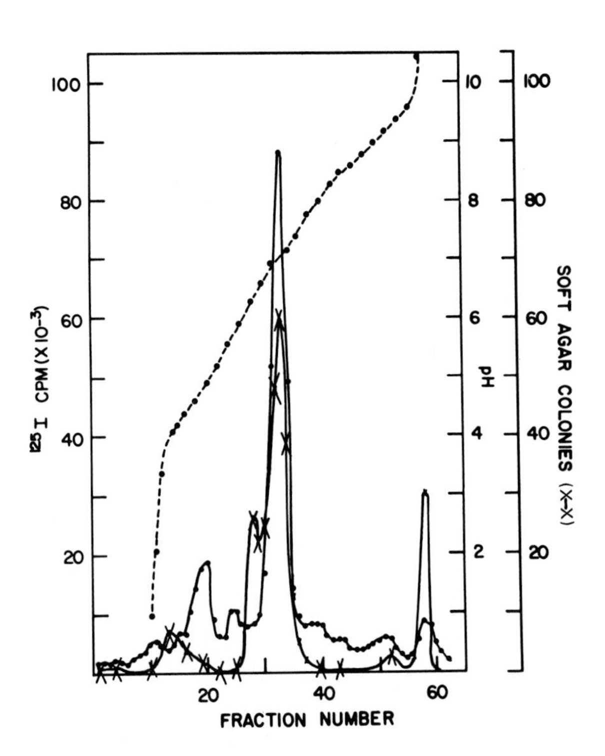

FIG. 4. Isoelectric focusing of biological activity from crude SGF. ^{125}I-labeled SGF eluted from formalin-fixed A431 cells, as described in Fig. 3B, was applied to an isoelectric focusing column with 100 μg crude SGF from the 10,000 molecular weight region of a Bio Gel P-60 column. Isoelectric focusing was performed as described in the legend to Fig. 3. A 100 μl aliquot of each indicated column fraction was diluted in 1 M acetic acid, lyophilized and tested for soft agar growth as described previously.

TABLE 4. Purification of ^{125}I-SGF using binding and elution from an EGF receptor-rich cell

Material	Specific Binding / Nonspecific Binding	Percent of Input dpm Bound
Bio-Gel P-60 Pool II	0.18	0.12
After the First Cycle	5.6	2.8
After the Second Cycle	18.7	23.9

Unlabeled SGF from Bio-Gel P-60 pool II was cycled on formaldehyde. Fixed A431 cells and the resultant fractions were tested for their ability to stimulate a NRK fibroblastic clone to grow in soft agar (Table 5). The lyophilized SGF (180 µg) was dissolved in 15 ml of sterile binding buffer, a 5 ml aliquot was set aside as untreated and the remainder was cycled over the fixed A431 cells. The ligand to be "cycled" was diluted in binding buffer and placed on fixed cells for the stated period of time at the stated temperature. To correct for nonspecific binding when radiolabeled ligands were used, 50 µg/ml of unlabeled EGF were added to the labeled ligand. After binding continued for the stated period, the unbound peptides were removed, and the cells were washed twice with binding buffer and twice with PBS in order to remove the unbound materials. The bound ligands were eluted from the fixed cells by washing the "monolayers" three times with 0.1% acetic acid. The eluted peptides were then concentrated by lyophilization. The cycled material was divided into two sections; that portion which did not bind to the cells and remained in the supernate after the second cycle of binding, and the material which bound to the cells and was eluted with dilute acetic acid. The untreated material was the most potent stimulator of cell division on a per volume basis. After dilution, the agar assay contained 1-2 µg of pool II SGF per ml. The portion which was bound to and eluted from the fixed cells contained over 90% of the activity found in the untreated SGF preparation. If one assumes the percentage of unlabeled peptides which bound to the fixed A431 cells is similar to that for the radiolabeled peptides (Table 4), then there was considerably less than 5 ng per ml of peptide from SGF pool II in the soft agar assay of the bound and eluted material. In the controls, which consisted of binding buffer cycled over fixed A431 cells, neither the "unbound" nor the "bound" materials had any soft agar growth stimulating activity.

TABLE 5. Soft agar growth stimulation by cycled Bio-Gel P-60 pool II SGF preparations

Preparation	Soft Agar Colonies Greater Than 50 Cells Per 100 Cells Seeded[a]
Untreated pool II at a final concentration of 1.2 μg/ml	60
Pool II unbound to fixed A431 cells	4
Pool II bound to and eluted from fixed A431 cells	55
Binding buffer unbound to fixed A431 cells	0
Binding buffer bound to and eluted from fixed A431 cells	0

[a]Ten ml of either binding buffer or binding buffer containing 12 μg per ml of pool II SGF was bound to fixed A431 cells for 45 min. at room temperature. The unbound material was then transferred to another dish of fixed A431 cells and bound for an additional 45 min. The plates were each washed twice with serum-free Waymouth's medium and PBS before eluting the "bound materials" from the fixed cells. The eluted materials were lyophilized and redissolved in a volume of binding buffer equal to that of the unbound material. Soft agar assays were set up using the above materials at a 1:10 dilution and read after 13 days.

Murine sarcoma virus-transformed cells lack available receptors for EGF. We have shown this altered phenotype is the result of the endogenous production of growth factors by the MSV-transformed cells themselves. There is no evidence that SGF acts as a complete carcinogen itself, producing permanent cell transformation; its properties resemble classical chemical promoters of carcinogenesis, like 12-0-tetradecanoylphorbol-13-acetate (4,29,37) (TPA), the highly active component of croton oil. While TPA is an exogenous plant derivative acting on an animal or a cell, SGF is an endogenous, virally induced growth promoter.

Retinoids (vitamin A and synthetic analogs) (30) block the action in vivo of exogenous and endogenous promoters, preventing carcinogens from producing new tumors, but do not reverse the growth of many established tumors (5,6,30,36). Retinoids prevent cancer of the lung (27,30), skin (6), bladder (31), and mammary gland (23) in experimental animals, block cell transformation induced by chemicals (22) and radiation (18,22) in culture, and reverse the anchorage-independent growth of transformed mouse fibroblasts (11). If SGF is part of the natural tumor-promoting system and retinoids are part of the natural defense against that system, then one should be able to demonstrate a direct antagonism in cell culture.

We have used a subclone (536-7) of a rat fibroblast cell clone (NRK 49F) (9) that showed pronounced morphologic transformation and anchorage-independent growth when treated with SGF, forming multiple cell layers and criss-crossing each other in an apparently random fashion. Although the effects are all reversible, these treated cells resemble MSV-transformed cells in their phenotype (8). The cells that were treated with both SGF and retinoids did not have a disordered growth pattern. Retinyl acetate, at 6 ng/ml, almost abolished the growth stimulatory effect of SGF, as determined by the final cell density the monolayer cultures reached ten days after the experiment began. The effect of retinoids on SGF-induced morphologic alterations was evident within a few days after treatment.

The retinoid concentrations (1-2 x 10^{-8} M) neither reversed the phenotype of virally transformed cells, nor blocked cell transformation produced by transforming viruses, such as the Moloney strain of MSV or SV40. Mouse 3T3 cells and rat fibroblasts were tested for susceptibility to transformation by MSV and by SV40. Neither retinyl acetate nor retinoic acid, up to 2 x 10^{-6} M, could be demonstrated to block either the initiation or the maintenance of virally induced transformation when efficient transforming viruses, like MSV or SV40 were used. In the same experiment, however, the SGF-induced morphologic transformation was inhibited. Retinyl acetate did not inhibit normal cell growth or the cloning efficiency of the rat fibroblast cell clones in petri dishes, but did have a pronounced effect on the final cell ("saturation") density of cells treated with SGF. SGF-induced cell growth was blocked and normal growth properties were essentially retained.

Table 6 shows that at concentrations well below those that show any evidence of toxicity, retinoids prevent SGF-induced colony formation in soft agar. Colonies that did form were smaller and contained fewer cells than those treated with SGF alone. Retinoic acid, retinyl acetate, retinyl methyl ether and retinylidene dimedone were all effective. Cells not treated with SGF and plated in agar remained as single cells with occasional 2-4 cell colonies. The clone used (536-7) has a spontaneous transformation rate, as determined by agar colony growth, of less than 1 in 10 cells plated. The preparation of SGF used, at 10 μg/ml, produced 40-50 large colonies per 10^{4} treated cells and many smaller colonies with between 4 and 20 cells as well. The inhibiting effect of the retinoids was less evident or absent when more active SGF preparations, or higher concentrations of SGF were used. Retinylidene dimedone, of the compounds tested, was the most efficient inhibitor of SGF-induced phenotypic transformation (Table 6, experiment 3). As a control against selective toxicity of retinoids to transformed cells, as compared to normal cells, MSV-transformed mouse and rat cells that grow well in agar without adding SGF were plated in soft agar in the presence of retinyl acetate at 2×10^{-6} M. No reduction in colony forming ability was seen.

TABLE 6. Effect of SGF and various retinoids on the colony forming ability of rat fibroblasts plated in soft agar[a]

Treatment	Colonies/plate Expt. 1	Expt. 2	Expt. 3
Untreated controls	0	0	0
+ retinyl acetate (1.9×10^{-8} M)	0	0	0
+ retinoic acid (2.0×10^{-8} M)	NT[b]	0	NT
+ retinylidene dimedone (1.5×10^{-8} M)	NT	NT	0
+ retinyl methyl ether (2.0×10^{-8} M)	NT	NT	0
SGF-treated (10 μg/ml)	44.5	39.0	49.5
+ retinyl acetate (1.9×10^{-8} M)	2.5	1.5	8.0
+ retinoic acid (2.0×10^{-8} M)	NT	3.2	NT
+ retinylidene dimedone (1.5×10^{-8} M)	NT	NT	0.5
+ retinyl methyl ether (2.0×10^{-8} M)	NT	NT	14.5

[a]On day 0, 1×10^5 rat fibroblast cells, clone 536-7, were treated in monolayer cultures using DMEM with 1% fetal calf serum. On day 2, they were seeded at 1×10^4 cells per plate in 0.3% soft agar containing the additions shown as previously described (8). All cells not treated with SGF (whether treated with retinoid or not) remained as single cells with occasional (<10%) small colonies of 2-4 cells. Colonies with greater than 20 cells after two weeks in agar were scored as positive.
[b]NT = Not tested.

These experiments establish that, in the system used here, retinoids block the transforming effect of the polypeptide hormone, SGF. Only one concentration of each retinoid, well below the level that shows any cell toxicity, was used, and both the growth promoter and the antagonists were added to the cells at the same time. This system is now available for further studies where the concentrations, duration of treatment and the nature of the interaction between each of the three components (promoter, antagonist and responding cell clone) can be varied in a systematic manner.

The general transformation model we are proposing has these features: Viruses and chemical carcinogens act by inducing cells to produce normally repressed or inactive growth promoting factors. These factors, which may be endogenous or exogenous to given cells, could be important in embryonic development, but if inappropriately expressed later in life could lead to transformation. Tumor viruses either provide transforming genes directly or activate cellular genes; chemical carcinogens do only the latter. These growth promoting and transforming factors may be produced during early embryogenesis and then "switched off". The endogenous viruses, with their capacity to recombine with cellular genes, have the ability to transfer information between cells and presumably within a cell like bacterial insertion sequences. They may well be vehicles that allow expression of the endogenous growth promoter structural genes. In this model the promoters, be they endogenous (SGF) or exogenous (TPA), act as proximal effectors of transformation.

The virogene-oncogene hypothesis (19) points out the possibly erroneous assumption that virally induced tumors would have to arise through external infection by emphasizing that virus-coded or virus-associated genes are present in several animal species already. These genes, rather than the environmentally transmissible agents, are more likely to be involved in the origin of natural cancers. The tumor viruses, although unnatural, in that they had often been selected for producing rapid disease, have provided extremely powerful tools to dissect out and understand the molecular mechanisms involved. Genetically transmitted viral genes and transforming genes are now accepted as being part of the normal genetic makeup of many organisms and of being activated by agents such as chemical carcinogens, hormones and radiation (1,33). In parallel with this is the frequently made assertion that chemical carcinogenesis and environmental carcinogenesis, or even industrial carcinogenesis, are almost interchangeable with one another. The finding that SGF, produced by animal cells themselves, is an extremely potent promoter in cell culture systems suggests that endogenous growth promoters may be significant factors in naturally occurring cancers.

In view of the efficacy with which an endogenous factor such as SGF promotes transformation, natural mechanisms for suppression of transformation must exist. The data presented here, as well as previous _in vivo_ studies, indicate that suppressors of transformation, such as retinoids, offer a potential approach to prevent the expression of the transformed phenotype.

We have examined cells transformed by a temperature-sensitive sarcoma virus mutant that has been produced by Donald Blair called 6M2 (3). The sarcoma virus-transformed nonproducer cells, derived from NRK, have a transformed phenotype at 32°C and grow very well in agar, while at 39°C they are almost indistinguishable from the parental normal rat kidney, NRK, cells and do not grow in agar or form multiple cell layers under normal conditions. When we tested these cells for EGF receptors it was found that at 32°C virtually no EGF receptors are detected, while at 39°C the levels of receptors are comparable to those found in the untransformed NRK cells (Table 7). One can shift the temperature back and forth and at the nonpermissive temperature the EGF receptors appear. At the permissive temperature they can no longer be detected. Supernatant fluids were collected from these cells and tested for growth stimulating activity. At 32°C it was found that the sarcoma virus-transformed cells produced a growth factor with a major peak of activity at 12,000, that like the wild type (wt) MSV-transformed cell factor, induced fibroblasts to grow well in agar. At 39°C no such activity could be detected. Similarly, the supernatants at 32°C showed EGF-competing activity while those grown at 39°C did not. As a control, using wild type, transformed cells, SGF production at 32oC and 39°C was roughly comparable, and the level of EGF receptors at the two temperatures remained low. We could further ask whether the cells at the nonpermissive temperature, 39°C, were capable of responding to SGF. The answer is that they were; SGF-treated 6M2 cells at 39°C formed colonies in agar as well and as efficiently as their untransformed counterparts. Thus, they are capable of responding to the production of the factor. We interpret these results to mean that the EGF receptors are produced at both temperatures. At the lower temperature, however, SGF is produced which interacts with and blocks expression of the EGF receptors. At the nonpermissive temperature, the cells fail to produce a competent SGF and available EGF receptors can be detected once again. The production of SGF, then, correlates both with the absence of EGF receptors and with the transformed cell phenotype. This set of experiments shows that MSV transformation does not irreversibly destroy the ability to produce EGF receptors and that the absence of available EGF receptors is a temperature dependent process.

TABLE 7. Properties of a cell clone transformed by a temperature-sensitive mutant of MSV

Properties	Temperature 39°C	39°C
Transformed phenotype	+	-
EGF receptors	-	+
SGF Production	+	-
Response to EGF	NT[a]	+

[a]Not tested.

A second system of comparing wild type murine sarcoma viruses with temperature-sensitive mutants uses Kirsten murine sarcoma virus (Ki-MSV) and a mutant, ts371 clone 5 derived from it. The cells transformed by the mutant virus also show rapid alterations in their available cell surface receptors when shifted from the permissive to the nonpermissive temperature. The cells were grown for several days at the permissive temperature, 35°C, and then shifted for a 24-hour period to 32°C, 37°C, and 39°C, the latter being nonpermissive for the expression of the transformed phenotype. After 24 hours, the cells were assayed for EGF receptors. As shown in Table 8, the cells transformed by the wild type virus show markedly reduced available receptors, characteristic of sarcoma virus transformation, at any of the temperatures. On the other hand, the cells transformed by the ts virus show a greatly increased number of receptors at 39°C, comparable to those found on the untransformed parental cells. At 32°C, they show only 10 to 20% the number of receptors shown at 39°C and the cells maintained at 37°C have been reproducibly found to possess intermediate levels of EGF receptors. The transformed cells then have the cell surface EGF receptors which are not detectable, by ^{125}I-EGF binding, at the permissive temperature presumably because some viral product interferes with the availability of membrane receptors to exogenously added ligands. A shift to the nonpermissive temperature, then, would cause an inactivation or disappearance of this product, presumably a protein and possibly the virus-coded guanine nucleotide binding protein described by Scolnick et al. (28), which blocks EGF receptor availability.

TABLE 8. The effect of temperature of incubation of the cell lines on the levels of ^{125}I-EGF binding

	Previous Day at	^{125}I-EGF Bound (cpm)
ts371 cl5	32°C	1,470
	37°C	3,940
	39°C	10,970
Untransformed NRK	32°C	8,700
	37°C	9,500
	39°C	8,300
Ki-MSV-Transformed NRK	32°C	350
	37°C	200
	39°C	260

Direct experiments testing the temperature-sensitive nature of the growth factors involved heating the concentrate at 56° C for varying periods of time and then measuring the residual activity as estimated by the induction of DNA synthesis in the indicator cells. Under the conditions used, upon heating at 56°C for varying periods of time, growth factors produced by the parental Ki-MSV-transformed rat cells are stable, showing no loss in growth stimulatory activity upon treatment for up to 2 hr. In contrast, the ts371 cl5 growth factor showed a 70% loss in growth stimulatory activity over the 2-hr. test period. The loss of activity upon heating can be shown for all three functional parameters: the ability to compete for binding to membrane EGF receptors, to stimulate thymidine incorporation by growth arrested monolayer fibroblast cultures and the ability to induce normal rat fibroblasts to proliferate in semi-solid agar.

We then examined whether the membrane EGF receptors available on the surface of the ts virus-transformed cells at 39°C were functional in their ability to respond to SGF. The untransformed normal rat kidney cells and the ts transformed cells were plated on soft agar in the presence and absence of SGF. In control incubations in the absence of SGF at 32°C and 37°C the ts transformed cells formed >500 colonies while the untransformed cells failed to produce any detectable colonies on soft agar at either temperature (Table 9). At 39°C both the normal parent and the ts transformed cells formed a large number of colonies in the presence of 20 μg/ml SGF but not when plated without the SGF. The lesion in the ts transformed cells at the nonpermissive temperature thus does not involve the ability of membrane EGF receptors to respond to SGF.

TABLE 9. The stimulation of soft agar colony formation by untransformed NRK cells and ts mutant-transformed cells in response to SGF treatment

Cell line	Assay Temperature	Additions	Agar Colony Formation
NRK	32°C	-	0
	37°C	-	0
	39°C	-	0
	39°C	SGF (20 μg/ml)	70
ts371 cl5	32°C	-	>500
	37°C	-	>500
	39°C	-	5
	39°C	SGF (20 μg/ml)	150

The ability of the cells, transformed by the mutant virus, to release SGF-like factor(s) was compared to that of the wild type transformed cells. Conditioned media from either cell line growing at 34°C, (NRK or ts371 cl5) was dialyzed against 0.1 M acetic acid and lyophilized. The concentrates were then tested for the ability to induce DNA synthesis, to compete for EGF receptors and to induce soft agar colony formation of the indicator cell line, the rat fibroblastic clone 49F (9). The results show that even at the permissive temperature the ts371 cl5 produced 3.0×10^3 agar growth units/mg protein while conditioned media from a parallel culture of wt Ki-MSV-transformed NRK cells contained 8.4×10^4 agar growth units/mg protein. The ts lesion in the transforming virus thus results in a greater than twenty-fivefold reduction in the release of SGF activity in the culture fluid under the standard conditions.

The above experiments indicated that a protein produced by the ts371-transformed cells is temperature-sensitive and may account at least in part for the temperature-sensitive nature of the transformed phenotype. The thermolability of the activity produced by the ts virus-transformed cell line relative to that obtained from cells transformed by the wild type parent virus, suggest that it may be the sarcoma virus gene product involved in the transformation of cells. The continued production of a temperature-sensitive protein is required for the maintenance of the transformed state. At the nonpermissive temperature the protein production may be decreased and the stability of the protein already produced would be diminished. As a consequence, the cells would revert to normal morphology. This experiment provides direct evidence that the temperature-sensitive mutant virus gene product and the growth stimulatory factor are related to one another; it suggests that the growth factor produced by Ki-MSV-transformed cells is produced by the mutant genome itself, rather that acting indirectly via induction of host cell genes.

A model consistent with the above observations is the production and release by MSV-transformed cells of an EGF-like peptide factor or factors which are able to bind to and block cellular EGF receptors, and act as mitogens either to the cells producing them or to other cells having functional receptors capable of binding these factors. This could account for both the lack of measurable EGF receptors on the MSV-transformed cells as well as for the lowered serum requirement for MSV-transformed cells.

The data presented here are consistent with the above model in that MSV-transformed cells produce at least one EGF-like peptide (SGF) which is capable of binding to and blocking EGF receptors. Using the EGF receptor to purify this peptide(s) the biological activity copurifies with the EGF receptor-binding activity. This factor from MSV-transformed cells has both similarities and differences when its properties are compared to mouse submaxillary gland EGF. It binds to EGF receptors with an affinity similar to EGF, and they both stimulate cell division in serum-depleted cells. They differ in that EGF has a more acidic pI (4.4) than SGF (6.8), they migrate differently in both SDS polyacrylamide gel electrophoresis and acetic acid (gel permeation) chromatography; and EGF will not stimulate anchorage independent growth while SGF will.

Clearly SGF does not mediate <u>all</u> of the effects of sarcoma virus transformation. It does, however, represent the major activity released by sarcoma virus-transformed cells and is not seen in similarly prepared samples processed from the fluids of normal, DNA virus, or some chemical carcinogen transformed cells. SGF then acts as one of the proximal effectors of cell transformation produced by sarcoma viruses.

References

1. Aaronson, S.A., and Stephenson, J.R. (1976): Biochim. Biophys. Acta, 458:323-354.
2. Bassin, R.M., Tuttle, N., and Fischinger, P.J. (1970): Int. J. Cancer, 6:95-107.
3. Blair, D.G., Hull, M.A., and Finch, E.A. (1979): Virology, 95:303-316.
4. Boutwell, R.K. (1974): CRC Crit. Revs. Toxicol., 2:419-443.
5. Bollag, W. (1971): Cancer Chemother. Rep., 55:53-55.
6. Bollag, W. (1972): Eur. J. Cancer, 8:689-693.
7. Cohen, S., Taylor, J. M., Murakami, K., Michelakis, A. M., and Inagami, T. (1972): Biochemistry, 11:4286-4292.
8. De Larco, J.E., and Todaro, G.J. (1978): Proc. Natl. Acad. Sci. U.S.A., 75:4001-4005.
9. De Larco, J.E., and Todaro, G.J. (1978): J. Cell. Physiol., 84:335-342.
10. De Larco, J. E., and G. J. Todaro (1978): Nature, 272:356-358.
11. Dion, L.D., Blalock, J.E., and Gifford, G.E. (1979): Exp. Cell Res., (in press).
12. Duc-Nguyen, H., Rosenblum, E.N., and Zeigel, R.F. (1966): J. Bacteriol., 92:1133-1140.
13. Dulak, N. C., and Temin, H. M. (1973): J. Cell. Physiol., 81:153-170.
14. Dulbecco, R., and Freeman, G. (1959): Virology, 8:396-397.
15. Fabricant, R.N., De Larco, J.E., and Todaro, G.J. (1977): Proc. Natl. Acad. Sci. U.S.A., 74:565-569.
16. Gospodarowicz, D., Greene, G., and Moran, J. S. (1975): Biochem. Biophys. Res. Commun., 65:778-787.
17. Haigler, H., Ash, J.F., Singer, S.J., and Cohen, S. (1978): Proc. Natl. Acad. U.S.A., 75:3317-3321.
18. Harisiadis, L., Miller, R.C., Hall, E.J., and Borek, C., (1978): Nature, 274:486-487.
19. Huebner, R.J., and Todaro, G.J. (1969): Proc. Natl. Acad. Sci. U.S.A., 64:1087-1094.
20. Kitos, P. A., Sinclair, R., and Waymouth, C. (1962): Exp. Cell Res., 27:335-342.
21. Lowry, O. H., Rosenbrough, N.J., Farr, A.L., and Randall, R.J. (1951): J. Biol. Chem., 193:265-275.
22. Merriman, R.L., and Bertram, J.S. (1979): Cancer Res., (in press).
23. Moon, R.C., Grubbs, C.J., Sporn, M.B., and Goodman, D.G., (1977): Nature, 267:620-621.
24. Nissley, S. P., and Rechler, M.M. (1976): National Cancer Institute Monograph, 40:167-172.
25. Pruss, R.M., and Herschman, H.R. (1977): Proc. Natl. Acad. Sci. U.S.A., 74:3918-3921.

26. Savage, C. R. Jr., and Cohen, S. (1972): J. Biol. Chem., 247:7609-7611.
27. Saffiotti, U., Montesano, R., Sellakumar, A.R., and Borg, S.A. (1967): Cancer, 20:857-864.
28. Scolnick, E.M., Papageorge, A.G., and Shih, T.Y. (1979): Proc. Natl. Acad. Sci. U.S.A., 76:5355-5359.
29. Slaga, T.J., Sivak, A., and Boutwell, R.K., editors (1978): Mechanisms of Tumor Promotion and Carcinogenesis. Raven Press, New York.
30. Sporn, M.B., Dunlop, N.M., Newton, D.L., and Smith, J.M. (1976): Fed. Proc., 35:1332-1338.
31. Sporn, M.B., Squire, R.A., Brown, C.C., Smith, J.M., Wenk, M.L., and Springer, S. (1977): Science, 195:487-489.
32. Taylor, J.M., Cohen, S., and Mitchell, W.M. (1970): Proc. Natl. Acad. Sci. U.S.A., 67:164-171.
33. Todaro, G.J., Callahan, R., Sherr, C.J., Benveniste, R.E., and De Larco, J.E. (1978): In: Persistent Viruses, ICN-UCLA Symposia on Molecular Biology, Volume 11, pp 133-145. Academic Press, New York.
34. Todaro, G.J., De Larco, J.E., and Cohen, S. (1976): Nature, 264:26-31.
35. Todaro, G.J., Lazar, G.K., and Green, H. (1965): J. Cell. Comp. Physiol., 66:325-334.
36. Verma, A.K., and Boutwell, R.K. (1977): Cancer Res., 37: 2196-2201.
37. Weinstein, I.B., and Wigler, M. (1977): Nature, 270:659-660.

Control Mechanisms in Animal Cells,
edited by L. Jimenez de Asua et al.
Raven Press, New York © 1980.

A Progression in the Production of Growth Factors with Progression in Malignant Transformation

Robert B. Bürk

Friedrich Miescher-Institute, CH-4002 Basel, Switzerland

SV28 cells in culture produce and release two growth factors into serum-free culture medium. The factors stimulate cell migration, cell proliferation and produce a transformed culture morphology. The factors have been purified and separated using cell migration as an assay. Each acts in the presence of saturating amounts of the other. BHK21/13 releases only one factor that acts in the assay in the presence of one SV28 factor but not the other. 2° and 3° cultures of fibroblasts from kidneys of baby hamsters produce no migration stimulating activity. Fetal calf serum contains two activities that act respectively in the presence of saturating amounts of the individual SV28 factors but not in the presence of both.

An hypothesis of malignant transformation being due to auto-stimulation by growth factors produced ectopically in their target cell is proposed. Although the DNA of all cells in a mammal codes for the growth factors needed by the cells of a particular tissue, synthesis of the factors is normally repressed in the target cells. The factors are synthesized in some other cell types and transported in the blood to their target, hence the factors are found in serum. In the progression of malignant transformation these factors come to be synthesized in their target cells. Specifically in the BHK21/13 cell line one factor is derepressed by mutation or teratogeny and the cells have a reduced serum requirement and partial autonomy of growth. In SV28 a further factor is derepressed and by auto-stimulation the cells have almost no serum requirement and almost full independence of the factors normally regulating their growth.

INTRODUCTION

Malignant cancer typically involves the proliferation of cells that infiltrate the neighbouring tissue. In the history of tissue culture it was early reported to be easier to culture tumour cells than normal cells. It seemed that there were unidentified growth factors in the serum that were required by normal and not tumour cells. Studies of cells transformed by viruses showed a decreased dependence on serum for growth (2, 10,14,15). The tumour viruses when they transform BHK21/13 also induce a disorientation of the normal colony (11) or monolayer morphology (12). It was proposed that transformation induced a cell to make its own growth factors (3,7,8) and that these factors induce the syndrome of transformation. I have pursued the growth factor that also stimulates cell migration (4,5) produced by a cell line SV28 that produces invasive metastasing tumours. It is now reported that SV28 appears to produce two growth factors and that its pretransformation counterpart BHK21/13 produces one growth factor and the normal cells produce none. These observations lead to a speculative hypothesis of ectopic auto-stimulation in malignant transformation.

METHODS

Cells

SV28 and BHK21/13 cells were cultured as described previously for SV28 (4). Secondary and tertiary cultures of fibroblasts from the kidneys of baby hamsters were produced from the kidneys of seven day old Syrian hamsters by standard methods (13) and cultured in Dulbecco's medium with and without a 10% supplement of tryptose phosphate broth with daily medium changes.

Migration assay

Migration assay was as described before (4), except that 35 mm dishes containing 2.5 ml medium and 2×10^5 3T3-B cells (our derivative of Balb/c 3T3 clone A31) were used and the razor blade was broken in half. Activity is the number of cells crossing 1.2 mm of the start line in 22 hours.

Purification

Purification of the migration factors was similar to our previous methods (5). Activity was concentrated by precipitation with 500 g/l $(NH_4)_2SO_4$ from SV28 or BHK21/13 serum-free conditioned medium. The activity was adsorbed to Dowex AG50W at pH 7.6 and was eluted at pH 12. It was dialysed, lyophilised and applied to Sephadex G100 in 0.01 M formic acid. The active fractions were pooled and made up to 55 ml with 0.6 ml pH 9-11 ampholines (LKB) (and sometimes 0.1 ml ethanolamine, see text) and water. This pool was combined with 55 ml 50% sucrose containing 1.9 ml pH 9-11 ampholines to form a linear gradient in an LKB 110 ml isoelectric focussing column. Cathode was 0.1 M NaOH and anode 10 mM formic acid. The voltage was set to be a max. 1000 V and the current a max. 4 mA and after 66 hours at 6° the column was emptied at 1.0 ml per min collecting 2 min fractions.

Analytical isoelectric focussing

The range of the standard LKB preformed isoelectric focussing gels was extended by soaking a 1.5x24 cm strip of Whatman No.3 paper with 1.0 ml undiluted pH 9-11 ampholines and placing it on the gel in the location squares 1.0-2.5 cm cathode side. The current was set at $\leqslant$ 50 mA and voltage at $\leqslant$ 1000 V. After 25 min the ampholine strip and the application strips were removed and focussing continued for 45 min. The gel was placed for 20 min in the LKB fixing solution at 60°, then 20 min at 60° Vesterberg Coomassie G250 staining solution followed by repeated changes of LKB destain solution according to the LKB Instruction Leaflet.

RESULTS

Fig. 1 shows the result of the isoelectric focussing step in the purification of migration activity from medium conditioned by SV28 cells and by BHK21/13 cells. Clearly, there were two peaks in the migration activity produced by the SV28 cells and apparently only one from the BHK21/13 cells. Fig. 2 shows Coomassie stained gels of various steps of the purification of the SV28 materials. The fractions are not yet completely purified. The recovery between different

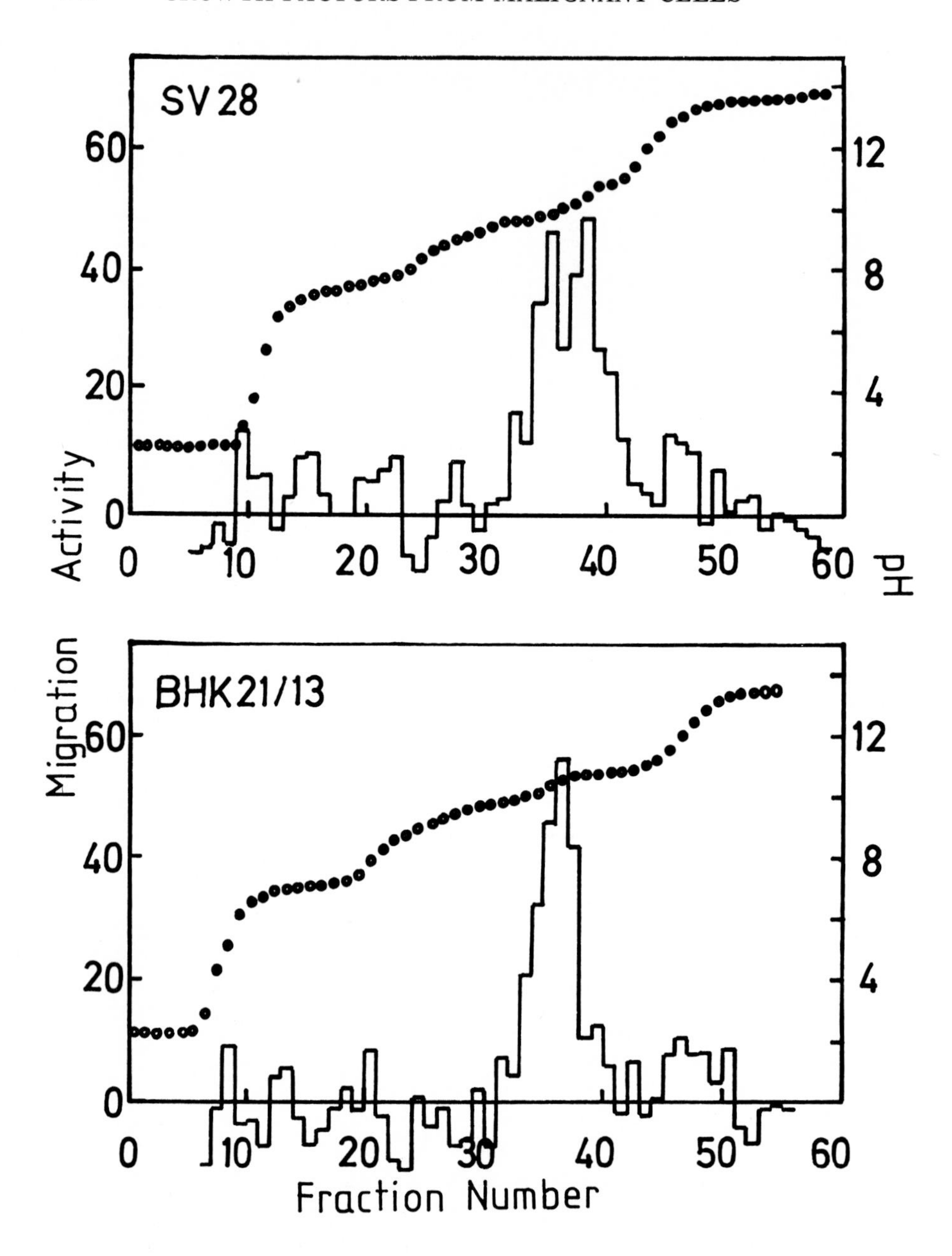

FIG. 1. Iso-electric focussing of migration activity from a) SV28 cells and b) BHK21/13 cells. Cathode on the right.

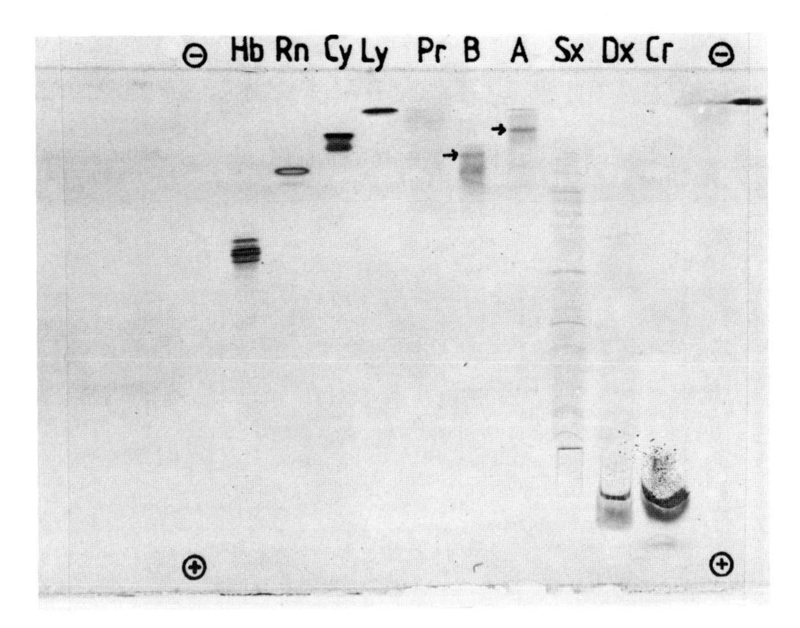

FIG. 2. Analytical isoelectric focussing in a polyacrylamide gel of various fractions during purification of the migration activity from SV28 cells. Approximately same amount of protein in each channel. Cr, crude medium; Dx, inactive fraction from Dowex; Sx, pooled active fractions from Sephadex; A, pooled active fraction A from focussing, B, pooled active fraction B from focussing; Pr, protamine; Ly, lysozyme; Cy, horse cytochrome c; Rn, ribonuclease; Hb horse haemoglobin. Arrows indicate where activity was recovered in parallel channels of A and of B.

runs varied between 0.2% and 2.0% and the purification between 200- and 1000-fold. The migration dose-response curves for A and for B reach a plateau at different levels. The actual level seems to depend on the batch of cells used for the assay. When A and B were added together, then the migration was greater than with A or B alone but usually less than the sum of A and B. By adding ethanolamine to the isoelectric focussing column the pH range could be extended and (Fig. 3) thus the two components from SV28 cells could be separated but at the cost of a heavy reduction in yield. The activities of A and B from SV28 were now additive at saturating values (Table 1). This additivity shows that A and B have been separated and that they are different substances acting on different targets. Further, A can

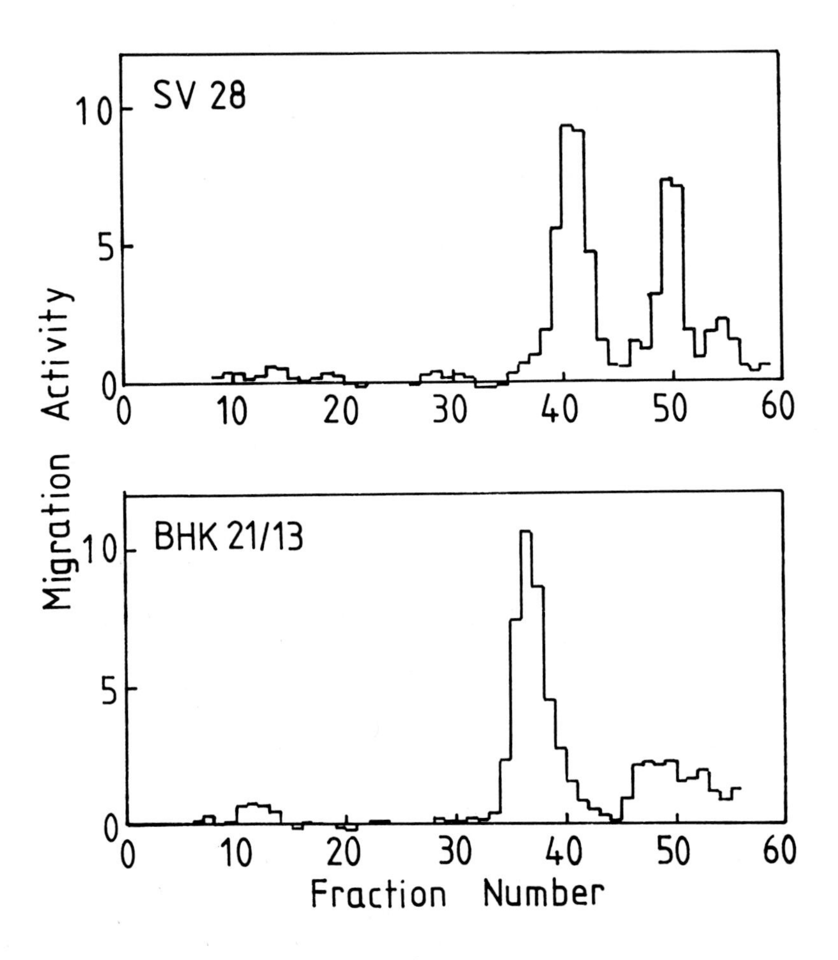

FIG. 3. Isoelectric focussing in a gradient supplemented with 0.1% ethanolamine of migration activity from a) SV28 cells and b) BHK21/13 cells. Cathode on the right.

TABLE 1. Interaction of purified A and B in migration assay. Mean of three counts on each of two plates is shown in cells per mm of starting-line

	10 μl	20 μl	50 μl
A		24.7	20.7
B		32.9	35.9
A + B	42.9	50.3	63.0

be measured as the increment over the plateau value of B and vice versa. This increment could be used to identify the activity from BHK21/13 cells. The crude media conditioned by BHK21/13 cells and by SV28 cells were dialysed against 10 mM formic acid and assayed alone and in the presence of saturating amounts of A and of B (Table 2).

TABLE 2. Assay of migration activity of BHK21/13 fractions alone or in presence of saturating A or B from SV28 (see Table 1)

	μl	---	A	B
			41.4	80.5
BHK crude	10	5.5	41.8	75.0
	50	25.8	71.3	75.7
BHK Sx	2	15.7		
	10	40.4	74.5	70.2
	50	64.2	97.7	68.5
SV28 crude	50		65.5	110.8

SV28 conditioned medium produces an increment in migration activity in the presence of both saturating A and of saturating B. BHK21/13 medium has no further migration activity in the presence of saturating B, but it produces the same increment in activity in the presence of saturating A, as it has activity alone. Further, the purified BHK21/13 activity also is active in the pres-

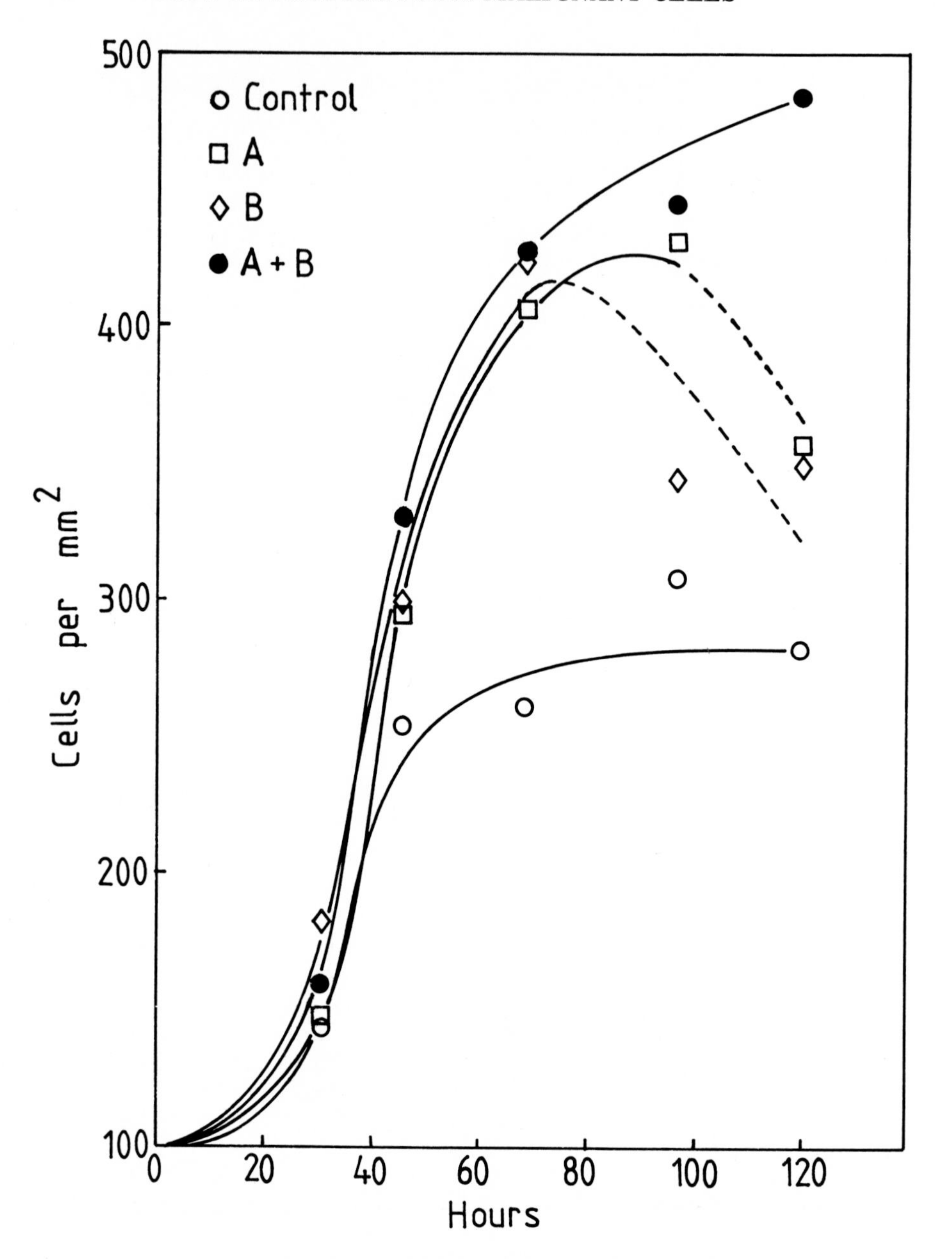

FIG. 4. Growth stimulating effect on 3T3 cells in 10% serum of factors A and B from SV28 cells

ence of saturating A but not at all the presence of saturating B. Therefore, BHK21/13 cells produce B and not A. The purification, although the efficiency is low, seems to truly reflect the position in the crude medium.

Both factors A and B can stimulate the growth of cultures of 3T3-B cells to above their saturation density (Fig. 4). It is conspicuous that A at all concentrations at which it produces migration produces a change in the culture morphology like that seen when 3T3 are transformed (1). On the other hand, B produces very little change in morphology (not more than the effect of adding 10% normal whole serum to the cultures) until there is about 50-fold excess above the amount needed to produce saturating migration activity (Fig. 5).

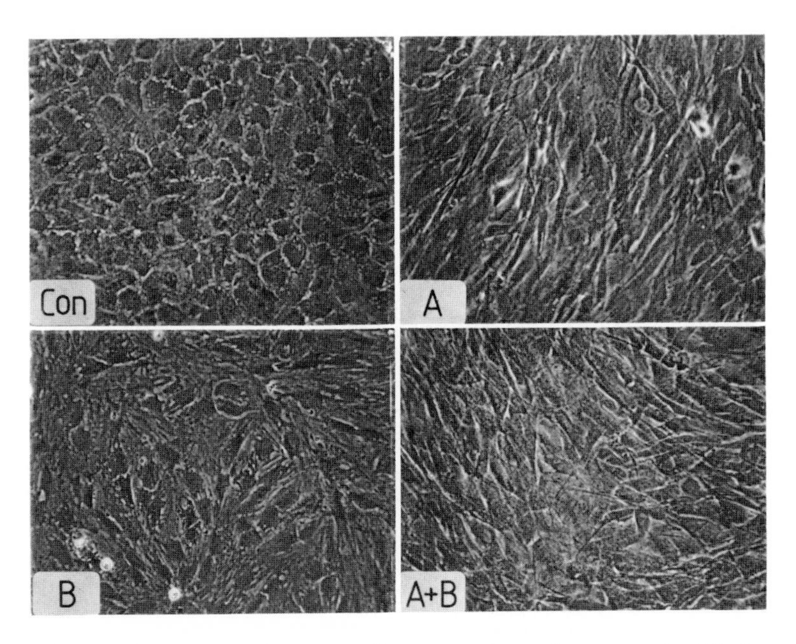

FIG. 5. Photograph of cultures in Fig. 4 at 70 hours. White rectangle was 105 μ long.

Fetal calf serum was found in the mixing experiments to be partially active in the presence of saturating A or in the presence of saturating B. Therefore, the serum has an activity which is not A and one which is not B. Serum had little if any activity in the presence of saturating A plus saturating B, therefore serum contains an A-like activity and a B-like activity.

When medium was conditioned by secondary or tertiary cultures of fibroblasts from kidneys of baby hamsters with or without supplements of 10% tryptose phosphate broth for 1,2,3,4 or 5 days, it had no significant migration activity.

DISCUSSION

We have shown here that SV28 cells release into serum-free tissue culture medium growth factors (A and B) that we assay by their migration activity and which we can purify and separate. In contrast, BHK21/13 cells only produce one growth factor (B) and secondary and tertiary fibroblasts from kidneys of baby Syrian hamsters produce none. A problem with BHK21/13 cells is that they spontaneously transform and then seem to produce an A-like activity. We have also shown that factor A applied to cultures of 3T3 cells makes them look transformed. Historically, BHK21/13 is a cell line derived from fibroblasts of kidneys of baby Syrian hamsters. The cultures retain the oriented morphology of normal fibroblast cultures but have a reduced requirement for serum growth factors. BHK21/13 forms tumours in hamsters when between 10^4 and 10^5 cells are injected. The tumours are benign in that they remain exclusively at the site of injection. SV28 is a line of cells derived from BHK21/13 by transformation with SV40 virus. The cells show an often rounded, irregular, rather neoplastic morphology in cultures and will grow indefinitely (up to 2,500 hours without subculture) without serum. Thirty cells of SV28 injected into hamsters produce a tumour at the site of injection that in about 70% of the animals metastasize, in particular to the lymph nodes, kidneys and lungs. We wonder whether the progression observed in the production of no, of one of two growth factors is in fact the cause of the progression from normal fibroblasts, to the benign BHK21/13, to the malignant SV28. Is it that genes for A and B exist in normal fibroblasts and that when gene B is derepressed the resulting phenotype is that of BHK21/13 and when genes B and A are derepressed, the resulting phenotype is that of SV28? Is it that with transformation the cells become auto-stimulating by the ectopic production of the growth factors of which they are the targets? Is it that the normal cells receive their growth stimulus via the blood (serum growth factors) and that with progressive and transformation they become more self-stimulating and hence more independent of the normal growth regulating stimuli (serum growth factors)?

In a meeting on Hormones and Cancer it seems legitimate to speculate on the role of growth factor production in cancer. Fig. 6 attempts to present pictorially an hypothesis that arose from the above speculations (6). In the top row there are three types of

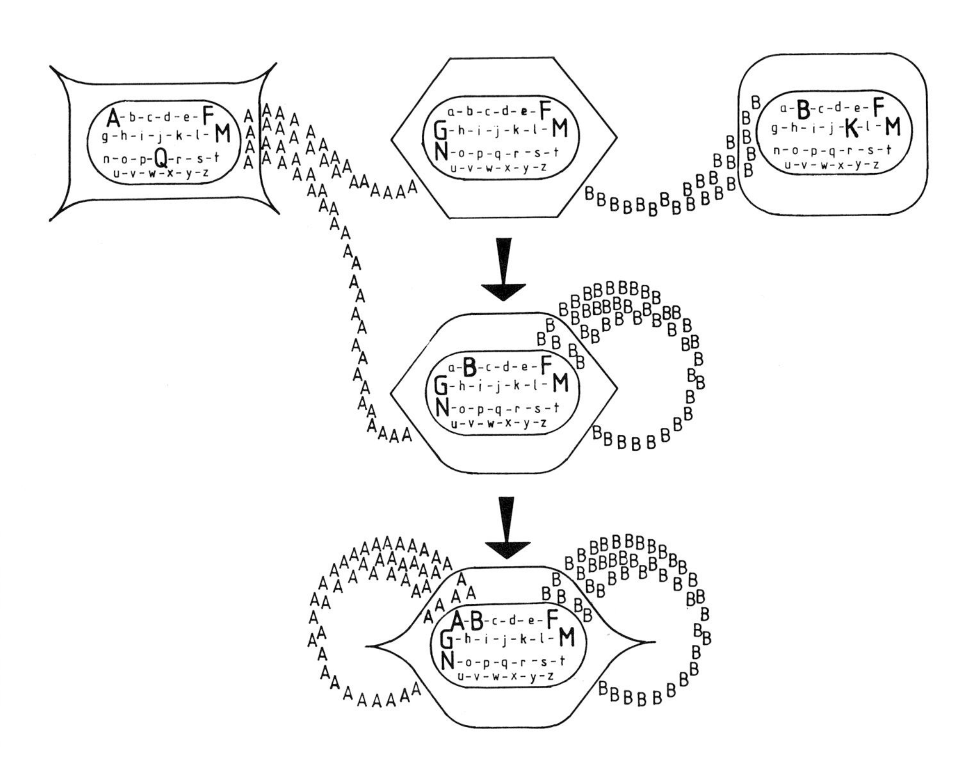

FIG. 6. Ectopic auto-stimulation. For explanation see Discussion.

cells from an organism with genes A to Z. In all types of cell genes F and M are expressed. In the left hand cell genes A and Q are also expressed, in the right hand B and K are expressed and in the middle cell G and N are expressed. A and B are exported and reach the hexagonal target cell through the circulation. (There is a sub-hypothesis where the genes are not expressed differentially in space but in time so that A, for instance, is an embryonic growth factor that comes to be expressed in an adult cell.) The hexagonal cell is supposed to be altered mutagenically or teratogenically so that gene B is additionally (and ectopically) expressed. By a further alteration gene A is expressed.

This hypothesis leads to a very pessimistic view of the possibility of curing cancer because, essentially, a normal growth factor is acting on its proper receptor in the proper target cell which is responding properly, proliferating and migrating (as it should perhaps in embryogenesis). The abnormality is the expression of a growth factor gene in a specific wrong place, namely in its target cell. Perhaps in order to be fail-safe a cell requires stimulation by several growth factors so that if only one is abnormally expressed then the target cell is still largely under normal growth control. The multi-hit nature of cancer mortality curves emphasized by Doll (9) would occur if the expression of many growth factor genes (including regulator genes and genes concerned with post-translational events) was needed for the cell to be fully malignantly transformed. Unfortunately, with our present knowledge of the control of gene expression there is no foreseeable practical way of correcting such abnormalities. The good side to the hypothesis is that transformed cells can be used as a source of specific growth factors as Todaro and De Larco (16) and we ourselves are doing.

ACKNOWLEDGEMENTS

I thank Mrs. A.-M. Rothen and G. Steck for their able technical assistance.

REFERENCES

1. Aaronson, S.A., and Todaro, G.J. (1968): J. Cell. Physiol., 72:141-148.
2. Bürk, R.R. (1966): Nature, 212:1261-1262.
3. Bürk, R.R., and Williams, C.A. (1971): In: Ciba Foundation Symposium on Growth Control in Cell Cultures, edited by G.E.W. Wolstenholme and J. Knight, pp. 107-125. Churchill Livingstone, London.
4. Bürk, R.R. (1973): Proc. Natl. Acad. Sci. USA, 70: 369-372.
5. Bürk, R.R. (1976): Exp. Cell Res., 101:293-298.
6. Bürk, R.R. (1977): In: The Transformed Cell, R. Pollack, R. Hynes and L.B. Chen, p. 20. Cold Spring Harbor Lab., N.Y.
7. Comings, D.E. (1973): Proc. Natl. Acad. Sci. USA, 70:3324-3328.
8. De Larco, J.E., and Todaro, G.J. (1978): Proc. Natl. Acad. Sci. USA, 75:4001-4005.
9. Doll, R. (1978): Cancer Research, 38:3573-3583.
10. Holley, R.W., and Kiernan, J.A. (1968): Proc. Natl. Acad. Sci. USA, 60:300-304.
11. Macpherson, I. (1965): Science, 148:1731-1733.
12. Macpherson, I., and Stoker, M. (1962): Virology, 16:147-151.
13. Paul, J. (1963): Cell and Tissue Culture. Livingstone, Edinburgh.
14. Stanners, C.P., Till, J.E., and Siminovitch, L. (1963): Virology, 21:448-463.
15. Temin, H.M. (1966): J. Natl. Canc. Inst., 37:167-175.
16. Todaro, G.J., and De Larco, J.E. (1978): Cancer Res., 38:4147-4154.

Control Mechanisms in Animal Cells,
edited by L. Jimenez de Asua et al.
Raven Press, New York © 1980.

Two Different Growth Factors from an SV40 Transformed BHK Cell Line Stimulate Both Migration and DNA Synthesis

*Peter Leuthard, *Germaine Steck, Robert R. Bürk, and Angela Otto

Friedrich Miescher-Institute, CH-4002 Basel, Switzerland

INTRODUCTION

Two characteristic properties of malignant cells are their ability to grow very rapidly and to invade the surrounding tissues, colonizing sites distant to the initial tumor. The former property is explained by a loss of the ability to respond to external regulatory mechanism(s), while the latter is basically related to their capability to migrate (1). In tissue culture the behavior of cells transformed by oncogenic viruses or carcinogens resembles that of malignant cells in vivo. Transformed cells replicate independently of growth factors in the culture medium and do not conform with the proper spatial relationship among the cells (7,8). In contrast, growth of normal cells is highly dependent on the presence of growth factors in the medium and is sensitive to the cellular organization in the monolayer.

Among the mechanisms postulated to be responsible for transformation is the production and secretion of growth factors by the transformed cells leading to continuous autostimulation (4,5,6). The isolation and characterization of such growth factors from various tumors and virally transformed cell lines has been pursued and the biological activity of these factors assayed in "normal" cell lines. In one line of studies it has been reported that SV28 cells, an SV40 transformed

*Present address: Research Department, Pharmaceutical Division, Ciba-Geigy Ltd., CH-4002 Basel, Switzerland

BHK cell line, which can be grown in serum-free medium, release factors into the medium, which are able to stimulate both migration and DNA synthesis in normal mouse 3T3 cells (3). We have been able to isolate and purify two biologically active factors, A and B. Here we provide evidence that they are chemically and biologically different.

MATERIALS AND METHODS

Cell Cultures

Stock cultures of SV28, Balb/c 3T3-B, and Swiss 3T3-K were maintained in Dulbecco's modified Eagle's medium (DEM) containing penicillin and streptomycin and supplemented with 10% fetal calf serum.

Migration Assay

Balb/c 3T3-B cells were plated at 2×10^5/35 mm dish in 2.5 ml DEM as for incorporation of ^{3}H-thymidine, and migration was assayed as described (2). Migration activity is represented by the number of cells having crossed 1.2 mm of the wound edge after 22 hr.

^{3}H-Thymidine-Incorporation

Balb/c 3T3-B cells were plated at 2×10^5/35 mm dish in 2.5 ml of DEM with 10% fetal calf serum and were allowed to become quiescent for 5 days without medium change. At the same time as the addition of factor, cells were labeled with 3 μM (1 μCi/ml) (methyl-^{3}H)-thymidine. After 24 hr duplicate cultures were processed for scintillation counting (9).

Radioactive Labeling of Nuclei

Swiss 3T3 cells were plated at 1.5×10^5/35 mm dish in 2 ml of DEM supplemented with 6.5% fetal calf serum and low molecular weight components (12). Cultures were allowed to become quiescent 4 days after an intermediate medium change. Additions were made directly to the culture medium and cells labeled with 1 μM (3 μCi/ml) (methyl-^{3}H)-thymidine for 28 hr (9).

Production and Purification of Migration Stimulating Growth Factor

To obtain conditioned medium SV28 cells were cultivated in serum-free DEM containing 10% tryptose phosphate broth. The purification procedure is summarized in Fig. 1. Details will be published elsewhere (11).

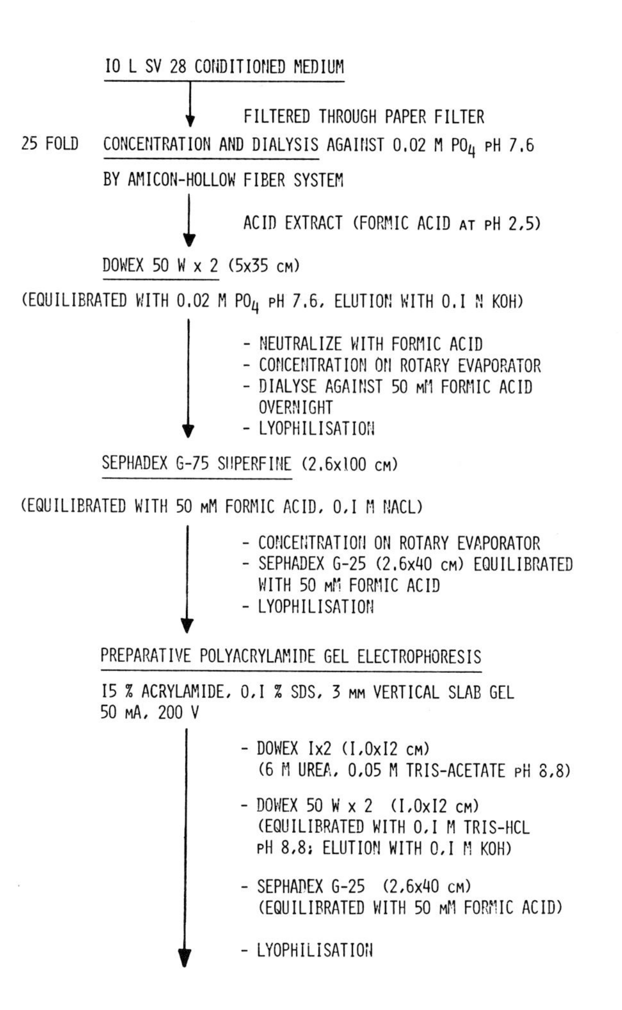

FIG. 1. Purification scheme of two factors from SV28 conditioned medium

RESULTS

The purification of conditioned, serum-free medium from SV28 cells (Fig. 1) resulted in two active fractions for stimulation of migration and DNA synthesis. Fraction A appears to be a protein with a M_r of 23,000 dalton and has been purified to homogeneity in SDS polyacrylamide gel electrophoresis. Fraction B with a M_r of 46,000 dalton, which could appear to be a dimeric form of A, however, gives as yet several bands in SDS gel electrophoresis and, unlike fraction A, cannot be recovered in a biologically active form from the SDS gel. The following biological assays will provide further evidence for two different factors.

Fig. 2 shows the dose-response curve for stimulation of cell migration using increasing amounts of fraction A or fraction B. The saturating amount of the fraction A reached a plateau that corresponded to

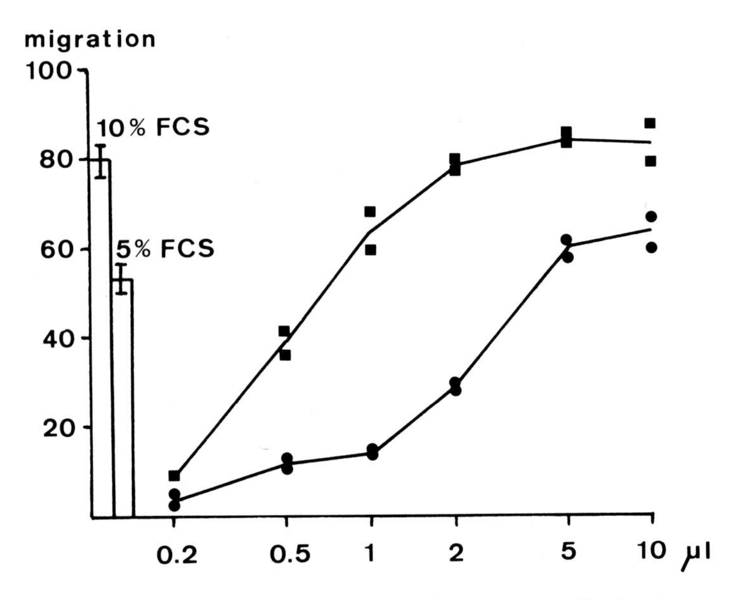

FIG. 2. Stimulation of migration in Balb/c 3T3 cells by different amounts of a stock solution of fraction A (236 µg protein/ml) (●) and of fraction B (117 µg protein/ml) (■). Migration is expressed in number of cells having crossed the wounding line. FCS: fetal calf serum.

the migration activity of 5% fetal calf serum. Addition of 189 ng protein of fraction B per ml culture medium resulted in the same migration as 10% fetal calf serum, equal to 8 mg protein/ml in the culture. This means that 37,000 x less protein was required to induce migration by fraction B than by fetal calf serum. The difference in the plateau and in the dose-response curve at low amounts of each fraction suggests that fractions A and B contain different factors for migration of 3T3 cells. Also, fractions A and B differ in their ability to induce morphological changes: cells treated with low amounts of fraction A became extremely elongated, had a criss-cross arrangement and piled up. Fraction B can induce this morphology only at high concentrations, i.e. about 10 times the dose required for half maximal stimulation of migration (see Bürk, this volume).

The dose-response curves for ^{3}H-thymidine incorporation are shown in Fig. 3. Increasing amounts of frac-

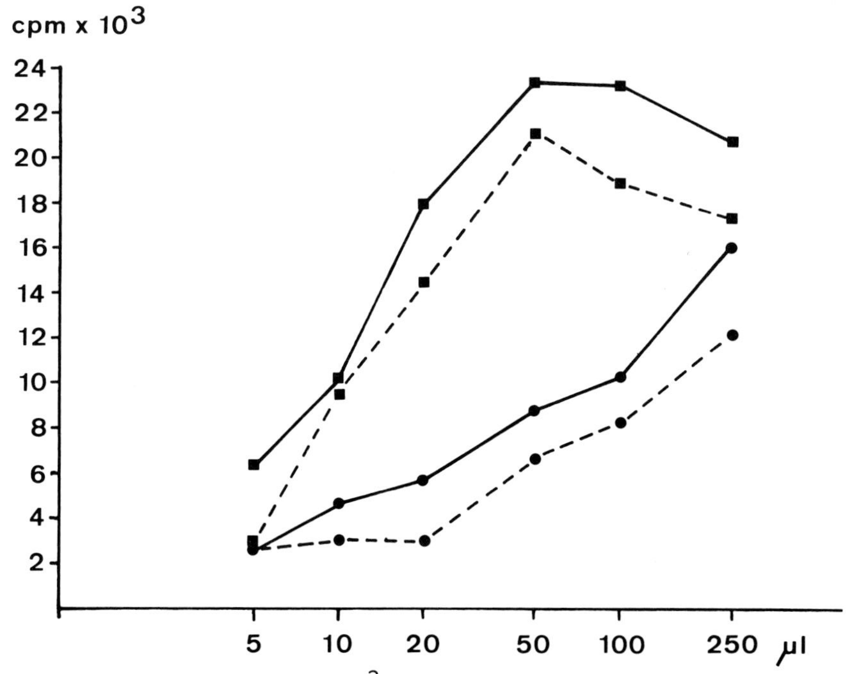

FIG. 3. Stimulation of ^{3}H-thymidine incorporation in Balb/c-3T3 cells by different amounts of a stock solution of fraction A (510 µg protein/ml) (●) and of fraction B (530 µg protein/ml) (■). Additions were made either directly to the culture (——) or after changing to serum-free medium (---).

tions A and B were added to quiescent 3T3 cells either directly to the culture medium or after changing to serum-free medium. Addition of fraction A or B to serum-free medium resulted in slightly lower ^{3}H-thymidine incorporation than additions directly to the culture medium. In either case, increasing amounts of fraction A resulted in a gradual and relative constant increase in ^{3}H-thymidine incorporation without reaching a plateau in the range tested. In contrast, increasing amounts of fraction B resulted in a sharp increase in ^{3}H-thymidine incorporation reaching a plateau at 4 times the concentration required for half maximum stimulation. In several experiments using different solutions of fractions A and B, it was also observed that higher amounts of each fraction were required to stimulate ^{3}H-thymidine incorporation than were needed for migration.

The percentage of radioactively labeled nuclei is a direct assay for the fraction of cells having initiated DNA synthesis in a cell population. Fig. 4 shows the dose requirement of fractions A and B for stimulation of quiescent cells to proliferate. Fractions A and B had very similar dose-response curves, both fractions stimulating 95% of the cells to initiate DNA synthesis

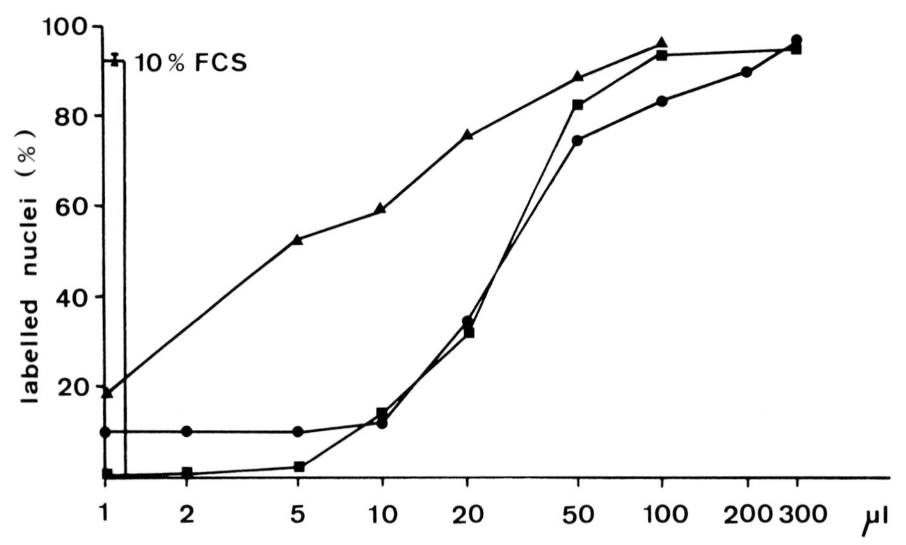

FIG. 4. Percentage of labeled nuclei stimulated by different amounts of a stock solution of fraction A and of fraction B in Swiss 3T3 cells. Both stock solutions were adjusted to 2 mg protein/ml. (●) Fraction A, (■) fraction B, (▲) equal amounts of fractions A and B.

after 28 hr at the saturating amount. This is the same percentage obtained with 10% fetal calf serum. Adding equal amounts of fractions A and B together resulted in a synergistic effect as shown in the shift of the dose response to lower concentrations. This result is a further indication that fractions A and B contain different biologically active factors.

Hormones and other compounds, which under certain defined conditions do not stimulate DNA synthesis by themselves, are known to have either a synergistic or inhibitory effect on the initiation of DNA synthesis stimulated by different growth factors (10). The effect of insulin as well as hydrocortisone has been tested on the labeling index stimulated by either fraction A or B.

As shown in Fig. 5A, insulin at physiological con-

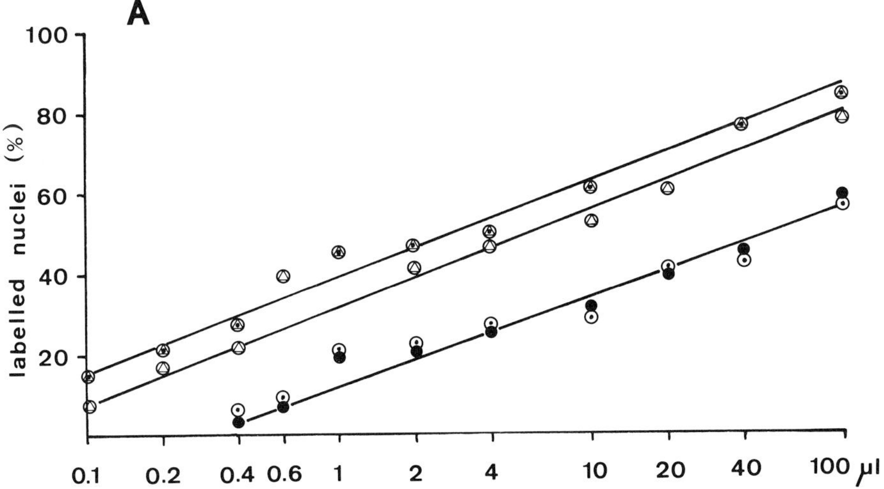

FIG. 5. Effect of insulin and hydrocortisone on the labeling index stimulated by fractions A and B.
A. Different amounts of a stock solution of fraction A (50 µg protein/ml) alone (●), with hydrocortisone (⊙), with insulin (◬), with hydrocortisone and insulin (◬).

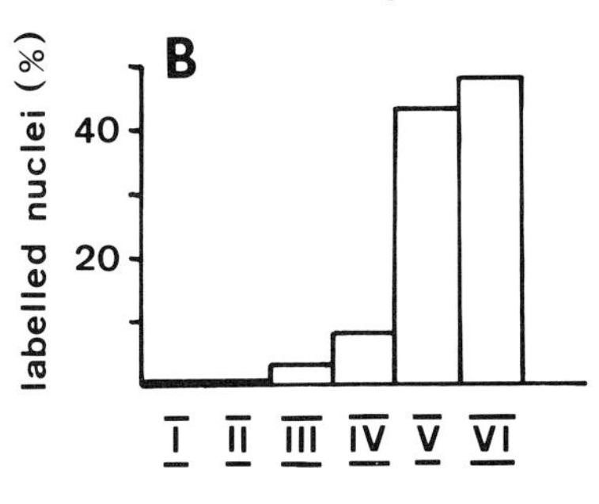

B.I: Hydrocortisone, II: insulin, III: fraction B (10 µg protein/ml culture medium), IV: fraction B plus hydrocortisone, V: fraction B plus insulin, VI: fraction B plus hydrocortisone and insulin. Concentrations used in the culture medium were for hydrocortisone 20 ng/ml and for insulin 50 ng/ml.

centrations (50 ng/ml) enhances the stimulatory effect of different amounts of fraction A, while hydrocortisone (20 ng/ml) had no effect. Addition of both insulin and hydrocortisone with fraction A seems to give a further increase in the labeling index. Similar results were obtained with fraction B (Fig. 5B), however, a higher purification will be required to exclude the interaction with proteins as yet remaining in this fraction. Nevertheless, the stimulatory effect of fractions A and B on the initiation of DNA synthesis can be enhanced by insulin analogous to results with defined growth factors such as $PGF_{2\alpha}$, FGF, EGF (10).

DISCUSSION

The production of growth factors by a number of cancer cells and transformed cell lines provides these cells with an external mean to grow independently of serum growth factors. The SV40 transformed BHK cell line SV28 releases two factors, A and B, into serum-free medium, each one stimulating both migration and proliferation of "normal" cells. The comparison of their chemical and biological properties suggests that they may possibly be two different proteins.

Each of these two factors can stimulate the initiation of DNA synthesis in 95% of a 3T3 culture within 28 hr - the same as addition of 10% FCS. This property is a basic difference with the mode of action of other growth factors such as EGF, FGF or $PGF_{2\alpha}$, which can only stimulate 25% of the cells to initiate DNA synthesis under the same conditions. Insulin, which at our experimental conditions at physiological concentrations has no effect, exerts a synergistic effect on each of these three growth factors resulting in 40-50% labeled nuclei (10). Adding insulin with factors A and B also has a synergistic effect by decreasing the amount of factor required for stimulation. Hydrocortisone has either an enhancing or an inhibitory effect depending on the growth factor; the stimulation of DNA synthesis by $PGF_{2\alpha}$ and EGF is inhibited, whereas hydrocortisone exerts a synergistic effect on FGF (see Jimenez de Asua, this volume). With low concentrations of factors A and B hydrocortisone seems to have only a marginal effect. These results suggest that different mechanisms are involved in the stimulation by factors A and B than by $PGF_{2\alpha}$ and EGF or FGF [13].

The fact that factors A and B together have a synergistic effect guarantees the cell population a high degree of stimulation even when these factors are synthesized and released in very low, subsaturating amounts. Furthermore, insulin is synergistic with factors A and B, which could suggest that hormones may play an active role in the enhancement of growth stimulation by factors produced by transformed or cancer cells. The elucidation of the interaction of such factors and hormones may lead to the understanding of the mechanisms involved in the constitutive proliferation of cancer cells.

ACKNOWLEDGEMENTS

The authors are deeply indebted to Dr. Luis Jimenez de Asua for support and stimulating discussion. We are also grateful to Dawn King for helping in parts of the experiments and to Drs. D. Monard and K.M.V. Richmond for critical reading of the manuscript.

REFERENCES

1. Abercrombie, M., and Ambrose, E.J. (1962): Cancer Research, 22:525-548.
2. Bürk, R.R. (1973): Proc. Natl. Acad. Sci. USA, 70: 369-372.
3. Bürk, R.R. (1976): Exp. Cell Res., 101:293-298.
4. Bürk, R.R., and Williams, C.A. (1971): In Ciba Foundation Symposium on Growth Control in Cell Cultures, edited by G.E.W. Wolstenholme and J. Knight, pp. 107-125. Churchill Livingstone, London.
5. Comings, D.E. (1973): Proc. Natl. Acad. Sci. USA, 70:3324-3328.
6. De Larco, J.E., and Todaro, G.J. (1978): Proc. Natl. Acad. Sci. USA, 75:4001-4005.
7. Dulbecco, R. (1969): Science, 166:962-968.
8. Dulbecco, R. (1970): Nature, 277:802-806.
9. Jimenez de Asua, L., Clingan, D., and Rudland, P.S. (1975): Proc. Natl. Acad. Sci. USA, 72:2724-2728.
10. Jimenez de Asua, L., Richmond, K.M.V., Otto, A.M., Kubler, A.-M., O'Farrell, M.K., and Rudland, P.S. (1979): In Hormones and Cell Culture, edited by G.H. Sato and R. Ross, pp.403-424. Cold Spring Harbor Laboratory.

11. Leuthard, P., Steck, G., and Bürk, R.R. (1980): Manuscript in preparation.
12. O'Farrell, M.K., Clingan, D., Rudland, P.S., and Jimenez de Asua, L. (1979): Exp. Cell Res., 118: 311-321.
13. Otto, A.M., Leuthard, P., Bürk, R.R., and Jimenez de Asua, L. (1980): Manuscript in preparation.

Control Mechanisms in Animal Cells,
edited by L. Jimenez de Asua et al.
Raven Press, New York © 1980.

Transformation Enhancing Factors in the Plasma of Cancer Patients

S. Barlati and P. Mignatti

Laboratory of Genetic Biochemistry and Evolution of the C.N.R. and Institute of Genetics, University of Pavia, 27100 Pavia, Italy

Transformed cells secrete into the culture medium several factors with specific biological and enzymic activities. Among these: overgrowth stimulating factor (16, 17), migration factor (6, 7), sarcoma growth factor (10, 18, 19), plasminogen activator (9, 11, 15, 20, 21) and transformation enhancing factors (2, 12, 13). An activity similar to the latter has been identified also in the plasma cryoprecipitate of patients affected with different neoplastic diseases (4, 14). It has not yet been demonstrated whether the TEF activity from in vitro transformed cells is or is not identical to TEF from the blood plasma of cancer patients; TEF's are therefore only operationally defined as a class of apparently large MW proteins (4, 13) which accelerate and/or favour the expression of cell transformation under appropriate experimental conditions. In the present paper we briefly summarize previous reports on TEF from blood plasma of cancer patients and present evidence of a possible connection between TEF activity and clinical conditions. We finally present data on the partial characterization of these factors.

ASSAY OF TRANSFORMATION ENHANCING FACTORS

Chicken embryo fibroblasts (CEF), infected by temperature sensitive (ts) mutants for transformation of Rous Sarcoma Virus (RSV), express a transformed phenotype only when incubated at the permissive temperature (37°C) and not at higher temperature (41°C).

As previously reported (4, 13) TEF activity is assayed by infecting CEF with ≃ 100 FFU/plate of ts mutants of RSV, PA1 (5) or PA2 (8) and incubating them at 41°C so as to get cultures which do not express foci with transformed morphology (latent foci). After an appropriate period of incubation, series of at least duplicate plates are shifted down to 37°C in the presence of liquid medium containing 5% decomplemented calf serum, 10 µg/ml of cyclohexamide and 0.1% (v/v) of the plasma fractions to be tested. After 2 to 4 hours incubation the cultures are stained with Methylene Blue (MB) (1, 3). This staining technique allows to selectively visualize foci in different transformation stages: stained cells with transformed morphology (STP), stained cells with a normal morphology (SNP) and foci of unstained cells with normal morphology (latent or USNP foci). A scheme of these transformation stages and pictures of the corresponding foci are reported in Fig. 1.

Only STP foci (corresponding to the final stage of transformation) are then scored at the microscope.

The presence of TEF activity, added in the culture medium, increases the number of foci reappearing in test dishes as compared to those in control dishes in which no TEF activity is present. The ratio of the number of foci in test dishes versus that

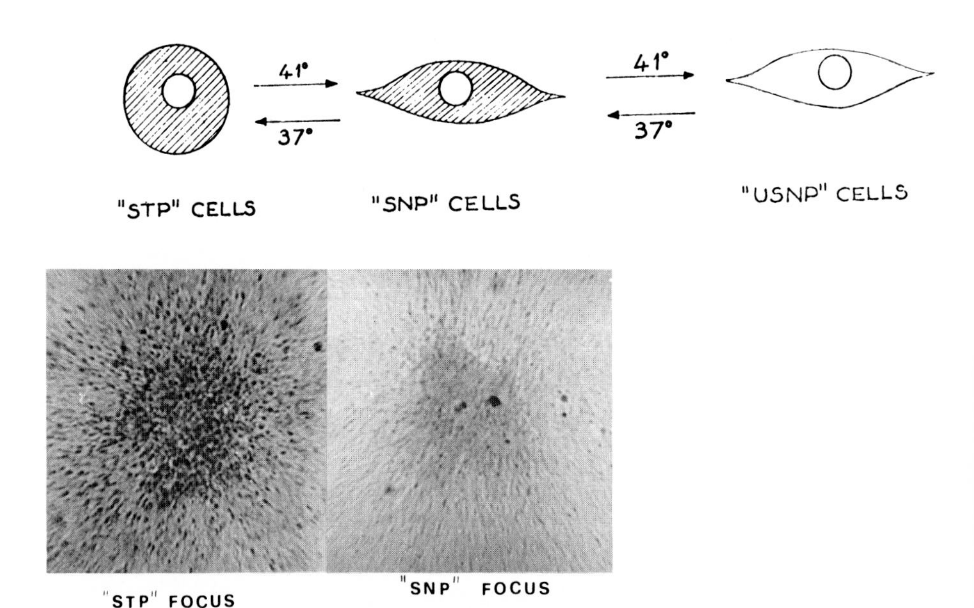

FIG. 1. Scheme of different transformation stages, visualized by MB staining, as a function of incubation temperature. Pictures of the correponding foci are also shown; USNP foci cannot be directly visualized (1, 3).

in control dishes, defined as Factor of Enhancement (FE), directly measures TEF activity. The FE resulting can be significantly higher than 1 or not significantly different from 1 depending on the presence or the absence of TEF activity; it can also be significantly lower than 1 indicating the presence of transformation inhibiting activities (4). A scheme of the assay is reported in Fig. 2.

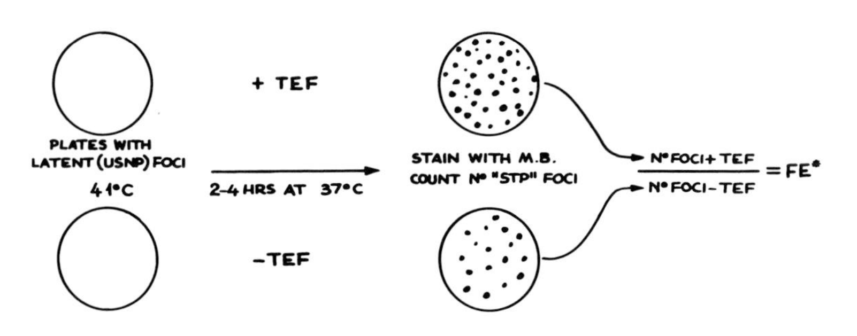

FIG. 2. Scheme of TEF assay. The dark spots in the plates, after the temperature shift-down (41°C → 37°C), correspond to the STP foci.

TEF ACTIVITY IN BLOOD PLASMA FRACTIONS

The assay of whole plasma taken from cancer patients before therapy was started, revealed the presence of TEF activity in some of the plasmas tested. No activity could be revealed in control plasmas. The fact that in several cases an inhibitory effect on transformation could be observed suggested the presence of transformation inhibitors which might have masked the expression of TEF activity. Plasma was therefore submitted to fractionation in an attempt to separate TEF from antagonistic plasma activities. This was achieved by cryoprecipitating plasma (4) and redissolving the precipitate in PBS, 1/5 of original plasma volume (cryo). Whole plasma, cryo and cryo supernatant (cryo sup) were tested in parallel; the results obtained are reported in Table 1. As shown all the cryos from the

TABLE 1. TEF activity in different blood plasma fractions.

Malignant disease	FE			Controls	FE		
	Plasma	Cryo sup	Cryo		Plasma	Cryo sup	Cryo
Hodgkin	1.6	0.3	3.5	Healthy	1.1	1.2	1.0
Lymphoma	2.1[a]	1.0	3.6	"	0.8	1.2	1.4
AML	2.5	9.1	2.9	"	0.9	0.5	1.1
AMML	0.5	1.1	3.5	"	0.1	0.1	0.0
AMML	0.0	0.0	2.2	"	1.1	1.0	1.1
Stem Cell L.	0.2	0.2	2.0	Pregnancy	0.1	0.5	1.5
Bronchog. carcinoma	2.0	0.0	4.2	"	0.2	0.4	0.7
Hepato carcinoma	1.5	1.5	3.0	Viral hepatitis	0.2	0.2	1.1

[a] Underlined values are TEF positive, i.e. significantly higher than 1 ($p \leq 0.05$).

Abbreviations used: AML: Acute Myeloblastic Leukemia; AMML: Acute Myelomonoblastic Leukemia; Stem Cell L.: undifferentiated Leukemia.

cancer patients and none from the controls reported, revealed a TEF activity. This suggested cryoprecipitation to be an efficient and simple method for revealing TEF activities in the blood plasma of cancer patients. This conclusion can now be extended to approximately 60 cancer patients and 60 controls (healthy subjects or affected by non malignant diseases) (unpublished data).

The histogram reported in Fig. 3 shows the score of the number of foci elicited in the presence of cryo from other 12 cancer patients and 12 controls, as compared to control dishes, in different TEF assays.

Relative FE's can be calculated by the ratios of dashed to empty columns.

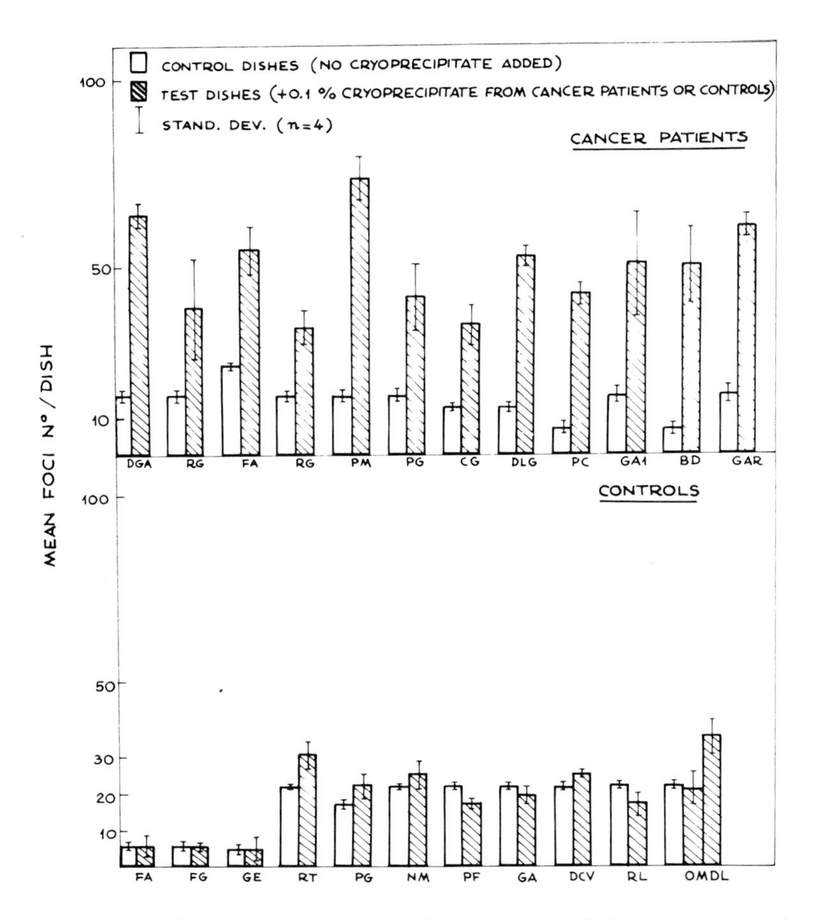

FIG. 3. The histograms report the scores of the number of foci elicited in several TEF assays of cryo precipitates from cancer patients and controls.

LOSS OF TEF ACTIVITY DURING REMISSION

The fact that most of the cancer patients tested, affected by a wide spectrum of malignant diseases, were found to be TEF positive, prompted us to investigate whether this activity might be related to the patient's clinical conditions and therefore used as a prognostic marker. We started this study by following up patients affected by lymphomas and leukemias and submitted to polychemotherapy. Cryos obtained from plasma samples at the time of onset of the disease and during remission, were stored frozen and tested simultaneously in the same TEF assay.

The results obtained by the assay of the five patients are reported in Table 2. All the patients were TEF positive (underlined values) before therapy and only one of them remained positive while considered in clinical remission. This suggests that a correlation may exist between the presence or the absence of TEF activity, and the clinical conditions.

TABLE 2. Loss of TEF activity during remission.

Malignant disease	FE I° Onset of disease	pI°[a)]	FE II° During remission	pII°[a)]
Lymphoma	3.6±0.5	0.005	1.1±0.2	n.s.
AML	2.5±0.2	0.01	2.3±0.4	0.02
AMML	3.1±1.0	0.05	0.9±0.4	n.s.
AMML	4.8±1.5	0.05	0.8±0.3	n.s.
Hairy Cell L.	2.4±0.5	0.02	1.0±0.2	n.s.

a) For legend and abbreviations, see footnote of Table 1. pI° and pII° indicate statistical significance of differences between respectively FE I° and FE II° and their own controls. Differences between FE I° and FE II° (p not reported) are all significative ($p \leq 0.05$) unless for AML.

PARTIAL CHARACTERIZATION OF TEF ACTIVITY

Plasma cryos from patients affected by acute leukemias

have been submitted to gel filtration. By this separation procedure four major protein peaks could be eluted and TEF activity was found associated with two peaks with MWs of 2.2 - 2.5 and 0.5 - 1.2x10^5 daltons (4).

Work is in progress to verify possible correlations with other known growth factor or proteins related to cell transformation. In Fig. 4 we report the results obtained by the assay of one of these factors: the Plasminogen Activator (PA), in the fractions of the cryo from a leukemic patient and a healthy control eluted from an Agarose Bio Gel A-5 m column. PA was assayed as previously described (20) using aliquots of column fractions diluted in 0.1 M TRIS-HCl buffer pH 8.1 in the presence and in the absence of 2 µg/ml purified Plasminogen (Plg). One peak of Plg dependent fibrinolytic activity was found associated with the 2.2 - 2.5x10^5 MW protein peak superimposing with the larger MW TEF activity in the eluate of the leukemic patients cryo. A much lower level of PA, on the contrary, could be detected in the correspondent fractions eluted from the cryo

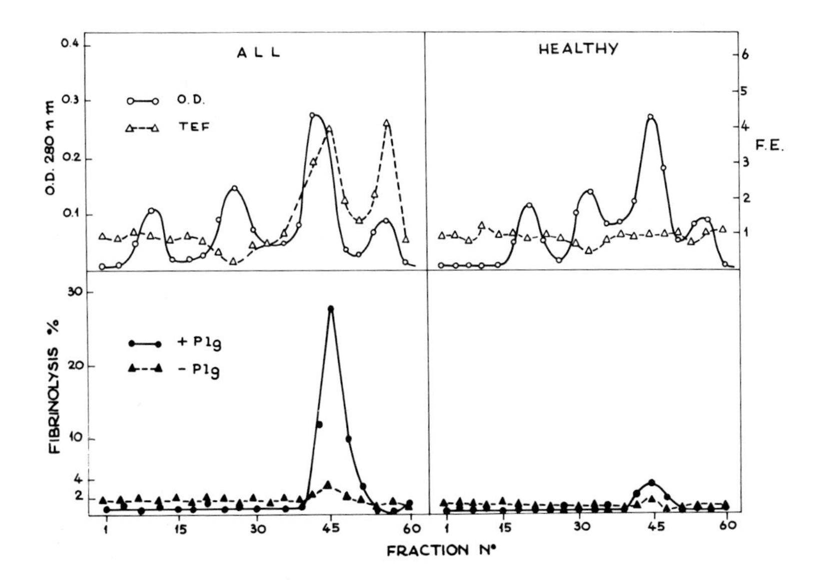

FIG. 4. Agarose A-5 m Bio Gel fractionation of the cryos' from the plasma of a patient affected by acute lymphoblastic leukemia (ALL) and of a healthy control.

of the healthy control, where no TEF activity was detectable.

The results indicate that TEF activity can be associated with PA activity but not identified only with it since no PA was detectable associated with the lower MW peak ($0.5\text{-}1.2 \times 10^5$ daltons) of TEF activity.

CONCLUSIONS

Transformation enhancing factors can be found in the culture medium of in vitro transformed cells (2, 12, 13) and in vivo in the blood plasma of cancer patients (4, 14). Surprisingly TEF activity has been found in the plasma of a wide variety of patients affected with various types of leukemias, lymphomas and solid tumors. We do not have yet sufficient data to conclude that TEF might be a general marker for transformation; however, the frequency of TEF positive, among cancer patients, of TEF negative, among controls, the frequent loss of TEF activity after successful therapy and the fact that the assay used is directly related with parameters of transformation, strongly suggest the possibility of a direct relationship between the presence of TEF and malignancies.

The partial characterization of TEF indicates, that in the case of patients affected by acute leukemias, the activity is found associated with two peaks of proteins of apparently large MW. Furthermore, the absence of PA from the lower MW peak indicate that TEF should not be solely due to PA; however the presence of PA activity, present at levels much higher than in controls, in the high MW peak of TEF activity, suggests that PA may possess or may be somehow related to TEF as previously proposed (4).

While continuing on the characterization of the TEF activity and on the comparison with other known transformation related factors, we are also verifying the limits of the potential diagnostic and prognostic value of the detection of TEF in human blood plasma.

ACKNOWLEDGEMENT

The authors thank Ms. M. Bensi and Mr. F. Tredici for their skillful technical assistance. This work was supported by the C.N.R. Progetto Finalizzato Virus, contract n. 204121/84/ /81817 and 81819.

REFERENCES

1. Barlati, S., Kryceve, C., and Vigier, P. (1974): Intervirology, 4:23-30.
2. Barlati, S., Kryceve, C., and Vigier, P. (1974): Abstracts XIth Int. Cancer Congress, Florence, 2:86.
3. Barlati, S., Mignatti, P., Kryceve, C., and Vigier, P. (1975/76): Intervirology, 6:25-31.
4. Barlati, S., Mignatti, P., Brega, A., De Petro, G., and Ascari, E. (1978): In: Avian RNA Tumor Viruses, edited by S. Barlati and C. De Giuli, pp. 331-348. Piccin Medical Books, Padova.
5. Biquard, J.M., and Vigier, P. (1972): Virology, 47:444-455.
6. Burk, R.R. (1973): Proc. Natl. Acad. Sci. USA, 70:369-372.
7. Burk, R.R. (1976): Exp. Cell Res., 101:293-298.
8. Calothy, G., and Pessac, B. (1976): Virology, 71:336-345.
9. Christman, J.K., and Acs, G. (1974): Biochim. Biophys. Acta, 340:339-347.
10. De Larco, J.E., and Todaro, G.J. (1978): Proc. Natl. Acad. Sci. USA, 75:4001-4005.
11. Goldberg, A.R. (1974): Cell, 2:95-102.
12. Kryceve, C., Vigier, P., and Barlati, S. (1976): Int. J. Cancer, 17:370-379.
13. Kryceve, C., Biquard, J.M., Barlati, S., Lawrence, D., and Vigier, P. (1978): In: Avian RNA Tumor Viruses, edited by S. Barlati and C. De Giuli, pp. 319-330. Piccin Medical Books, Padova.
14. Mignatti, P., Brega, A., Ascari, E., Sitar, G., and Barlati, S. (1979): In: Therapy of acute leukemias, edited by F. Mandelli, pp. 876-883. Lombardo Editor, Rome.
15. Quigley, J.P. (1976): J. Cell Biol., 71:472-486.
16. Rubin, H. (1970): Science, 167:1271-1272.
17. Rubin, H. (1970): Proc. Natl. Acad. Sci. USA, 67:1256-1263.
18. Todaro, G.J., De Larco, J.E., and Cohen, S. (1976): Nature, 264:26-31.
19. Todaro, G.J., De Larco, J.E., Nissley, S.P., and Rechler, M.M. (1977): Nature, 267:526-528.
20. Unkeless, J.C., Tobia, A., Ossowski, L., Quigley, J.P., Rifkin, D.B., and Reich, E. (1973): J. Exp. Med., 138:85-111.
21. Unkeless, J., Danos, K., Kellerman, G.M., and Reich, E. (1974): J. Biol. Chem., 249:4295-4305.

Control Mechanisms in Animal Cells,
edited by L. Jimenez de Asua et al.
Raven Press, New York © 1980.

Growth Factors Localized in Tumor Nuclei

K. Nishikawa, Y. Yoshitake, C. Okitsu, and K. Adachi

Department of Biochemistry, Kanazawa Medical University, Ishikawa 920-02, Japan

Recently, some workers have demonstrated the presence of growth factors for cultured cells in the conditioned medium of cultured tumor cells (2, 3, 5, 6, 11, 14). We found at least two different growth factors in cytosol fraction of a Rhodamine (Rd) fibrosarcoma of rat; one for the completion of BALB/3T3 cell division (cell division factor, CDF) and another for the stimulation of DNA synthesis of the cells (DNA synthesis factor, DSF) (13). On the other hand, growth factor-like activity has been shown in histone or nonhistone fractions (7, 10, 15).

In this paper, we represent DSF activity for BALB/3T3 cells in the acid extract of chromatin obtained from the nuclei of a Rd fibrosarcoma.

MATERIALS AND METHODS

Cells and Culture

The BALB/3T3-3K cell line, a clone of BALB/3T3 that had been maintained in our laboratory (13), was isolated. Under normal subculture conditions for 3T3 cells (1), these cells maintain their high requirement for serum for longer than 3 months. The cells were maintained in the standard medium as described previously (13).

Animals and Tumor

The rats and Rd fibrosarcoma used were of the same kinds as those in the previous study (13).

Preparation of Cytosol and Histone Fractions

Cytosol fraction (crude tumor extract) was prepared from Rd fibrosarcoma according to the method described previously (13). Nuclei were isolated from the tumor or normal rat liver according to the method of Chauveau et al (4) modified by Miyazaki et al (9). Chromatin was prepared from the nuclei by the method of Nakamura et al (12) with some modifications. The nuclei were suspended in 9 volumes (V/W) of 0.2 mM EDTA (pH 7.2) and the suspension was sonicated for 1 min, followed by centrifugation at 12,000g for 30 min. The resulting supernatant was supplemented with such an amount of NaCl that the salt concentration would be 0.15 M. The mixture was centrifuged at 25,000g for 20 min. The resulting precipitate was suspended in water. Histone fraction was prepared by extracting from the chromatin with 0.1 M sulfuric acid and precipitating by adding ethanol according to the method of Miyazaki et al (8). The dried material was dissolved in phosphate buffered saline (PBS) or 0.2 M acetic acid and added in the medium for the assay.

Assay of DNA Synthesis

Subconfluent BALB/3T3-3K cultures grown in standard growth medium was used. The trypsinized cells were plated at the density of 10^5 in 60-mm Falcon plates with 5 ml Dulbecco's modified Eagle's medium (DME) containing 3 % calf serum (CS). Five hours later, the medium was removed, the attached cells were washed with 5 ml PBS, and the plates were transferred into DME containing 0.2 % CS. Twenty-four hours later, the medium was removed and replaced with fresh DME containing 0.2 % CS, followed by adding test sample. Sixteen hours after the addition, 50 µl of 2 x 10^{-5} M, 20 µC/ml [^{3}H]thymidine solution was added. After 3 more hours of incubation, the plates were washed with PBS and trichloro-

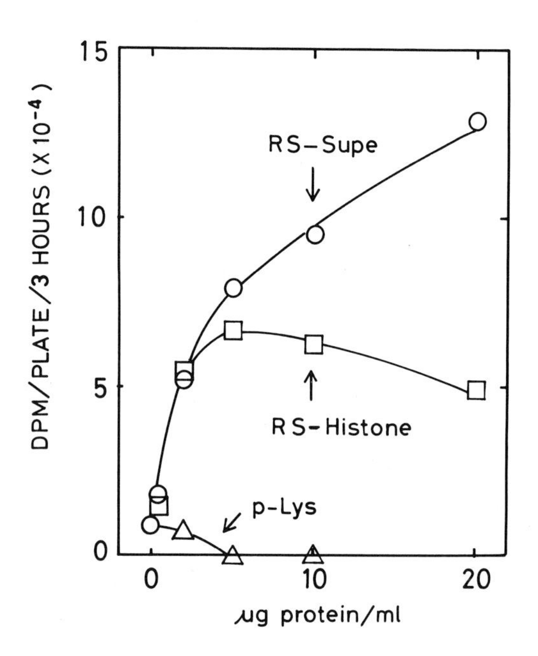

FIG. 1. Stimulation of DNA synthesis in response to cytosol and histone fractions from a Rd fibrosarcoma. The experimental conditions are described in the text, except that various concentrations of the samples were added as indicated. ○ (RS-Supe), Plus cytosol fraction; □ (RS-Histone), plus histone fraction; △ (p-Lys), plus poly-L-lysine (molecular weight, higher than 70,000).

acetic acid (TCA). DNA was extracted with 0.5 M NaOH and then precipitated by adding excess TCA. The precipitate was collected on Millipore filters (HA), washed with TCA and ethanol and then dried. The filters were dissolved in methyl cellosolve and the resulting solutions were mixed with toluene scintillation fluid. Radioactivity incorporated into DNA was counted by means of Aloka Liquid Scintillation Spectrometer (LSC-653). All points were done in duplicate.

Chemicals

DME was the product of Flow Laboratory. CS was purchased from GIBCO. Calf thymus histone (Type II) and poly-L-lysine (Type I-B) were the products of SIGMA.

RESULTS AND DISCUSSION

We found DSF activity not only in the cytosol fraction from a Rd fibrosarcoma, but also in the acid extract of chromatin prepared from the nuclei of the tumor (FIG. 1). DNA synthesis, as determined by [^{3}H]thymidine incorporation into DNA of BALB/3T3-3K cells was stimulated in proportion to the concentrations of the cytosol (supernatant) fraction; the maximal level corresponded to that achieved by adding high concentration of CS. Nuclei were isolated from a Rd fibrosarcoma. It was microscopically confirmed that the preparation was pure nuclei nearly free from cytoplasmic tags. Chromatin preparation which was extracted from the nuclei had a protein versus DNA ratio of 2.0 - 2.3. Acid extract (histone fraction) prepared from chromatin

TABLE 1. Intracellular distribution of DSF activity in Rd fibrosarcoma

Fractions	DSF activity[a] (units/g tissue[b])
Cytosol	420
Histone	110

[a]One unit of the activity is defined as the amount equivalent to 1 mg CS proteins in its ability to cause the stimulation of [^{3}H] thymidine uptake into DNA. The units were calculated from the activities at concentrations lower than 3 μg protein/ml. In the case of histone fraction, the unit was calibrated by measuring the amount of DNA/g tissue and estimating a histone versus DNA ratio of 1.

[b]Wet weight.

was analyzed to contain five major histone classes almost free from other proteins from the electrophoretic patterns on polyacrylamide gels in 2.5 M urea and 0.9 N acetic acid or in sodiumdodecyl sulfate. When thus prepared histone fraction was added, DNA synthesis was also stimulated; maximal activity was achieved at about 5 µg protein/ml and the stimulatory effect decreased at higher concentrations. The maximal activity at optimal concentration of histone fraction was about half that in the presence of excess concentration of the cytosol fraction. Poly-L-lysine inhibited the DNA synthesis. Moreover, DNA synthesis in the presence of DSF was inhibited by the polycation in a competitive manner (data not shown). Distribution of DSF activity in Rd fibrosarcoma cell is shown in TABLE I. Total activity in the histone fraction was about one fourth that in the cytosol fraction. Although the active substance in the cytosol fraction could be extracted with 0.2 M formic acid or acetic acid, it could not be extracted with 0.1 M sulfuric acid, whereas histonefraction was obtained as the sulfuric acid extract.

DSF in the cvtosol fraction has partially purified (K. Nishikawa et al, in prep.). The activity was precipitated between 10 and 80 % saturation ammonium sulfate at pH 7 and then ext-

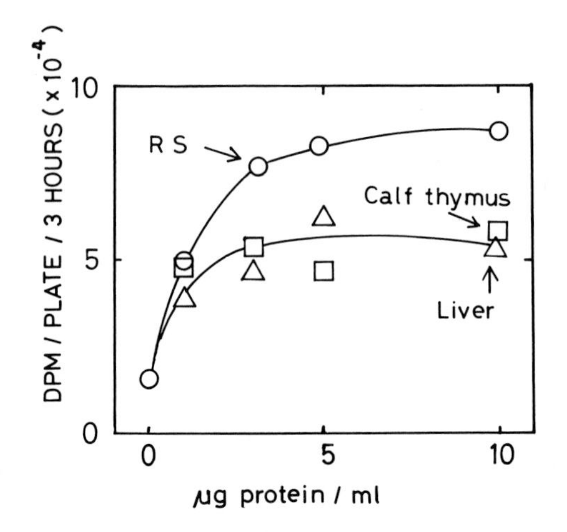

FIG. 2. Stimulation of DNA synthesis in response to histone fractions from various tissues. The experimental conditions were the same as those for FIG. 1, except that various concentrations of histone fractions from various sources were added as indicated. ○(RS), Plus histone fraction from Rd fibrosarcoma; △(Liver), plus histone fraction from normal rat liver; □(Calf Thymus), plus whole histones (Type II) from calf thymus purchased from SIGMA.

racted from the precipitate with 0.2 M formic acid. Further purification was achieved by molecular sieve fractionation on Sephadex G-75 in 0.1 M formic acid containing 1 M ammonium sulfate. The activity was stable and self- or coaggregation was avoided under the conditions. The major activity corresponded to protein fraction having molecular weight of about 12,000. This fraction was further purified by isoelectric fractionation with Ampholine carrier ampholytes of pH range from 9 to 11. Major active material had a pI value of around 12. This preparation stimulated DNA synthesis at 50 ng/ml or less.

DSF in the histone fraction has partially purified (Y. Yoshitake et al, in prep.). The active fraction had a molecular weight range from 10,000 to 25,000 by molecular sieve fractionation on Sephadex G-75 in acid. By means of isoelectric fractionation with higher pH range, the active protein fraction corresponded to a pI of around 11. The histone fraction from the tumor was further fractionated by gel chromatography on Biogel P60 in 0.02 N HCl and 0.05 M NaCl in order to separate each histone. Among the histone classes, H1 fraction had the highest specific activity. This fraction showed good stimulatory activity for the DNA synthesis at 500 ng/ml or less.

Both histone fractions from normal rat liver and calf thymus had also DSF activity (FIG. 2). However, these stimulatory effects at optimal concentrations were lower than that of the histone fraction from the tumor.

It is under investigation to ascertain whether the active substance(s) in the tumor histone fraction is histone itself or another protein. Although the relationship between the DSF's localized in cytosol and chromatin of tumor cells is unclear, there may be a possibility that DSF in cytosol is modified polyleptide of DSF in chromatin or vice versa. Our finding of DSF activity in nuclei may suggest that some of growth factors act in nuclei after internalization.

ACKNOWLEDGMENT

This study has been in part supported by grant from the Research Laboratories, Toyo Jozo Co., Ltd., Shizuoka, Japan.

REFERENCES

1. Aaronson, S.A. and Todaro, G.J. (1968):Science., 162:1024-1026.
2. Bourne, H.R. and Rosengurt, E. (1976):Proc. Natl. Acad. Sci., 73:4555-4559.
3. Burk, R.R. (1976):Exp. Cell Res., 101:293-298.
4. Chauveau, J., Moule, Y. and Rouiller, C.H. (1956):Exp. Cell Res., 11:317-321.
5. DeLarco, J.E. and Todaro, G.J. (1978):Proc. Natl. Acad. Sci., 75:4001-4005.

6. DeLarco, J.E. and Todaro, G.J. (1978):Nature.,272:356-358.
7. Hahn, E.C. (1974):Biochem. Biophys. Res. Commun., 57:635-640.
8. Miyazaki, K., Nagao, Y., Matsumoto, K., Nishikawa, K. and Horio, T. (1973):GANN., 64:449-463.
9. Miyazaki, K., Hagiwara, H., Nagao, Y., Matsuo, Y. and Horio, T. (1978):J. Biochem., 84:135-143.
10. Murakami, O. and Yamane,I. (1976): Cell Structure and Function., 1:285-290.
11. Nair, B.K. and DeOme, K.B. (1973): Cancer Res., 33:2754-2760.
12. Nakamura, T., Hosoi, K., Nishikawa, K. and Horio, T. (1972): GANN., 63:239-250.
13. Nishikawa, K. and Okitsu, C. (1979): In: Control of Proliferation in Animal Cells: Hormone and Cell Culture., edited by R. Ross and G. Sato, Vol.6:pp.441-452, Cold Spring Harbor Press, New York.
14. Rubin, H. (1970):Science., 167:1271-1272.
15. Tuan, D., Smith, S., Folkman, J. and Merler, E. (1973): Biochemistry., 12:3159-3165.

Control Mechanisms in Animal Cells,
edited by L. Jimenez de Asua et al.
Raven Press, New York © 1980.

A New Theory of Neovascularisation Based on Identification of an Angiogenic Factor and Its Effect on Cultured Endothelial Cells

B. R. McAuslan

Molecular and Cellular Biology Unit, C.S.I.R.O., North Ryde, N.S.W. 2113, Australia

INTRODUCTION

Several decades ago it was observed that extensive angiogenesis occurs around some tumor transplants (11,14). A striking demonstration of this is the extensive vascularisation of the tumor formed in rats within days of injecting Walker carcinoma cells.

The interdependence between angiogenesis and tumor growth was publicised mainly by Folkman in a series of papers and reviews in which, among other observations, he showed (i) that transplanted neoplastic tissue is limited in growth and does not become a clinically evident tumor unless the host tissue can provide it with blood vessels (8), (ii) that when neoplastic cells are present at an appropriate site, the surrounding endothelium is stimulated so that endothelial cells appear hypertrophic and may have an increased mitotic index (4).

Gimbrone and co-workers (10),showed that if extracts of carcinoma cells were used to impregnate an inert polymer, fragments of which were then implanted into an avascular site such as the anterior eye chamber, blood vessels were induced to grow out from the nearby tissue (the iris in the case of the eye chamber) and to vascularise the polymer fragment. It was suggested (6) that carcinomas produce a tumor angiogenic factor, TAF, that stimulated neovascularisation and subsequent rapid growth of the carcinoma. TAF appeared to be a 100,000 MW endothelial cell specific mitogen (9).

Unfortunately the report of increased endothelial cell mitosis in vivo in response to crude TAF led other investigators to assume that the enhanced proliferation of cultured endothelial cells in response to crude extracts of carcinoma cells (5) or of Balb/c3T3 cells (3) was due to angiogenic factor and that angiogenic factor was therefore, a mitogen. We believe this conclusion is erroneous and indeed, it has been shown that the primary response to angiogenically active crude extracts appears to be a chemotactic attraction of endothelial cells with mitosis occurring only secondarily (2). The following is an outline of work leading to the identification of the active moiety of an angiogenic factor and on the basis of this and from its effects on cultured endothelial cells I propose a new theory on the primary event in neovascularisation.

ASSAY OF ACTIVITIES INDUCING NEOVASCULARISATION

There are several methods for demonstrating angiogenic activity of test fractions. Two that I think are the most informative are (i) the renal assay in which mice are injected with test material and sections of kidney subsequently examined by electron microscopy for changes in the capillaries of medulla or cortex (13,16) and (ii) the pellet vascularisation assay described above (10). This method has been frequently used to demonstrate tumor angiogenic factor activity.

With all methods one observes essentially the same phenomenon, neovascularisation. All suffer from at least two major drawbacks; they are impractical for quantitative estimates of the response and they are cumbersome. As discussed later, I shall propose a new assay for angiogenic activity based upon the response of endothelial cells in culture.

ENDOTHELIUM STIMULATING FACTOR FROM WALKER CARCINOMA

Following early reports on changes in the vascular system in response to parotid extracts (15), a highly active endothelium stimulating factor, ESF, was clearly demonstrated to exist in extract of bovine parotid or mouse sub-mandibular glands (13). This ESF produced a response in the renal endothelium (13) that was exactly like that produced in animals by crude tumor angiogenic factor TAF prepared from Walker carcinoma cells, i.e. it caused endothelial cell hypertrophy and neovascularisation. However, we could find no evidence of increased endothelial cell mitotic index at foci where there was an obvious expansion of the capillary bed.

Since Walker cell extracts were claimed (4) to induce neovascularisation and in addition an increase in the mitotic index of endothelium it seemed likely that Walker carcinoma cells might contain both an endothelium stimulating factor responsible for neovascularisation as well as a mitogen causing proliferation of endothelial cells. Using the renal assay we showed that Walker cells contained an ESF of very low MW (210) that induced neovascularisation in vivo, but did not promote endothelial cell proliferation in vivo or in vitro (16). This low MW ESF was clearly separable from a high MW mitogen that did act on cultured endothelial cells and we concluded that ESF at least contributes to those effects ascribed to crude TAF.

ENDOTHELIUM STIMULATING FACTOR FROM BALB/c3T3 CELLS

Balb/c3T3 cells have angiogenic activity (7). It was claimed (3) that Balb/c3T3 cells produce a mitogen with some specificity for cultured bovine aortal endothelial cells since avascular (corneal) endothelial cells were not stimulated. Despite the lack of evidence that TAF is an endothelial cell mitogen (4,16) it was suggested that the 3T3 cell mitogen might be the angiogenic factor. Using the renal assay we have isolated from Balb/c3T3 cells an ESF activity that in all properties (MW, heat stability, chemical properties) appears to be similar or identical to that isolated from Walker carcinoma (18). ESF from Balb/c3T3 did not stimulate proliferation of cultured aortal endothelial cells. It could be separated from a mitogen in 3T3 cell extracts that did. The 3T3 cell mitogen is a MW 120,000 polypeptide that acts on both aortal and corneal bovine endothelial cells.

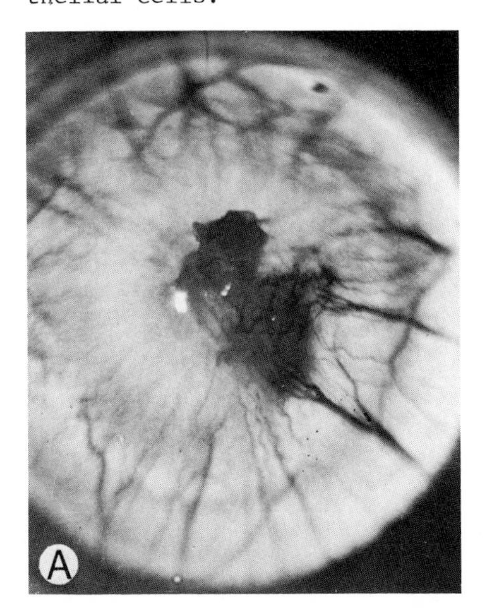

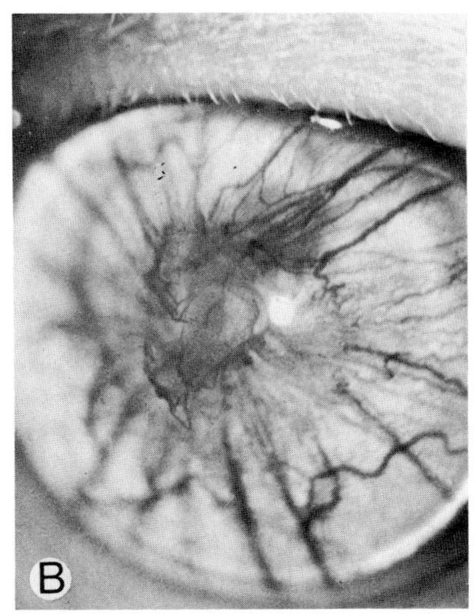

Figure 1. Angiogenic activity of crude 3T3 cell extract or of an ESF-active fraction. Cell fractions were incorporated into sterile polymer. Fragments of this were inserted into the anterior eye chamber of rats and viewed by means of an opthalmic slit lamp 10 days later. A. Polymer plus crude extract of 3T3 cells, B. Polymer plus ESF (M.W. 210), Vascularisation of the pellet was found only with ESF-positive fractions. Magnification 14X. (18).

Could the Balb/c3T3 cell derived low MW ESF activity account for the angiogenic factor activity of such cells? To decide this, we assayed ESF by the procedure commonly used to assay TAF. ESF active material (MW 3000 or 210) from 3T3 cells was incorporated into a copolymer of ethylene-vinyl acetate and fragments of polymer inserted into the anterior eye chamber of rats. As shown in Fig. 1 ESF caused vessels to grow out from the iris and invade the polymer head. No response was produced by polymer alone or by polymer containing non-ESF active fractions. Preheating ESF to 340°C. did not inactivate its activity in the renal test or the pellet vascularisation test. (18). Since angiogenic activity (pellet vascularisation assay) and ESF activity (renal assay) are found in fractions which have the same molecular weight, the same thermo-stability and are sufficiently similar in chemical properties to copurify, there are no basic differences between ESF and TAF activities. Contrary to previous ideas (9,7) it is clear that a small heat stable molecule, non-mitogenic for cultured endothelial cells has angiogenic factor activity.

ISOLATION AND IDENTIFICATION OF ESF FROM BOVINE PAROTID GLANDS

Although preliminary studies on bovine ESF showed that activity occurred as at least two MW species (86,000 and 3000), we found (17) that when crude or partially purified parotid extract was fractionated by preparative isoelectric focusing, all ESF activity was detected in only a single region corresponding to an isoelectric point of 7.8. The MW of the activity is about 3000 and this can be further dissociated to a large inactive carrier and a low MW (210) ESF activity that, as in the case of Balb/c3T3 cells or Walker carcinoma cells, is remarkably heat stable. On the premise that ESF activity was due to an inorganic substance we submitted the heated (340°C 60') MW210 ESF-active material to semi-quantitative analysis by spark source mass spectrometry. Apart from those common elements expected in biological material only two trace elements were present at high concentration. These were zinc (2000 ppm) and copper (8000 ppm).

When sterile aqueous solutions (over the range of 10^{-5} to 10^{-4} M) of simple salts of Cu^{2}, Zn^{2}, Mg^{2} Mn^{2}, Ni^{2}, Cr^{2}, Co^{2}, Mo^{2}, Fe^{3}, or Al^{3} were used to inject mice Cu^{2} salts alone induced renal neovascularisation just as crude bovine parotid extract or low MW ESF prepared from all sources did (17). We conclude from this and related experiments that ESF activity of crude bovine extract or the MW210 material isolated from it is due to copper ions and that the purification data is consistent with the idea that copper is bound ionically to a carrier molecule of MW approximately 3000, which in turn can form high molecular weight complexes.

ANGIOGENIC ACTIVITY OF Cu^{2} IONS

The most convincing way of demonstrating the angiogenic activity of Cu^{2} salts is probably the pellet vascularisation assay. When such assays were conducted only those pellets containing either crude or purified bovine ESF or salts of Cu^{2} became vascularised (14) as demonstrated in Fig. 2. Salts of the related transition elements Ni or Zn or of Cr, Co, Mo, Mn, Mg and Al were inactive when tested by a range of concentration from 10^{-5} to 5×10^{-4}M. Therefore bovine ESF activity and angiogenic factor activity are indistinguishable by this test and can be ascribed to copper ions.

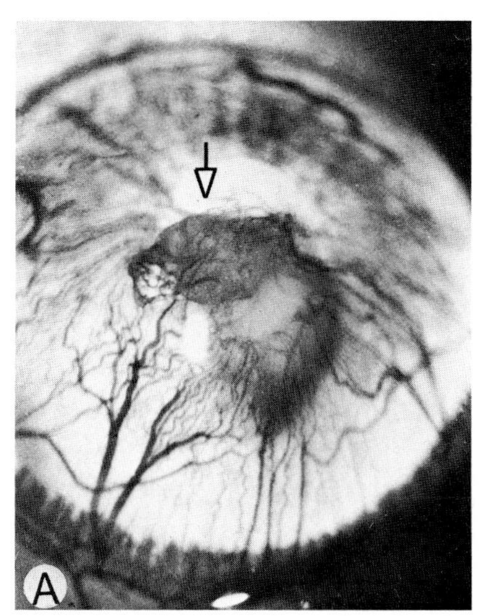

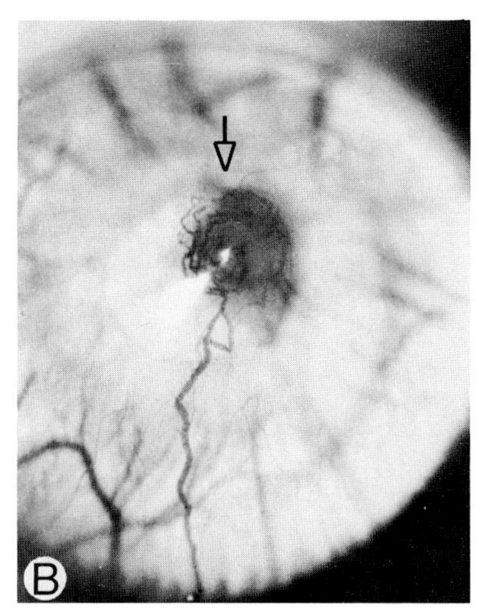

Figure 2. Angiogenic activity of bovine ESF or of Copper salt. Aqueous solutions of substances to be tested were incorporated into ethylene-vinyl acetate copolymer (10). Sterile fragments of polymer containing test material were inserted into the anterior eye chamber of anaesthetised rats and viewed by means of an opthalmic slit lamp 10-14 days later. A. Polymer plus bovine ESF (M.W.210) B. Polymer plus $CuSO_4$ (A.R.) final concentration 10^{-4}M. Arrows point to the polymer fragments. Magnification 14X. No vascularisation of pellet was found in controls which consisted of polymer plus physiological saline (17).

THE RESPONSE OF CULTURED ENDOTHELIAL CELLS TO COPPER IONS

Copper ions elicit neovascularisation (17). Since the primary response to angiogenic factor appears to be a chemotactic attraction of endothelial cells (2) and if copper ion is the active principle of angiogenic factor, then cultured endothelial cells should be stimulated to migrate in response to copper ions. This is exactly what we found (20). Cell motility is readily demonstrated by the Albrecht-Buehler technique (1) in which phagokinetic tracks are formed by cells migrating across a surface of colloidal gold. A line of bovine aortal endothelial cells,BAE, has been isolated and tested. Cultured BAE cells have almost negligible motility in normal growth medium. However, when Cu^{2} or Cu^{1} salt (optimal final concentration 2×10^{-6}M) is added to medium, cells become highly motile, producing tracks of the order of 500μm or more in 48 hours. (Fig. 3). A variety of salts of Zn, Cr, Co, Mo, Fe, Mn or Al had no detectable effect on motility. Thus the high specificity for copper ions for cell motility is entirely in accord with our studies on the specificity of the angiogenic response to copper ions.

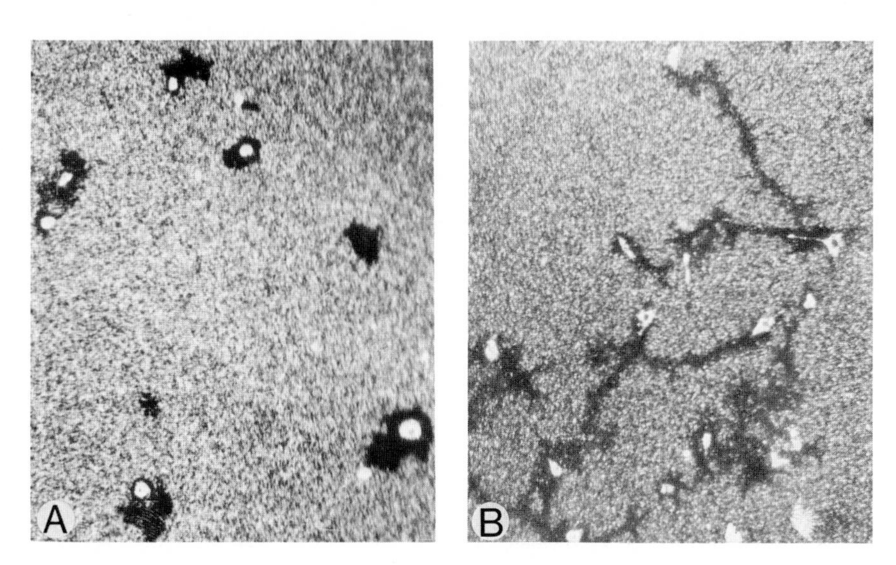

Figure 3. Endothelial Cell motility in response to Copper ions. Cultured bovine aortal endothelial cells, BAE, were seeded onto a gold coated surface so that one could visualise the phagokinetic tracks produced by motile cells (1). A. BAE cells, control, B. BAE cells, medium plus 10^{-6}M $CuSO_4$. All cultures were photographed 72 hours after addition of $CuSO_4$ (20).

A NEW ASSAY FOR ANGIOGENIC ACTIVITY

A recent review (12) discussed the possibility of using the angiogenic response of biopsy material as a tool to identify women at high risk of breast cancer but pointed out that the high cost and qualitative nature of current assays have limited clinical utilisation. If, as it now seems, copper is the active moiety, methods such as atomic absorption spectroscopy might seem to be the logical choice for detection of angiogenic potential. However, the intensity of the angiogenic response is influenced by the nature of the copper complex and the type of assay used. For example, the higher molecular weight species of ESF such as the 3000 MW complex, generally elicit a more marked response than simple copper salts; some naturally occurring copper complexes give no angiogenic response or at best a marginal effect. The eye chamber assay detects small amounts of low MW angiogenic factor such as $CuSO_4$ with greater reliability than obtained with the renal tissue assay. It is unlikely then, that quantitation of angiogenic activity could be achieved by these procedures either.

I suggest that endothelial cell motility provides the basis of a rapid, convenient assay of angiogenic activity that could be made quantitative. We have found that cultured cell mobilising activity and neovascularising activity coincide for MW fractions 80-100,000, 3000 and 210 from Walker carcinoma extracts (20). Fractions that were not active did not enhance motility. This observation applies also to sub-fractions from bovine parotid gland and Balb/c3T3 cells and to a wide range of synthetic copper complexes (20).

A NEW THEORY OF NEOVASCULARISATION

Given that angiogenic activity is not confined to tumors but found in a number of normal tissues (12) it seems likely that there could be a mechanism to induce neovascularisation that in its general aspects is common to carcinomas and to normal tissue at such times as vascularisation of the corpus luteum and the uterus during the reproductive cycle. On the basis of our data so far (17,20) I propose that for neovascularisation to occur (i) normal tissue acquires a mechanism for increasing angiogenic copper in its vicinity. This might occur by induction of a system for concentrating or modifying some copper complex at the site to be vascularised. In carcinomas this must be constitutive as a consequence of neoplastic transformation. (ii) The naturally occurring angiogenic precursor, say a Cu-carrier-protein of MW 80,000 or greater is converted to Cu-carrier (MW 3000 ESF?) which is more angiogenically active. (iii) The Cu-carrier presents Cu^{++} or Cu^{+} ions to the endothelial cell which is thus mobilised and exhibits positive tropism for a copper gradient. (iv) Subsequently endothelial cell mitosis occurs in response to specific mitogens well in arrear of the lead cells (2) followed by recruitment of other cell types to form a nascent capillary vessel.

At this stage of development of the theory it is important to show a naturally occurring copper complex that would serve as a source of angiogenic copper. One likely candidate for this role is ceruloplasmin, a cuproprotein (8 moles Cu/mole protein) occurring naturally in high amounts in serum. Ceruloplasmin Type III causes a marked increase in the motility of cultured aortal endothelial cells (Fig. 4) and from it we have isolated a fragment that not only enhances cell motility in vitro but that also induces angiogenesis in vivo.

ACKNOWLEDGEMENTS: The experimental work referred to was conducted in collaboration with several colleagues to whom I express my appreciation of their efforts and skills. In particular I thank Dr. Hoffman for his collaboration on the isolation work and for the tedious task of conducting all the vascularisation assays.

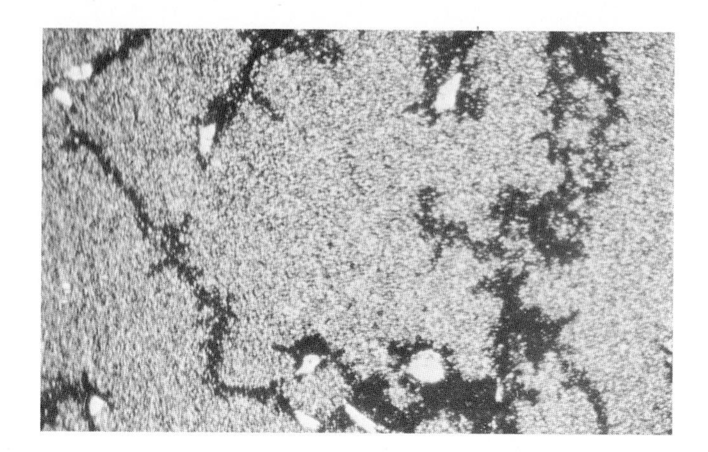

Figure 4. Endothelial cell motility in response to ceruloplasmin. The phagokinetic tracks shown were produced within 72 hours by BAE cells migrating in response to human ceruloplasmin type III added to the medium (30 ng ceruloplasmin/ml) (20).

REFERENCES

1. Albrecht-Buehler,G. (1977): Cell, 11: 395-404
2. Ausprunk,D.H. and Folkman,H. (1977): Microvasc.Res. 14:53-65
3. Birdwell,C.R., Gospodarowicz,D. and Nicholson,G.L. (1977): Nature 268: 528-531
4. Cavallo,T., Sade,R., Folkman,J and Cotran,R.S. (1972) J.Cell. Biol. 54: 408-420
5. Fenselau,A., Mello, R.J. (1976): Cancer Res. 36: 3267-3273
6. Folkman,J. (1974): Adv. Cancer Res. 19: 331-358
7. Folkman,J. (1976): IN: Cancer Vol.3,edited by R.L. Becker p.355. Plenum Press, N.Y.
8. Folkman,J., Cole,P., Zimmerman,S (1966): Ann. Surg. 164: 491-502
9. Folkman,J., Merler,E., Abernathy, C., Williams, G. (1971): J.Exp. Med. 133: 275-288
10. Gimbrone,M.A., Leapman,S.B., Cotran,R.S., Folkman,J. (1972): J. Exp. Med. 136: 261-276
11. Goldman,E. (1907): Lancet 2: 1236-1240
12. Gullino,P.M. (1978): J. Natl. Cancer Inst. 61: 639-643
13. Hoffman,H., McAuslan,B.R., Robertson,D, Burnett, E. (1976): Exp.Cell Res. 102: 269-275
14. Ide,A.G., Baker,N.H., Warren, S.L. (1939): A.J.R. 42: 891-899
15. Ito,Y. (1960): Ann. N.Y. Acad. Sci. 85: 228-310
16. McAuslan,B.R., Hoffman,H. (1979): Exp.Cell Res. 119, 181-190
17. McAuslan,B.R.,Hoffman,H.,Hamilton,E.(1979) unpublished
18. McAuslan,B.R.,Hoffman,H.,Hannan,G.N.(1979) unpublished
19. McAuslan,B.R.,Reilly,W. (1979): J.Cell.Physiol.(in press)
20. McAuslan,B.R.,Reilly,W. (1979). unpublished

Control Mechanisms in Animal Cells,
edited by L. Jimenez de Asua et al.
Raven Press, New York © 1980.

Estromedins: Uterine-Derived Growth Factors for Estrogen-Responsive Tumor Cells

David A. Sirbasku

Department of Biochemistry and Molecular Biology, University of Texas Medical School at Houston, Houston, Texas 77030, U.S.A.

The mechanism by which estrogens promote the growth of estrogen-responsive tumors remains the subject of extensive study. My laboratory has been examining the possibility that estrogens may promote the in vivo growth of some types of hormone-responsive tumors via an indirect or mediated mechanism. Our studies began with the development of permanent cell lines from three well known estrogen-responsive rodent tumors or cell populations. These lines were: the rat pituitary tumor GH3/C14 cell line, obtained as a secondary subclone (15) of the original GH3 cells established by Tasjian et al (17); the H-301 line (14) established from the estrogen-induced and estrogen-dependent Syrian hamster kidney tumor (7); and the MTW9/PL cell population (12), developed from the estrogen-responsive MTW9A rat mammary tumor (5).

With all three of these cell lines, my colleagues and I have found an unexpected yet reproducible result; estrogens are required for optimal growth in vivo, but have no direct effect on the rate of cell division in culture under a wide variety of experimental conditions (3,6,14). While these results could well be interpreted to mean that the lines have lost estrogen responsiveness in culture, in at least one case (GH3/C14 cells) the estrogen-receptor system was present, and shown to have properties comparable to normal tissue in vivo (16). These results indicated that a functional estrogen-receptor system is not sufficient to explain a growth response in a target cell. These observations, as well as others previously cited (13), have led to our proposal of a new mechanism which may be involved in hormone-responsive cell growth in vivo.

Beginning evidence for an indirect mechanism was obtained when we identified estrogen-inducible growth factor activities in extracts of rat uteri (11) and kidneys (13), and showed that these activities were relatively more potent mitogens for cells that formed hormone-responsive tumors than for normal cells or other types of tumor cells (11). Other preliminary studies (13) showed that the uterine activity had polypeptide properties by various criteria.

Nevertheless, the identification of a growth activity in uterus and kidney does not necessarily confirm their role *in vivo*; the demonstration of their presence in the general circulation is critical. With these observations in mind, and others presented below, we have proposed the mechanism of estrogen-induced mammary tumor cell growth presented in Figure 1.

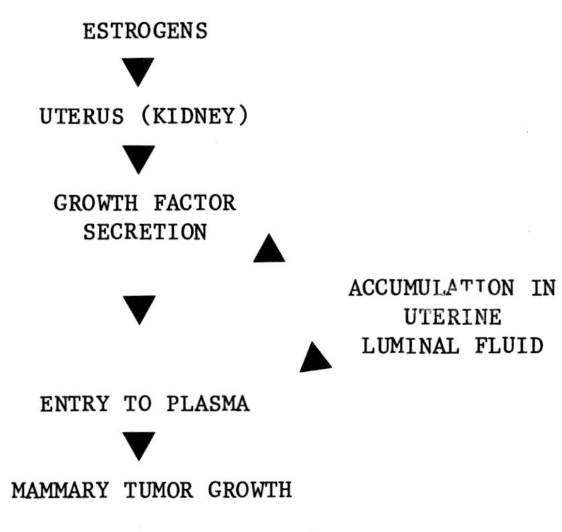

FIGURE 1. Estromedin mechanism for estrogen-responsive growth of mammary and pituitary tumor cells *in vivo*.

If the basic hypotheses is correct, one proof would be to demonstrate that the growth factor activity present in uterine extracts (11) is capable of reaching extracellular fluids. The most direct route for this activity release would probably be secretion directly into plasma, although we have considered other possibilities. It is well known (8,10) that estrogen treatment causes the accumulation of a considerable volume of luminal fluid in the uterine horns, and that this fluid contains estrogen and progresterone induced proteins that may be able to reach the general circulation (1,2,4,9). Hence, demonstration of similar activities in extracts of uterine tissue, estrogen-accumulated uterine luminal fluid, and/or plasma would be important. Beyond these observations, the demonstration that a similar growth factor could be extracted from mammary tumor cells growing in vivo (but not in vitro) would give further support to the estromedin mechanism (Figure 1). Such tumor associated activity would presumably be due to accumulation of the growth activity in the target tissue. We have begun attempts to identify activities in these tissues and fluids, and initially have selected four criteria for demonstrating similarity among the activities. These critieria are: (1) chromatography characteristics on carboxymethyl sephadex (CMS), (2) chromatography characteristics on Affi-gel blue resin, (3) molecular weight estimation by G-100 sephadex (or Biogel P-100), and (4) precipitation of the growth activity by sheep or goat antisera raised to purified rat serum albumin (RSA). A summary of our results is presented in Table 1.

One of the primary goals of our present studies is to purify and chemically characterize the uterine derived mammary and pituitary estromedin activity. During the initial course of these studies, we found that 80% or more of the activity in the crude uterine extracts would pass through a CMS column equilibrated with a low ionic strength buffer at pH 7.0, although 40% of the protein remained associated with the resin. The flow through fraction from CMS was then applied to an Affi-gel blue column, which was previously shown to remove serum albumin (18,19) from tissue extracts. Under the conditions employed (see the footnotes of Table 1) the flow through fraction contained 70% or more of the protein, by no growth activity. When the serum albumin fraction was eluted from Affi-gel with a single 1.4m NaCl step, the growth activity was recovered in this fraction. We had, by this time, determined that the molecular weight of the uterine activity was approximately the same as RSA or slightly larger, and that the isoelectric point of the growth activity was near that of albumin at 4.8 to 5.0. However, extensive testing of Cohn Fraction V purified RSA alone showed no mitogenic activity. These data all suggested that the growth factor activity may be a molecule which is associated with the residual RSA in the tissue extracts and that Cohn fractionation methods remove this activity. We had previously determined (D.A. Sirbasku, unpublished results) that the amount of albumin in the initial uterine extracts was approximately 5% of the total protein. At this point the albumin was precipitated from the uterine extract (prepared as in reference 11) by addition of sheep antiserum raised to highly purified rat serum albumin. The specificity of the antiserum was estimated by both the Ouchterlony double diffusion method, and immunoelectrophoresis. The antiserum formed precipitating antibodies with only one component in rat serum, rat uterine extracts, rat uterine luminal fluid, and MTW9/PL tumor extracts and the component was the same in all cases. With this knowledge in hand, we began to determine whether activities showing these similar chemical and immunological properties could be identified in rat uterine

TABLE 1. A Summary of the Properties of Mammary Growth Factor Activity(s) Derived From Uterus, Uterine Luminal Fluid, Plasma (Serum), and Mammary Tumor Extract.

	Origin of Extract or Fluid			
	Uterus	Luminal Fluid	Plasma (serum)	MTW9/PL Mammary Tumor
1. Carboxymethyl[a] Sephadex chromatography	flow through active	flow through active	flow through active	flow through active
2. Affi-gel blue[b] Chromatography	1.4M NaCl eluted active	1.4M NaCl eluted active	N.D.[e]	1.4M NaCl eluted active[c]
3. Molecular Weight estimate	70 to 80,000	N.D.	70 to 80,000	50 to 80,000
4. Precipitation by Antiserum raised to rat serum albumin[d]	80-100% precipitated	100% precipitated	N.D.	100% precipitated

[a]The carboxymethyl sephadex chromatography was run at pH 7.2 with 10mM sodium phosphate running buffer. The flow through fraction was that part which passed through in the running buffer without additional salt.

[b]Affi-gel blue was purchased from Biorad Laboratories and the running buffer was 10mM sodium phosphate pH 7.4. The activity in most cases was eluted with a single step of 1.4m NaCl added to running buffer.

[c]Activity separates equally between the flow through and the protein eluted with a single 1.4M NaCl step.

[d]Precipitation of the growth factor activities with sheep anti-RSA was carried out by standard immunological methods using increasing concentrations of antiserum to the final point of complete activity precipitaton after 18 hrs. at 4°C.

[e]N.D. indicates not determined.

luminal fluid, female rat serum or plasma, and in extracts of MTW9/PL rat mammary tumors growing in vivo. The rat uterine luminal fluid was collected as described before (13), and the extract of mammary tumors were prepared by the identical method used previously to prepare the uterine extract (11). The data in Table 1 show that estrogen induced accumulations of luminal fluid have a mammary cell mitogen(s) with properties similar if not identical to those in the uterine tissue extracts. Female rat plasma or serum contains a similar activity as judged by two criteria, although to date the Affi-gel blue chromatography has not been conducted. The attempt to precipitate the growth factor activity from plasma by addtion of antiserum to RSA has not been effective, since very considerable amounts of antiserum were required to precipitate the approximately 50% albumin present in serum. Under conditions employed to date, the added amount of antiserum begins to show very significant growth promotion alone.

Finally, the extracts of the mammary tumors growing in vivo did show a very potent growth factor activity for MTW9/PL cells in culture (see Table II). This activity has been separated into two forms, one form which flows through the Affi-gel column in running buffer, and a second form which binds to Affi-gel and is eluted with 1.4m NaCl. The preliminary data with antibody, however, suggest that both of these forms may be precipitable by sheep anti-RSA. The meaning of these two forms remains to be resolved. Nevertheless, the data from the carboxymethyl sephadex chromatography and the molecular weight estimation suggest the activity is similar to that of the uterus. In another series of experiments (Eastment and Sirbasku, manuscript in preparation) we were unable to identify mammary growth activity from the MTW9/PL cells in culture, further suggesting that this activity associated with the tumors may be the result of accumulations in vivo.

TABLE II. Growth of MTW9/PL Rat Mammary Tumor Cells In Culture In Response to Extracts of MTW9/PL Tumors

Amount of Tumor Extract added μg/ml	Number of MTW9/PL cell population doublings in six days
0	0
5	0.32
10	1.18
50	2.76
100	2.78
500	2.98

In summary, the data available this far suggest to us that the growth activity we have designated an "estromedin" is present in appropriate tissues and fluids in the tumor bearing animal, and that because of the similar biochemical and immunological properties, we believe these factors are the same, or very similar. These data further support the estromedin hypothesis, and provide a framework for future studies.

The author wishes to thank his colleagues, Mrs. Frances E. Leland, Dr. Robert H. Benson, and Dr. Caroline T. Eastment for their enthusiastic participation in these studies. This work was supported by an American Cancer Society grant BC-255A, and a National Cancer Institute grant 1-R01-CA26617.

1. Armstrong, D.T. (1968): Am. J. Physiol. 214:764-771.
2. Clemetson, C.A.B., Verma, U.L., and DeCarlo, S.J. (1977): J. Reprod. Fert. 49:183-187.
3. Eastment, C.T. and Sirbasku, D.A. (1978): J. Cell. Physiol. 97:17-28.
4. Kennedy, T.G. and Armstrong, D.T. (1975): Endocrinology 97:1379-1385.
5. Kim, U. and Depowski, M.J. (1975): Cancer Res., 35:2068-2077.
6. Kirkland, W.L., Sorrentino, J.M. and Sirbasku, D.A. (1976): J. Natl. Cancer Inst. 56:1159-1164.
7. Kirkman, H. (1959): Natl. Cancer Inst. Monogr., No. 1, pp. 1-58.
8. Meglioli, G. (1976): J. Reprod. Fert. 46:395-399.
9. Meglioli, G., Krakenbuhl, C. and Desaulles, P.A. (1969): Experientia 25:194-195.
10. Shih, H.E., Kennedy, J., and Huggins, C. (1940) Am. J. Physiol. 130:287-291.
11. Sirbasku, D.A. (1978): Proc. Natl. Acad. Sci. USA, 75:3786-3790.
12. Sirbasku, D.A. (1978): Cancer Res., 38:1154-1165.
13. Sirbasku, D.A. and Benson, R.H. (1979): In: Cold Spring Harbor Conferences on Cell Proliferation, Hormones and Cell Culture, Vol. 6, pp. 477-497.
14. Sirbasku, D.A. and Kirkland, W.L. (1976): Endocrinology, 98:1260-1272.
15. Sorrentino, J.M., Kirkland, W.L. and Sirbasku, D.A. (1976): J. Natl. Cancer Inst., 5:1149-1153.
16. Stancel, G.M., Heindel, J.J. and Sirbasku, D.A. (1979): Endocrinology (in press).
17. Tashjian, A.H., Yasamura, Y., Levine, L., Sato, G.H. and Parker, M.L. (1968): Endocrinology, 82:342-352.
18. Travis, J. and Pannell, R. (1973): Clin. Chim. Acta 49:49-52.
19. Willie, L.E. (1976): Clin. Chim. Acta 71:355-357.

Control Mechanisms in Animal Cells,
edited by L. Jimenez de Asua et al.
Raven Press, New York © 1980.

A Pituitary-Derived Growth Factor for Rat Mammary Tumor Cells: Phosphoethanolamine

Tamiko Kano-Sueoka and Janice E. Errick

Department of Molecular, Cellular and Developmental Biology, University of Colorado, Boulder, Colorado 80309, U.S.A.

INTRODUCTION

We have been studying growth and differentiation characteristics in culture of a clonal cell line, 64-24, which was established from a hormone-dependent rat mammary carcinoma MCCLX (3, 4, 5, 6). The MCCLX tumor grows optimally in animals supplemented with estradiol pellets and also can grow in successively pregnant animals. Pituitary glands of estrogen-supplemented animals are always enlarged. Hypophysectomy and subsequent prolactin replacement of the tumor-bearing animals indicated that the tumor requires some product of the pituitary gland, prolactin possibly being the major one (2).

The 64-24 cells proliferate well in the presence of fetal calf serum, but do not proliferate or do so only slowly in the presence of calf serum. Since in vivo the tumor cells require a hyperfunctioning pituitary for rapid growth, various pituitary hormones, including prolactin, were tested for growth-stimulatory effects on 64-24 cells when added to the calf serum-supplemented medium. None of the hormones gave a significant stimulation. Therefore, crude pituitary extract was examined for the possible presence of a growth-stimulatory material. Indeed, we discovered that the crude bovine pituitary extract contains the mitogen activity of 64-24 cells (4). Subsequently, this pituitary-derived growth factor for the mammary cells (MGF) was purified to homogeneity and the chemical structure of this molecule was identified as being phosphoethanolamine (5, 6). The biological activity of phosphoethanolamine has also been indicated.

PURIFICATION OF MGF

The entire purification procedure of MGF is summarized in Table 1. These procedures were developed according to the properties of MGF, such as small size, relative stability against heat and acid, and ninhydrin positiveness. The schematic

TABLE 1. Purification steps of MGF

Steps	Location of biological activity
1. Grinding of frozen bovine pituitaries in cold 0.075 N acetic acid	--
2. Centrifugations	Supernatant
3. A series of AMICON Diaflo ultrafiltration	PM10 filtrate: <10,000 daltons
4. Preparative isoelectric focusing (pH 3 to 5 or 6)	pH 3.8 to 4.0 fractions
5. Sephadex G25 column, eluted with distilled H_2O	$V_{0.65-0.7}$ fractions
6. Sephadex G15 column, eluted with distilled H_2O	$V_{0.4-0.5}$ fractions
7. High voltage paper electrophoresis, 4,000 volt, 150 mA, 1 hr, pH 1.9	5 to 5.5 cm toward the cathode from the origin

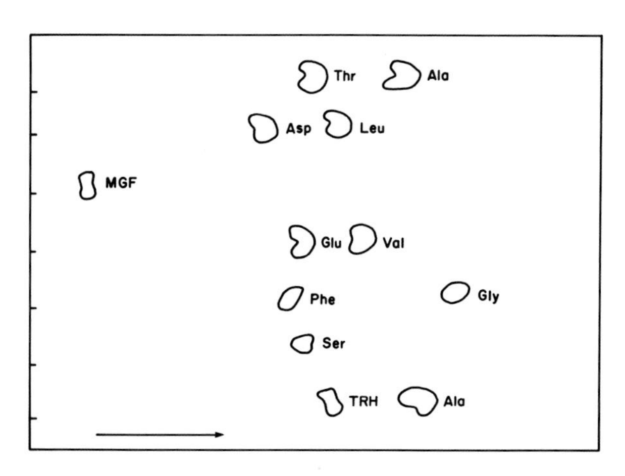

FIG. 1. High voltage electrophoresis of MGF. Electrophoresis was carried out as described in Table 1. The paper was sprayed with ninhydrin reagent to detect ninhydrin-positive spots. Standard amino acids and TRH (thyrotropin-releasing hormone) were electrophoresed as references (from 6).

presentation of the result of the final purification step, high voltage electrophoresis, is given in Fig. 1. Only the area where the ninhydrin-positive spot, designated as MGF, is located contains the biologically active material. The amount of MGF thus isolated was estimated by ninhydrin colorimetric assay (9) and expressed as the number of ninhydrin-positive residues. The final yield was generally 10 to 12 μmol/100 g wet weight pituitaries.

STRUCTURAL IDENTIFICATION OF MGF

The structural identification of MGF was carried out by using an automatic amino acid analyzer and two-dimensional chromatography following dansylation of intact and hydrolyzed MGF (5). Fig. 2A shows an elution profile of a purified sample (15 nmol) from the amino acid analyzer. A sharp ninhydrin-positive peak appeared soon after the elution began, and the rest of the elution profile was virtually free of ninhydrin-positive material except for ammonia and a minor peak eluting just before ammonia. Fig. 2B shows the elution profile of a standard amino acid mixture (10 nmol each) plus phosphoethanolamine (20 nmol). Comparison of Fig. 2A and 2B indicates that the sample peak eluted virtually at the same time as phosphoethanolamine did. If MGF is phosphoethanolamine, treatment with phosphatase or acid should be able to hydrolyze the molecule, yielding ethanolamine and orthophosphate. Therefore, a sample of MGF preparation was subjected to alkaline phosphatase digestion and examined by the amino acid analyzer (Fig. 1C). The ninhydrin peak corresponding to phosphoethanolamine now disappeared almost completely; instead a new peak appeared just before ammonia. As we predicted, this newly

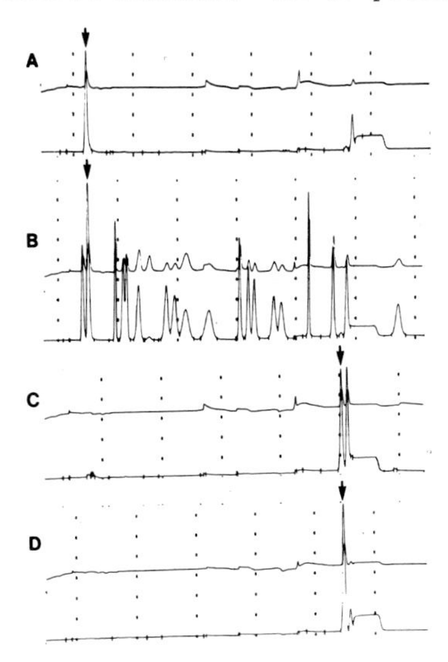

FIG. 2. Analysis of intact and phosphatase-digested MGF on amino acid analyzer. A. Elution profile of intact sample. The lower profile represents the absorption at A_{590} nm and the upper profile at A_{440} nm. The arrow indicates the elution position of MGF. B. Elution profile of standard amino acid and phosphoethanolamine. The arrow indicates the peak of phosphoethanolamine. C. Elution profile of alkaline phosphatase-digested MGF. The arrow shows the elution position of the digested product. The second prominent peak is ammonia. D. Elution profile of ethanolamine. The arrow indicates the elution position of ethanolamine (from 5).

generated peak coincides with that of ethanolamine, as indicated in Fig. 2D.

The dansyl chloride technique was also used for the identification of the intact, phosphatase-digested and acid-digested samples. As in the case of the analysis with the amino acid analyzer, the intact or the digested sample exhibited a single dansylated spot on two-dimension chromatograms (Fig. 3A, 3B). As expected, the intact sample migrated to the same position as dansylated phosphoethanolamine (Fig. 3C, spot 1) and phosphatase-digested sample migrated with ethanolamine (Fig. 3C, spot 2). The acid hydrolysis (6 N HCl, 105°C for 18 hr) gave exactly the same result as that of phosphatase digestion.

BIOLOGICAL ACTIVITY OF PHOSPHOETHANOLAMINE

The results presented above indicate strongly that MGF is phosphoethanolamine. Further proof was provided by showing the biological activity of phosphoethanolamine. Using 64-24 cells, the dosage response curves of MGF and phosphoethanolamine were compared. The results are presented in Fig. 4. Phosphoethanolamine indeed has mitogenic activity and the dosage response curves of phosphoethanolamine and MGF are virtually the same at lower concentration range. However, above 5 nmol/ml, phosphoethanolamine does not seem to exert further growth stimulatory effects, and similar results were obtained when different lots of commercially available phosphoethanolamine were tested. The

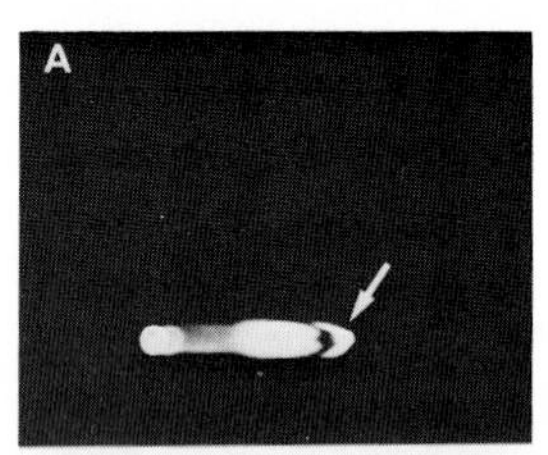

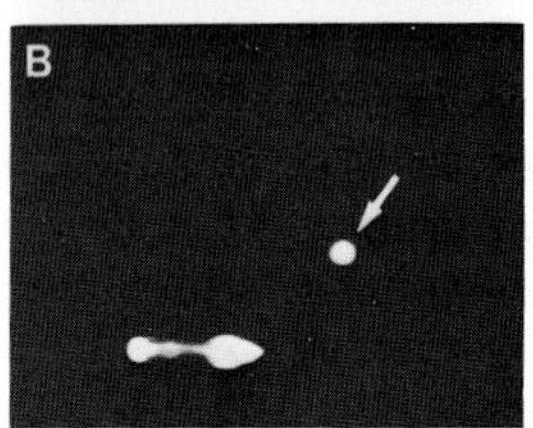

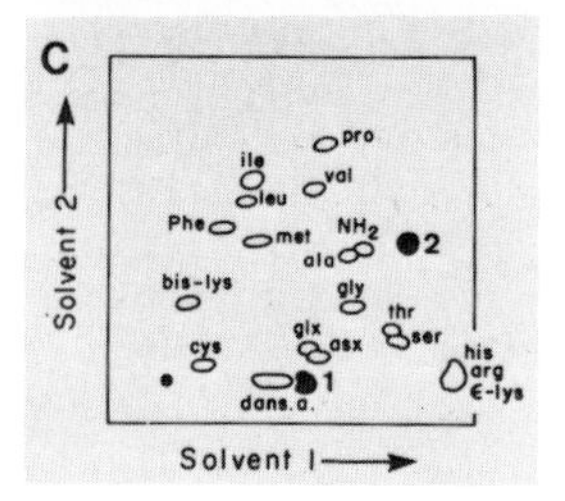

FIG. 3. Chromatograms of dansylated-intact and -hydrolyzed MGF. A. Chromatogram of dansylated-intact MGF. The arrow indicates the dansylated MGF, and a large streaked spot next to the MGF spot is free dansylic acid. B. Chromatogram of phosphatase-digested dansylated MGF. C. Chromatogram tracing of dansylated products of standard amino acids, phosphoethanolamine (1), and ethanolamine (2) (from 5).

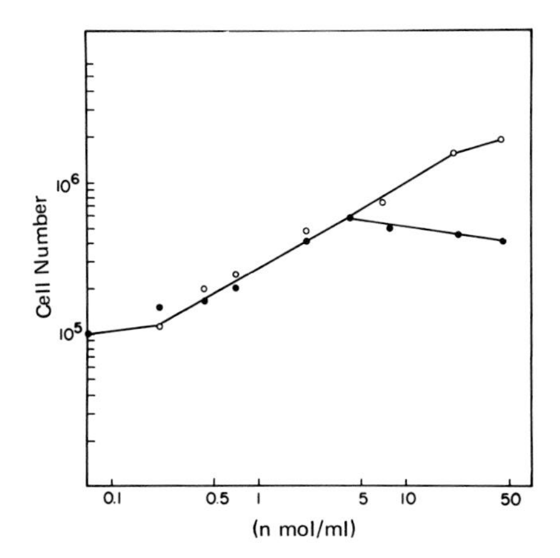

FIG. 4. Dosage response curves of MGF and phosphoethanolamine. 3×10^4 cells/plate (60 x 10 mm) of 64-24 cells were plated in a calf serum (5%) supplemented medium with increasing amounts of MGF or phosphoethanolamine. The cells were counted six days after plating. Each point represents an average of the number of cells from three plates. (•), phosphoethanolamine; (o) MGF.

discrepancy between the two at higher concentrations may be due to the fact that phosphoethanolamine preparations contain some contaminants having a growth inhibitory effect or that the MGF preparation contains something other than phosphoethanolamine which, at higher concentrations, further stimulates the growth of the cells. This problem remains to be clarified.

DISCUSSION

Many agents that are known to stimulate the proliferation of animal cells in culture are either polypeptides or steroid hormones. Two recently discovered growth factors specific for mammary epithelial cells seem also to be polypeptides (10, 11). The results presented here show convincingly that MGF is phosphoethanolamine. This is the first case describing the mitogenic activity of phosphoethanolamine or any related compound.

Phosphoethanolamine is present in literally every tissue and organ in animals (8) and is a precursor for phosphatidylethanolamine, which is one of the major components of membrane lipids. The metabolic effect of phosphoethanolamine, however, is virtually unknown. Since phosphoethanolamine is known to be particularly rich in nervous tissue (1), it is reasonable that we found pituitaries to be a good source. However, it is not known whether pituitary glands contain more phosphoethanolamine than, e.g., cerebrum, or whether the synthesis of phosphoethanolamine is hormonally controlled.

An important question to which we have no definite answer yet is whether phosphoethanolamine acts as a hormone or a nutrient.

The dose-response experiment suggests that it exerts the effect more like a hormone than a nutrient, since less than 1 nmol/ml in a culture medium can be biologically active. The required concentration of an ordinary nutrient is usually on the order of 10^3 to 1 nmol/ml.

Another unanswered problem is functional specificity of phosphoethanolamine with regard to the kind of cells and tissues that respond. Hormone-dependent mammary carcinoma cell lines other than 64-24 line, autonomous mammary carcinoma cell lines, normal mammary epithelial cells, and tissues other than mammary epithelium are to be tested to define the spectra of cells that respond to this mitogen. *In vivo* effects of phosphoethanolamine on a given tissue, particularly on mammary gland, will also be examined. So far, we have explored this question only to a limited extent. Several established cell lines, rat hepatoma cells; mouse 3T3 cells; a rat fibroblast line; a rat mammary tumor line, WRK-1; and a hormone-independent variant of 64-24 cells were tested, and none of them showed a significant response. On the other hand, normal mammary epithelial cells seem to respond to phosphoethanolamine. Although the results are still preliminary, phosphoethanolamine stimulated the incorporation of ^{3}H-thymidine into primary cultured cells of mid-pregnant rat mammary epithelium (Table 2). The stimulation was found only when a group of hormones are added in addition to phosphoethanolamine in the medium. Interestingly, when the effect of phosphoethanolamine was tested in a serum-deprived medium, the growth of 64-24 cells was also stimulated only when phosphoethanolamine was present in the medium along with prolactin, insulin, and hydrocortisone. The mechanism of action of phosphoethanolamine is an obviously important subject which is being pursued with the use of radiolabeled compounds.

Table 2. Incorporation of ^{3}H-thymidine in primary cultures of mid-pregnant rat mammary gland[a]

Supplements	cpm/mg protein	percent of control
None	5,346	100
5H[b]	22,463	420
5H + 1 ng/ml PEA[c]	24,426	456
5H + 10 ng/ml PEA	30,202	565
5H + 100 ng/ml PEA	32,141	601
5H + 1 μg/ml PEA	39,243	634

[a]Collagenase-digested glands (2 mg/ml in 199 + 5% FCS for 24 hr at 37°C) were plated in 199 + 5% FCS. On second day, medium was changed to 199 + 1% FCS and various supplements were added. After 24 hr, ^{3}H-thymidine incorporation into TCA-precipitable material during a 2-hr pulse was determined. Protein was measured using the method of Lowry (7).

[b]Insulin (5 μg/ml); hydrocortisone (1 μg/ml); prolactin (5 μg/ml); transferrin (5 μg/ml); T_3 (10^{-10} M).

[c]PEA: phosphoethanolamine.

REFERENCES

1. Altman, P.L., and Dittmer, D.S., editors (1973): Biology Data Book. Vol. II. Fed. Am. Soc. Exp. Biol., Bethesda, Maryland, p. 1211-1226.
2. Horn, T., and Kano-Sueoka, T. (1979): Cancer Res. (in press).
3. Kano-Sueoka, T., and Hsieh, P. (1973): Proc. Natl. Acad. Sci. USA, 70:1922-1926.
4. Kano-Sueoka, T., Campbell, G.R., and Gerber, M. (1977): J. Cell Physiol., 93:417-424.
5. Kano-Sueoka, T., Cohen, D.M., Yamaizumi, Z., Nishimura, S., Mori, M., and Fujiki, H. (1979): Proc. Natl. Acad. Sci. USA, (in press).
6. Kano-Sueoka, T., Errick, J.E., and Cohen, D.M. (1979): Cold Spring Harbor Conferences on Cell Proliferation, 6:499-512.
7. Lowry, O.H., Rosebrough, N.J., Farr, A.L., and Randall, R.J. (1951): J. Biol. Chem., 193:265-275.
8. Meister, A. (1965): Biochemistry of the Amino Acids. Vol. I. Academic Press, New York, London, p. 111.
9. Moore, S., and Stein, W.H. (1948): J. Biol. Chem., 176:367-388.
10. Ptashne, K., Hsueh, H.W., and Stockdale, F.E. (1979): Biochemistry, 18:3533-3539.
11. Sirbasku, D.A. (1978): Proc. Natl. Acad. Sci. USA, 75:3786-3790.

Control Mechanisms in Animal Cells,
edited by L. Jimenez de Asua et al.
Raven Press, New York © 1980.

Coordinate Expression of Erythroid Markers in Differentiating Friend Erythroleukaemic Cells

Jean-François Conscience and Werner Meier

Friedrich Miescher-Institute, CH-4002 Basel, Switzerland

We have been studying the expression of several erythroid marker enzymes in Friend erythroleukaemic cells (FLC). These cells are widely used as an *in vitro* model of murine erythropoiesis (for reviews, see ref. 13 and 17). Upon treatment with a variety of different chemicals, these permanent cell lines are induced to undergo changes that resemble those occurring during normal erythroid differentiation. In doing so, the cells cease dividing (12). Systematic investigations of a number of different erythroid markers have allowed a more complete characterization of the system and provided answers to the following questions: (1) Where are the uninduced FLC blocked along the erythroid pathway of differentiation? (2) How far do they progress toward a more advanced erythroid state, when submitted to induction? (3) How well do they reproduce *in vitro* the coordinate sequence of events occurring *in vivo*? This information is essential to estimate the relevance and the limitations of Friend cells, when used as a model of erythropoiesis.

Our studies included the following markers: Hemoglobin (Hb), carbonic anhydrase (CA), acetylcholinesterase (AChE), lactate dehydrogenase (LDH), 6-phosphogluconate dehydrogenase (6PGD), glucose-6-phosphate dehydrogenase (G6PD) and catalase (CTL). Most of the results reviewed here have been published *in extenso* elsewhere (3,4,6).

For each marker, we asked the following questions:

1) Is it a *marker of erythroid differentiation* in the FLC system?
2) What is the *constitutive level* of expression in different uninduced FLC clones and in non-erythroid cells?
3) What is the *induced level* of expression reached in different FLC clones after treatment with an inducer, and how does it compare with the levels expressed in erythrocytes?

4) What is the time course of the changes that take place during induction and how does it compare with the *in vivo* situation?

To answer these questions, we used a set of carefully chosen cell types (3,5), as summarized below:

1) Eleven FLC clones, each originating from one of the three major families of FLC (clones 745, FSD-1 and SFST-3) and differing in their degree of response to inducers.
2) Eight non-erythroid mouse cell lines, including the following epigenetic types:
   a) 2 L cell clones (transformed fibroblasts), A9 and B82
   b) 1 Hepatoma cell line expressing liver-specific functions (Hepa 1a)
   c) 1 Thymoma cell line (136.5)
   d) 1 Ig-producing myeloma cell line (RPC 5.4)
   e) 1 Bone marrow-derived attached cell line producing Friend virus (IsCl3)
   f) 1 Spleen-derived attached cell line producing Friend virus (GB 239)
   g) 1 Transformed embryo fibroblast cell line producing only the helper of the Friend virus complex (Eveline).
3) Forty clones of somatic hybrids each derived from one of the following three independent fusions: FLC x A9, FLC x B82 and FLC x Hepa 1a.
4) Erythrocytes obtained from normal DBA/2 mice.

The cell lines were kept under strictly controlled growth conditions, harvested in exponential growth phase, and extracted with the detergent NP-40 and sonicated (3). Induction was usually carried out with DMSO (1.3 to 2%), but sometimes also with 1 mM butyrate, 0.1 mM hemin, or the combination of hemin and DMSO. In the cell lines tested, we failed to find a difference between the effects of these inducers on any of the markers studied, except that DMSO, alone or in combination with hemin, was more potent than butyrate or hemin alone. The enzyme results were always expressed as activity per mg protein.

Hemoglobin

In vivo, Hb is a marker which appears late during erythropoiesis (7). As expected, non-erythroid cells are completely devoid of this marker. Uninduced Friend cells have a low constitutive level (19). This level, as well as a correspondingly low constitutive level of globin mRNA (2), is completely suppressed in the hybrid clones.

Upon induction, Hb increases sharply in some FLC clones, after a 2- to 3-day lag, to reach, after 6 to 7 days, levels which correspond to about 30 to 40% of the erythrocyte levels. This increase occurs only in a limited number of so-called inducible clones. Other clones do not respond (non-inducible) or respond by producing less Hb (moderately inducible).

Carbonic anhydrase

The validity of including CA among the erythroid markers of FLC differentiation is established by two findings. Firstly, like Hb, CA is suppressed in all the hybrid clones tested. Secondly, it increases during induction in all the FLC clones which are inducible for Hb. *In vivo*, CA activity increases progressively throughout erythropoiesis and is already elevated in cells that are just initiating Hb synthesis (7). In all FLC tested, CA is present at high constitutive levels in the uninduced cells. In non-erythroid cells, little or no activity is detected. Upon induction, CA activity increases 3- to 4-fold to reach about 60% of the erythrocyte level. The increase can be first observed 2-3 days after induction and only in those clones which are inducible for Hb.

Acetylcholinesterase

No data are available about the time course of AChE production during normal erythropoiesis. It is known that the enzyme is present in high amounts, bound to the erythrocyte membrane (10). Furthermore, erythrocyte AChE is a "true" AChE similar to the form present in nervous tissues and different from the numerous serum pseudocholinesterases, with respect to substrate specificity and sensitivity toward various inhibitors. In all uninduced FLC tested, we find a high level of a "true" AChE, while non-erythroid cells have no detectable activity or an activity which cannot be identified with a "true" AChE. Furthermore, in the hybrid clones, AChE is largely suppressed, like Hb and CA. Thus, AChE is a valid marker of FLC differentiation. Upon induction, however, the behaviour of AChE differs from that of Hb and CA. Firstly, it is increased not only in all clones which are inducible for Hb and CA, but also in many that are not. Secondly, the time course of the increase in AChE activity is different from that of Hb and CA: It can already be observed one day after addition of the inducer. Finally, the levels of AChE activity reached in the fully induced FLC are higher than the erythrocyte levels. This fact is probably attributable to rapid post-synthetic degradation of the enzyme in vivo (14). Thus, in FLC, AChE presents a pattern of expression which is very similar to the one described for spectrin, another erythrocyte membrane protein (8).

Lactate dehydrogenase

In vivo, during erythropoiesis, overall LDH activity decreases to reach a very low level in erythrocytes (20). Most of the decrease occurs before the time of the final cell division. In addition, in most mouse strains, the LDH B isozyme disappears completely, leaving only LDH A activity in the erythrocytes (21). Using isozyme electrophoresis on cellulose acetate gels, we have shown that all uninduced Friend cells tested have only LDH A activity. This activity, assayed spectrophotometrically, is similar in uninduced FLC and in non-erythroid cells, as well as in the hybrid clones. Upon induction, LDH A activity decreases in all FLC clones which are inducible for Hb and CA, and eventually reaches a level close to that of the erythrocyte. The decrease in activity also occurs with a 2-day lag, like the responses of Hb and CA.

6-Phosphogluconate dehydrogenase

6PGD activity decreases sharply, in vivo, at the time of the final cell division (7). In Friend cells, spectrophotometric assays have shown that the activity of 6PGD is similar in the FLC clones, non-erythroid cell lines and hybrid clones tested. Upon induction, most FLC clones maintain a constant level of 6PGD activity. Only one highly inducible clone, derived from clone 745, shows a large drop in activity, late during the induction period (day 5). Levels are reached which are similar to those found in erythrocytes.

Glucose-6-phosphate dehydrogenase

In vivo, a sharp decrease in G6PD activity takes place at the time of the final cell division (7). Uninduced FLC clones, non-erythroid cell lines and the hybrid clones share similar levels of G6PD activity. Treatment with an inducer does not affect G6PD activity which remains higher than in erythrocytes.

Catalase

CTL is a very late marker of erythropoiesis which, *in vivo*, does not increase before the reticulocyte stage (7). Uninduced FLC have levels of CTL activity which are indistinguishable from those found in non-erythroid cells and in the hybrid clones. No increase in activity is observed during induction and the levels remain lower than those found in erythrocytes.

CONCLUSIONS

The constitutive levels of Hb, CA and AChE, compared to those found in non-erythroid cells, are compatible with the idea that uninduced FLC maintain a rather advanced state of erythroid differentiation. Compared to the variations that occur *in vivo*, these constitutive levels allow one to place the FLC at a stage which precedes the final division of the erythroblast, just at the time Hb production is initiated. This model is supported by the fact that FLC also maintain low constitutive levels of spectrin (8). The model implies that the changes taking place during induction, like the late changes *in vivo*, are mostly quantitative, affecting the level of expression of the various markers, rather than the switching on or off of the corresponding genes. Recent molecular data confirm this point (18). Significant and persistent differences are found between the FLC clones studied, when the levels of expression of a given marker are compared prior to induction. However, these differences are usually small and they do not fit the kind of pattern that would be expected, if they were to reflect different degrees of erythroid maturation in the various clones. On the basis of our results, we conclude that all the clones studied, regardless of their origin, are blocked at essentially the same stage of erythroid differentiation.

Upon induction, several regulatory programs are turned on sequentially, each one involving a given set of markers. An early program, activated soon after the addition of the inducer, includes not only AChE, as described here, but also such membrane markers as spectrin (8), H2 antigen levels (1), lectin agglutinability (9), glucose transport (11), and transferrin receptor levels (15). A non-histone chromosomal protein (IP-25) is induced at the same time (16). A second program is turned on later and includes Hb, CA and LDH. Still later, a program involving a decrease in 6PGD activity is activated. _In vivo_, very late programs accompany the final maturation steps of the erythrocyte and include changes in CLT and G6PD activities. The various FLC clones vary in the extent to which they are able to activate these programs. The early program is turned on in almost all the clones studied, as shown by the increase in AChE activity. The program involving Ca, Hb and LDH is activated only in the so-called inducible clones. 6PGD decrease is observed in only one inducible clone and, finally, the programs involving CLT and G6PD are never turned on under our induction conditions. This last finding is in good agreement with the fact that we do not observe enucleation _in vitro_. Thus, the section of the erythroid pathway of differentiation which is reflected in the FLC during induction varies considerably in length between the various clones. In inducible clones, the coordinate expression of the various markers and their grouping into distinct programs is in good agreement with the _in vivo_ situation. This fact speaks for a high degree of similarity between FLC differentiation and normal erythropoiesis. Thus, FLC represent a system which can be used as a model to study erythroid differentiation.

The applicability of the system, however, is limited to the late stages of erythropoiesis, since uninduced FLC are blocked at an advanced stage and do not undergo the final maturation steps during induction. This is an intrinsic limitation of the FLC system which makes it unsuitable for the study of early events such as the commitment of the hemopoietic stem cell or the primary activation of erythroid genes. This limitation could only be overcome by the development of transformed erythroid cell lines blocked at earlier stages of differentiation.

REFERENCES

1. Arndt-Jovin, D.J., Ostertag, W., Eisen, H., Klimek, F., and Jovin, T.M. (1976): J. Histochem. Cytochem., 24:332.
2. Conkie, D., Affara, N., Harrison, P.R., Paul, J., and Jones, K. (1974): J. Cell Biol., 63:414.
3. Conscience, J.-F., and Meier, W. (1979): Exptl. Cell Res., (in press).
4. Conscience, J.-F., Miller, R.A., Henry, J., and Ruddle, F.H. (1977): Exptl. Cell Res., 105:401.
5. Conscience, J.-F., Ruddle, F.H., Skoultchi, A., and Darlington, G.J. (1977): Somatic Cell Genet., 3:157.
6. Davis, J.M., Conscience, J.-F., Deslex, S., Fischer, F., and Meier, W. (1980): Erythropoiesis and Differentiation in Friend Leukemic Cells, (G.B. Rossi, ed.)Elsevier/North Holland, Amsterdam.
7. Denton, M.J., Spencer, N., and Arnstein, H.R.V. (1975): Biochem. J., 146:205.
8. Eisen, H., Bach, R., and Emery, R. (1977): Proc. Natl. Acad. Sci. USA, 74:3898.
9. Eisen, H., Nasi, S., Georgopoulos, L.P., Arndt-Jovin, D., and Ostertag, W. (1977): Cell, 10:689.
10. Galehr, P., Plattner, F. (1928): Pflügers Arch. ges. Physiol., 218:488.
11. Germinario, R.J., Kleiman, C., Peters, S., and Oliveira, M. (1977): Exptl. Cell Res., 110:375.
12. Gusella, J., Geller, R., Clarke, B., Weeks, U., and Housman, D. (1976): Cell, 9:221.
13. Harrison, P.R. (1977): International Review of Biochemistry, Biochemistry of Cell Differentiation, II, 15:227.
14. Herz, F., and Kaplan, E. (1974): Brit. J. Haematol., 26:165.
15. Hu, H.-Y.Y., Gardner, J., Aisen, P., and Skoultchi, A.I. (1977): Science, 197:559.
16. Keppel, F., Allet, B., and Eisen, H. (1977): Proc. Natl. Acad. Sci. USA, 74:653.
17. Marks, P.A., and Rifkind, R.A. (1978): Ann. Rev. Biochem., 47:419.
18. Minty, A.J., Birnie, G.D., and Paul, J. (1978): Exptl. Cell Res., 115:1.
19. Reeves, R., and Cserjesi, P. (1979): Develpt. Biol., 69:576.
20. Setchenska, M.S., and Arnstein, H.R.V. (1978): Biochem. J., 170:193.
21. Shows, T.B., and Ruddle, F.H. (1968): Proc. Natl. Acad. Sci. USA, 61:574.

Control Mechanisms in Animal Cells,
edited by L. Jimenez de Asua et al.
Raven Press, New York © 1980.

Involvement of Calcium Ions in Neuroblastoma Neurite Extension

Rosemarie Hinnen and Denis Monard

Friedrich Miescher-Institute, CH-4002 Basel, Switzerland

INTRODUCTION

Cultured mouse neuroblastoma cells extend neurites when treated with serum-free media (15), agents which elevate intracellular cyclic AMP levels such as prostaglandin E_1 (2), phosphodiesterase inhibitors (9) and analogues of cyclic 3',5'-AMP (8). All these treatments are unphysiological, even at a low degree of morphological differentiation, and they all interfere with the growth rate of the cells. A macromolecular factor released from glial cells which induces morphological differentiation in neuroblastoma cells without interfering with their growth rate has been reported (6). The biochemical events leading to the formation of neurites, however, are not understood.

In the present study we present evidence for a possible role of calcium in the induction of neuroblastoma neurite extension by glial conditioned medium.

RESULTS AND DISCUSSION

Calcium Dependence of Neurite Outgrowth

Glia-induced morphological differentiation in neuroblastoma cells, clone NB_2A, was studied as a function of the calcium concentration in the medium. Figure 1 illustrates the correlation between neurite outgrowth in the presence or absence of glial conditioned medium

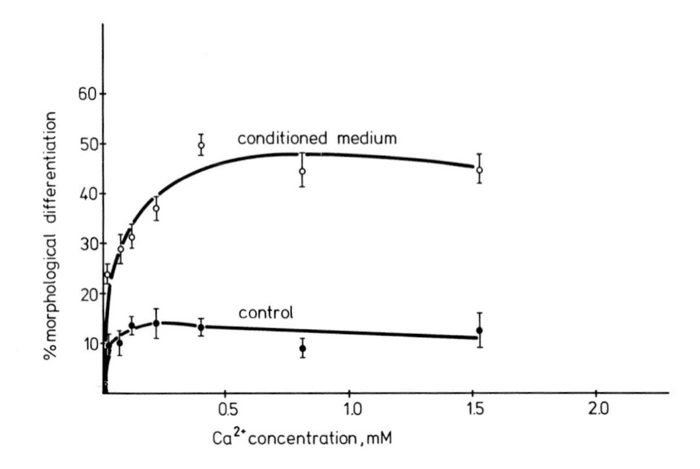

FIG. 1. Morphological differentiation of neuroblastoma cells as a function of the calcium concentration in the medium. Culture dishes (35 mm) were inoculated with 35,000 EDTA-dislodged mouse neuroblastoma cells, clone NB_2A, and incubated at 37°C in Dulbecco's modified Eagle's medium (DMEM) containing 10% fetal calf serum (FCS) (North American Biologicals Inc.). The medium was replaced 16-18 h later by DMEM containing 0.8% FCS (14). The bioassay for morphological differentiation was carried out 22-24 hr later in 2 ml DMEM containing different calcium-EGTA ratios. The free calcium concentration in the medium was determined with a calcium-sensitive electrode, Philips IS-Ca. The extent of morphological differentiation, expressed as the percentage of cells with neurites, was determined after 4 hr incubation at 37°C in the presence or absence of 10 μl serum-free conditioned medium (5) from C6 glioma cells (about 25 times concentrated and dialysed against 10 mM phosphate, pH 7.4). Points and error bars represent mean and standard deviation from triplicate values.

and the presence of calcium in the medium. Experimental details are described in the legends to the figures. Although adherence of the cells to the dish, a prerequisite to neurite extension, was dependent upon concentrations of calcium higher than 10 μM, optimal process formation required at least 300 μM calcium in the presence of glial conditioned medium. In the absence of glial conditioned medium calcium had very little effect on neurite outgrowth. Magnesium could replace calcium only at high concentrations (10 mM) but did not require the presence of glial factor(s) or calcium for the extension of neurites.

The action of calcium can be explained in at least two different ways. Calcium could either be needed for the interaction of the glial factor(s) with the cell surface or have an intracellular regulatory function involving neurite formation.

The latter possibility was studied using $^{45}Ca^{2+}$ to monitor calcium fluxes of neuroblastoma cells in the presence and absence of glial conditioned medium. No significant difference of $^{45}Ca^{2+}$ influx or efflux was found upon addition of glial factor(s). It is possible that small changes were difficult to detect due to limitations in the sensitivity of the technique. Plastic Petri dishes of different brands (Corning, Falcon, Nunc) bound large amounts of calcium which masked most of the $^{45}Ca^{2+}$ associated with the cells. This problem was partially overcome by using glass coverslips which bound much less calcium than plastic dishes. An additional problem was caused by the low cell densities required for morphological differentiation. Despite these technical problems, a one- to two-fold difference in $^{45}Ca^{2+}$ fluxes would have been significant. Our inability to detect any effect of glial factor(s) may infer that the $^{45}Ca^{2+}$ fluxes are either marginally or not at all affected by glial conditioned medium.

Effects of A23187 on Neurite Extension

The above results with $^{45}Ca^{2+}$ are insufficient to infer whether or not translocation of calcium across the membrane is necessary to effect neurite extension. Therefore, the ionophore A23187, known to act as a carrier for calcium (12), was tested on neuroblastoma process formation. Low concentrations of A23187 (0.02 μM) led efficiently to neurite extension, mimicking the action of conditioned medium (Fig. 2). The combination of 0.02 μM ionophore and conditioned medium resulted in an additional stimulation of neurite outgrowth over conditions with conditioned medium alone. At higher concentrations, the ionophore inhibited the neurite extension induced by glial factor(s) by either toxicity to the cells or a change in the internal calcium concentration which could interfere with tubulin

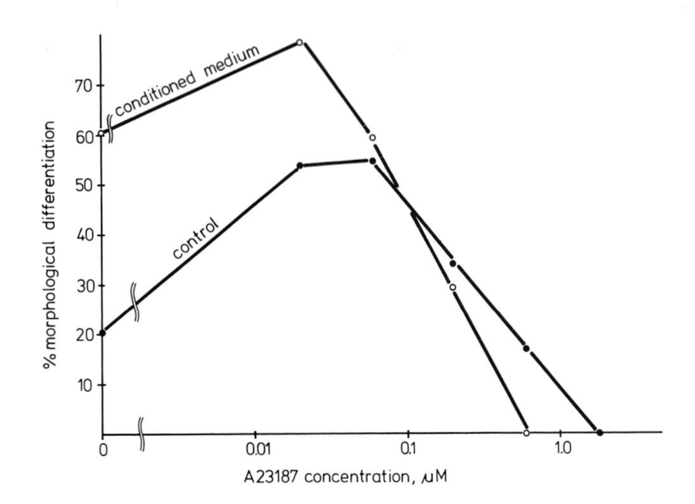

FIG. 2. Morphological differentiation as a function of A23187 concentration. The bioassay was carried out as described in Fig. 1, except that the external calcium concentration was kept constant at 1.8 mM and the ionophore concentration was varied. Additions of the ionophore were made in ethanol, the final concentration of ethanol not exceeding 0.2% v/v. Each point is the mean of duplicate values.

polymerisation (1,16). Moreover, the effect of 0.02 μM ionophore was dependent upon the presence of calcium in the medium. At 10 μM calcium the ionophore has little effect, whereas maximal stimulation of neurite extension is obtained with 1.0 mM calcium in the medium. These results suggest an increased displacement of calcium by the ionophore, implying that translocation of calcium plays a role in process formation.

Fluxes of $^{45}Ca^{2+}$ were monitored also in the presence of A23187. Cells were allowed to accumulate $^{45}Ca^{2+}$ for 30 minutes prior to addition of ionophore (Fig. 3). Both concentrations of ionophore caused a fast release of $^{45}Ca^{2+}$ followed by reuptake. More $^{45}Ca^{2+}$ was transported into the cells in the presence of 0.2 μM A23187 than 0.02 μM. These data suggest that the calcium concentration in the cytoplasm or some other cellular compartment has a regulatory function in the control of neurite extension.

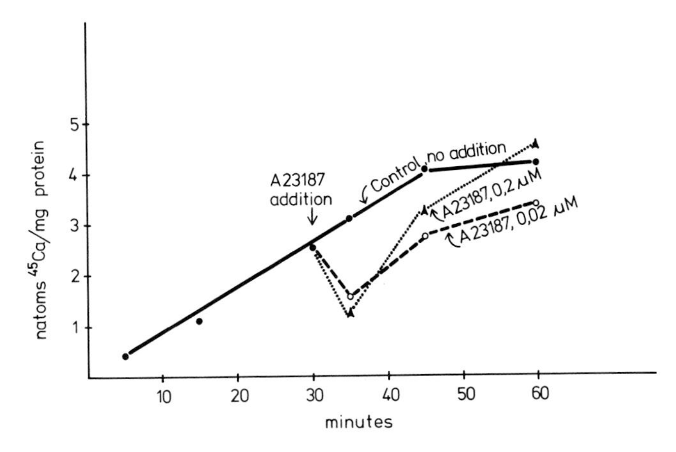

FIG. 3. Uptake of $^{45}Ca^{2+}$ in the presence of different A23187 concentrations. For $^{45}Ca^{2+}$ fluxes cells were seeded in DMEM containing 10% FCS at 400,000 cells per 35 mm dish. Each dish contained a 15 mm round glass coverslip. Cells were grown for 16-18 hr at 37°C. Prior to the assay, the cells were washed three times by dipping the coverslips into buffer containing 110 mM NaCl, 5 mM KCl, 0.8 mM Na phosphate, 0.5 mM $MgSO_4$, 10 mM glucose, 1.8 mM $CaCl_2$ and 40 mM Hepes, pH 7.4, 37°C. The assay was performed at 37°C in the same buffer but containing 1-3 Ci/mole $^{45}Ca^{2+}$. Cells were allowed to accumulate $^{45}Ca^{2+}$ for 30 minutes prior to addition of ionophore. The incubation was further continued and terminated at different times by washing the cells in ice cold buffer of the same composition without isotope. Coverslips with and without cells were counted for radioactivity in the presence of Bray's solution in a Mark II liquid scintillation counter. Each time point is the mean of the values of two coverslips.

Addition of glial conditioned medium instead of the ionophore in a similar experiment as is described for Fig. 3 did not lead to increased mobilization of $^{45}Ca^{2+}$. Glial factor(s) might, therefore, act via a different mechanism(s) than the ionophore. It (they) might trigger release and accumulation of calcium from internal stored or membrane-binding sites to which the isotope does not have ready access. Schubert et al. (13) argue that nerve growth factor (NGF) serves in the mobilization of calcium which leads to increased adhesion and neurite extension in PC12 cells. In their system, NGF and cyclic AMP stimulate $^{45}Ca^{2+}$ efflux slightly but are ineffective in $^{45}Ca^{2+}$ influx.

The ionophore A23187 produces a rapid release of K^+ concomitant with Ca^{2+} uptake in red blood cells (11). Our preliminary results indicated a small but consistent increase in the permeability of the plasma membrane to $^{86}Rb^+$ in the presence of either 0.02 μM A23187 or glial conditioned medium. Higher concentrations of the ionophore led to larger permeability changes. Although a change in Rb^+ permeability is an indirect way of assaying the involvement of calcium in neurite extension, nevertheless, it is compatible with a transient change in the cytoplasmic calcium concentration.

Effects of Membrane Potential Changes on Process Formation

Agents that depolarize the resting membrane potential of erythrocytes (3) were tested for their effect on neurite outgrowth in neuroblastoma cells. Valinomycin, a highly specific ionophore for K^+ (10), led to process formation in the presence of increasing K^+ concentrations (Table 1). The effect of valinomycin was potentiated by increasing the concentration of K^+ in the medium. The combination of valinomycin, K^+ and conditioned medium produced even higher cell numbers with neurites. Ouabain, also known to depolarize the membrane potential (7), stimulated morphological differentiation (Table 2).

Our results with K^+, valinomycin, and ouabain are in contrast to those obtained by Koike (4). Valinomycin and increasing concentrations of K^+ inhibited process formation in his clone of mouse neuroblastoma (N18). Differences in the two cell lines and especially differences in the procedure to obtain responsive cells for neurite outgrowth might be among the reasons for the different results.

The action of glial factor can be imitated by either the calcium ionophore A23187 or by conditions known to cause depolarization of the membrane potential. These data suggest an interaction of glial factor with the membrane leading to changes in ion conductance and membrane potential. In agreement with these results, membrane depolarization has been reported to cause in-

TABLE 1. Effect of valinomycin and K^+ on neuroblastoma neurite extension

Additions	% Morphological differentiation Exp. 1	Exp. 2
Control in DMEM (5 mM K^+, 110 mM Na^+)	11.5	18.9
Conditioned medium	52.7	56.2
Valinomycin 0.3 μM	21.5	33.5
Valinomycin + conditioned medium	65.2	64.3
Control in 15 mM K^+, 100 mM Na^+	19.4	23.2
Valinomycin	29.3	45.7
Valinomycin + conditioned medium	73.7	67.9
Control in 25 mM K^+, 90 mM Na^+	15.6	
Valinomycin	45.4	
Valinomycin + conditioned medium	77.7	
Control in 50 mM K^+, 65 mM Na^+	17.1	17.6
Valinomycin	55.5	65.4
Valinomycin + conditioned medium	70.6	66.5

The bioassay was carried out as described in Fig. 1, except that the concentration of K^+ in DMEM was increased and the Na^+ concentration decreased accordingly. Valinomycin was added in ethanol, the final concentration of solvent not exceeding 0.2% v/v. The results of each experiment are given by the means of duplicate values.

increased adhesion and neurite extension as well as an increased influx of calcium in PC12 cells (13).

Further studies will center on the effect of glial factor(s) on ion conductance and membrane potential.

TABLE 2. Effect of ouabain on neuroblastoma neurite extension

Additions	% Morphological differentiation
Control in DMEM	22.2
Conditioned medium	73.0
Ouabain, 100 μM	27.5
Ouabain, 200 μM	47.0

The bioassay for morphological differentiation was carried out in DMEM. Ouabain was added in DMSO, the final concentration of solvent not exceeding 0.2% v/v. The result represents the mean of duplicate values.

SUMMARY

A dose-response relationship has been found between the morphological differentiation of neuroblastoma cells induced by glial factor(s) and the calcium concentration in the medium.

Low concentrations of the calcium ionophore A23187 mimic the action of glial conditioned medium implying that the mobilization of calcium is a significant event leading to the extension of neurites.

Furthermore, process formation is initiated by agents known to depolarize the membrane potential. The interaction of glial conditioned medium with the cellular membrane could, therefore, cause changes in ionic conductance and membrane potential, which could have an effect in neuroblastoma neurite extension.

ACKNOWLEDGEMENTS

The authors express their gratitude to Dr. Evelyn Niday for constructive criticism of the manuscript and to Mrs. Magda Rentsch for the preparation of the tissue cultures.

REFERENCES

1. Fuller, G.M., and Brinkley, B.R. (1976): J. Supramol. Struct., 5:497-514.
2. Gilman, A.G., and Nirenberg, M. (1971): Nature, 234:356-357.
3. Hoffman, J.F., and Laris, P.C. (1974): J. Physiol. Lond., 239:519-552.
4. Koike, T. (1978): Biochim. Biophys. Acta, 509:429-439.
5. Monard, D., Rentsch, M., Schürch-Rathgeb, Y., and Lindsay, R.M. (1977): Proc. Natl. Acad. Sci. USA, 74:3893-3897.
6. Monard, D., Solomon, F., Rentsch, M., and Gysin, R. (1973): Proc. Natl. Acad. Sci. USA, 70:1894-1897.
7. Post, R.L., Kume, S., Tobin, T., Orcutt, B., and Sen, A.K. (1969): J. Gen. Physiol., 54:306S-326S.
8. Prasad, K.N., and Hsie, A.W. (1971): Nature New Biol., 233:141-142.
9. Prasad, K.N., and Kumar, S. (1974): In Control of Proliferation in Animal Cells, edited by Clarkson, B. and Baserga, R., pp. 581-594. Cold Spring Harbor Laboratory.
10. Pressman, B.C. (1976): Ann. Rev. Biochem., 45:501-530.
11. Reed, P.W. (1976): J. Biol. Chem., 251:3489-3494.
12. Reed, P.W., and Lardy, H.A. (1972): J. Biol. Chem., 247:6970-6977.
13. Schubert, D., La Corbière, M., Whitlock, C., and Stallcup, W. (1978): Nature, 273:718-723.
14. Schürch-Rathgeb, Y., and Monard, D. (1978): Nature, 273:308-309.
15. Seeds, N.W., Gilman, A.G., Amano, T., and Nirenberg, M. (1970): Proc. Natl. Acad. Sci. USA, 66:160-167.
16. Weisenberg, R.C. (1972): Science, 177:1104-1105.

Control Mechanisms in Animal Cells,
edited by L. Jimenez de Asua et al.
Raven Press, New York © 1980.

Factors Stimulating Neurite Outgrowth in Chick Embryo Ganglia

C.-O. Jacobson, T. Ebendal, K.-O. Hedlund, and G. Norrgren

Department of Zoology, Uppsala University, S-751 22 Uppsala, Sweden

The by now well known protein nerve growth factor (NGF) plays an essential role in differentiation and maintenance of sympathetic and sensory neurons. The effects of NGF on developing neurons has been extensively reviewed by Levi-Montalcini (11). Its obvious neurotropic effect on sympathetic nerve fibres, experimentally demonstrated in chick embryos and neonatal rodents by Levi-Montalcini et al. (12), is a promising step towards an answer to the question whether tissues are able to stimulate their own innervation in a specific way. In this report we give evidence for the existence of at least one class of factors, functionally and immunologically distinct from mouse NGF, that stimulates the outgrowth of neurites in chick embryo ganglia. These results are in line with those of a number of other reports on neurite-stimulating factors different from NGF (8, 9, 13, 14). The ability of NGF and these other trophic factors to stimulate various types of ganglionic neurons will also be discussed.

BIOASSAY

The nerve-growth promoting factors of different origin have been tested on various ganglia from chick embryos (White Leghorn). Spinal, sympathetic and ciliary ganglia are used in most experiments, but stimulative influences on trigeminal, nodose and Remak's ganglia have also been investigated.

In vivo neurites from the ganglia in many instances grow towards their targets in association with connective-tissue type extracellular matrices, the framework of which is a three-dimensional collagen network. In order to mimic this situation as much as possible in vitro the explanted ganglia were placed within 0.5-0.8 mm thick collagen gels. In such a three-dimensional lattice chemical concentration gradients are protected from convection as they may well also be in extracellular matrices.

The amount of neurites extending from the ganglia can be approximately quantified by counting neurites, or neurite bundles, intersecting cross-hair lines positioned perpendicularly to a ganglion radius along which outgrowth occurs (2, 4).

STIMULATIVE FACTORS FOUND IN DIFFERENT TISSUES

Chick Embryo Tissues

A number of chick embryo tissues were tested for their abilities to stimulate nerve growth. Small pieces of heart (ventricle), colon, liver, kidney, skin, skeletal muscle, spinal cord (4), choroid and retina (unpublished) were cocultured with sensory, sympathetic and parasympathetic ganglia. Three tissue explants were either surrounding a ganglion or facing it from one or two sides, the distance between ganglion and tissue explants set to about 1 mm. In some experiments different tissues surrounded the ganglion. It was found that tissues from the various organs influenced the neurite outgrowth from the ganglia to different extents. For example, heart and kidney stimulated the sympathetic ganglia more than did skin or skeletal muscle, whereas colon, heart and liver stimulated Remak's ganglion more than did skeletal muscle. Pieces of spinal cord had no stimulatory effect on the fibre extension from any of the ganglia tested.

The finding that Remak's ganglion was stimulated by most of the chick tissues was of certain interest, since it consists of cholinergic neurons. In order to investigate whether also other cholinergic parasympathetic ganglia, not responsive to NGF (11), are stimulated by chick embryo tissues, we used the ciliary ganglion in subsequent experiments. It was found that all tissues tested but those from the central nervous system, the neural retina included, evoked at least some neurite outgrowth response in spinal and sympathetic ganglia, as well as in ciliary ganglia. The heart ventricle was consistently found to be a reliable tissue for eliciting heavy neurite extension (FIGS. 1 and 3). Moreover we found, like Adler et al. (1), that the choroid, including the pigment epithelium, of the eye of Day 12 to 17 chick embryos distinctly stimulates the ciliary ganglion. However, this tissue stimulates the sympathetic ganglia as well (FIG. 4) and thus the cholinergic neuronotrophic factor proposed by Adler et al. (1) may function also in sympathetic neurons or there may be several trophic factors released from the choroid.

Antiserum to mouse salivary gland NGF did not block the neurite stimulation by the chick tissue in any of the ganglia, nor did antiserum to snake NGF (5, 6). Immunological and functional dissimilarities between the factor released from the chick embryo tissues and NGF are thus established.

Age-dependency for trophic activity.

The ability of heart tissue to elicit outgrowth of ganglionic neurites at various stages was tested (2). It was found that all ganglia tested (sympathetic, Remak's, ciliary and spinal) responded with the most heavy outgrowth when the heart tissue were taken from Day-16 to -18 embryos. This timing is interesting considering the invasion of sympathetic nerves in the ventricle wall around Day 16 to 21 in vivo (10), and also with the acquisition of the full number of parasympathetic heart ganglia of Day 15 to 20 (16). On the other hand, the ciliary ganglion was found to be most responsive to heart tissue already on Day 6, while the other tested ganglia were most sensitive on Day 8. The period of maximal responsiveness to the heart seems to be well correlated with the period of initial neurite outgrowth from the ganglia during normal development.

Purification of the chick heart factor.

The neurite inducing factor could be extracted from lyophilized Day-18 chick embryo hearts (7). After fractionation by gel filtration on Sephadex G-200 the activity was localized mainly in a peak corresponding to a molecular weight of about 40,000 dalton. A minor activity peak in a fraction corresponding to a considerably higher molecular weight was also found. Rechromatography of the active fractions did not change the positions of the activity. The gel filtration resulted in an increase in the specific activity of about ten times and the yield from the extracted supernatant of roughly 60%. Anti-mouse NGF did not block the activity. The stability of the heart factor will be a good help in further purification and characterization of this trophic factor.

Mammalian Tissues

Calf tissues.

The brain and heart were tested for their ability to produce neuronotrophic factors with the intention to find a source that would give large amounts of material for purification. Pieces of fresh brain tissue (6) cultured together with ganglia, as well as extract of lyophilized brain tissue (unpublished results), elicited markedly dense halos of fibres in ciliary ganglia (FIG. 6), but evoked no responses in spinal or sympathetic ganglia. The stimulative ability was not inhibited by anti-NGF and was found to be most expressed by the medulla oblongata. The heart experiments gave almost identical results: strong effects on the ciliary ganglia, weak on sympathetic and spinal ganglia.

Rat and mouse tissues.

A most interesting result was obtained from experiments with hearts of neonatal rats and adult male mice (unpublished results). When explants of these tissues were cocultured with the ganglia neurite outgrowths that could be inhibited by anti-NGF were elicited in spinal and sympathetic ganglia. In ciliary ganglia no outgrowth was evoked. Homogenized tissue, on the other hand, as well as frozen and thawed rat heart pieces, caused a heavy outgrowth in the ciliary ganglia (FIG.5) but also a slight neurite outgrowth in sympathetic and spinal ganglia. Anti-NGF did not influence these results.

At least two different trophic factors may thus be produced by the heart cells. A tentative interpretation is as follows: The factor that foremost stimulates the ciliary ganglion is present in the heart when this is taken out from the embryo but is released to the culture medium only after freezing and thawing the tissue. NGF, on the other hand, was not found to be present initially in the heart tissue but was obviously synthetized by the living heart explants under culture conditions.

The mouse brain tissue has an effect on the ganglia that to some extent resembles that found for calf brain. The weak stimulation of spinal and sympathetic ganglia that is found, beside the ciliary stimulation, is not blocked by anti-NGF.

Mouse salivary gland NGF.

The stimulatory effect of NGF on ganglionic neurons derived from epidermal placodes was studied in chick trigeminal ganglia, the ventro-lateral (VL) neurons of which are of epidermal origin (3). It was clear that the VL neurons responded to NGF in culture by showing a higher survival rate and a more differentiated ultrastructure than they do in vivo even if the effects on the neural-crest derived population of trigeminal neurons were more striking. Cultures using the nodose ganglion were also performed (Hedlund and Ebendal, in manuscript). The neurons of this vagus ganglion are all derived from an epidermal placode. In the presence of

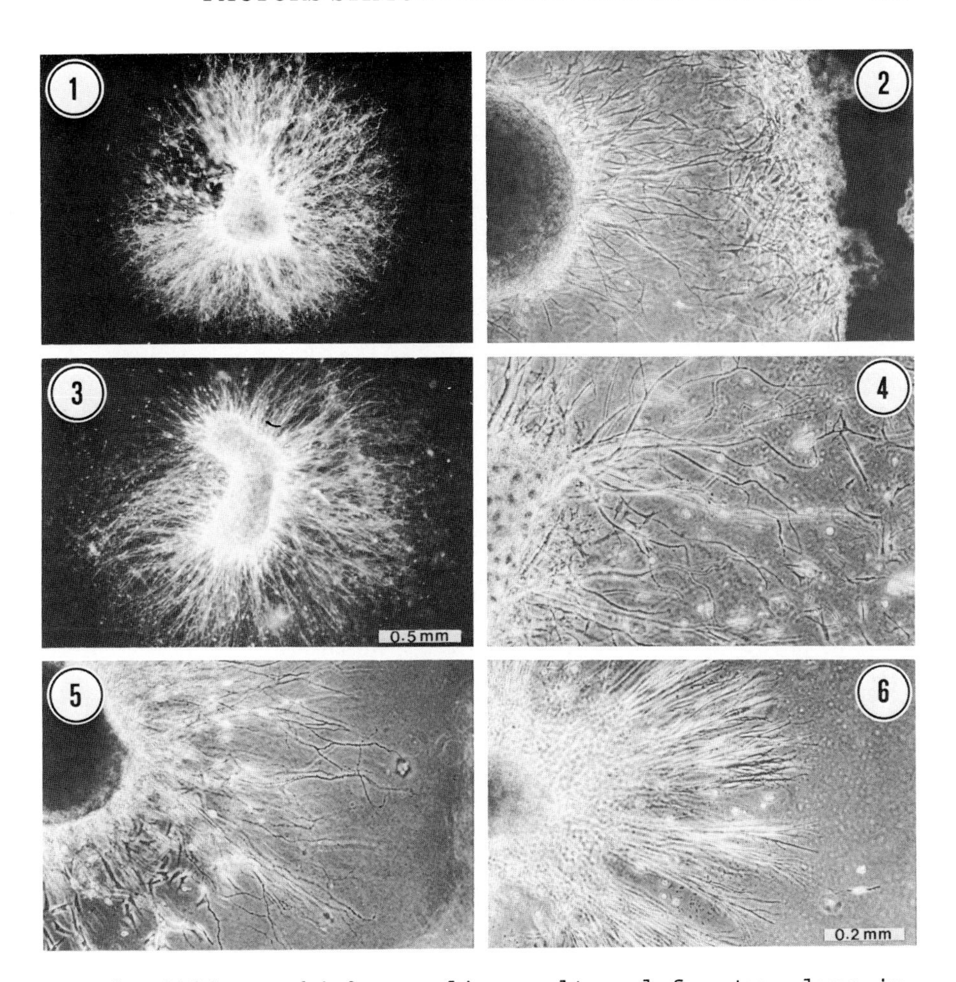

FIG.1. Ciliary chick ganglion cultured for two days in a collagen gel with explants of embryonic Day-18 chick hearts on both sides. A dense halo of neurites is elicited. Darkfield.

FIG.2. Ciliary ganglion (left) cultured with embryonic Day-12 chick eye choroid (right). A fairly dense outgrowth of neurites is evoked. Phase contrast.

FIG.3. Sympathetic ganglion as in FIG.1. A prominent halo of neurites is induced by the heart.

FIG.4. Sympathetic ganglion as in FIG.2. The choroid tissue evokes a modest neurite outgrowth.

FIG.5. Ciliary ganglion combined with frozen-thawed neonatal rat heart. A fairly dense halo of neurites is evoked by the non-living tissue.

FIG.6. Ciliary ganglion cultured with added extract of calf brain (about 0.2 mg of protein per ml). A dense outgrowth of neurites is elicited by the extract.

NGF the explanted ganglia were surrounded by a halo of coarse outgrowing neurites. Light and electron microscopy showed that NGF gave a better survival of the neurons. Also, the neurons increased their volume with about 50% and showed an increase in number of neurofilaments and microtubules in the presence of NGF. Antibodies to NGF blocked these effects.

Human glial cells.
Medium conditioned by cells from different lines of normal human glial cells alicits neurite outgrowth in spinal and sympathetic ganglia, an effect totally inhibited by anti-NGF (5). The characteristics of the glial factor were studied by gel filtration, ion exchange chromatography and preparative flat-bed electrofocusing (Norrgren et al., in preparation). By none of these techniques could the glial factor be established to differ from mouse NGF isolated as the 7S complex.

NEUROTROPISM

Do the factors that stimulate neurite outgrowth from chick embryo ganglia have the ability also to influence the growth direction of the fibres? A consistent observation is, that when a ganglion faces stimulative explants on only one side, the resulting neurite outgrowth is much denser and longer on this side while only few fibres extend on the away side of the ganglion. To investigate the mechanism creating this asymmetry in the outgrowth, a series of experiments were performed in which the heart explants were removed or additional heart explants were introduced after different times (Ebendal, in preparation). Furthermore, the position of ganglia was in some instances changed by transferring them to new gels after initial stimulation.

The results show that it is not necessary to have the heart present in the gel for the total culture period of two days to get a good directed response but that 6, 12 and 24 h of initial presence of the heart explants is sufficient to "condition" the gel to support partial outgrowth. A second wave of neurite outgrowth could also readily be evoked on the away side of the ganglion by positioning additional heart explants opposite to the first set of explants. Obviously the ganglion has no "memory" for being stimulated since when ganglia which responded by heavy outgrowth were transferred to new gels without heart, no neurites extended to replace the halo that had been ripped off. On the other hand, when they were transferred to a new gel with heart explants, once again a nicely directed outgrowth was evoked towards the heart tissue in 24 h.

These experiments thus clearly show that the neurite extension is directly dependent on the local presence of the heart factor in the gel. It is also clear from diffusion experiments using methylene blue that the back side of the ganglion is locally "shadowed" by the bulk of ganglionic tissue. The conclusion was reached that chemokinesis - the rate of neurite extension being positively correlated to the concentration of the trophic factor - is sufficient to explain the asymmetric outgrowth zones but that chemotaxis in neurites - the neurite orientating by sensing the slope of the diffusion gradient of the trophic factor - is not ruled out by the results. The findings thus present some evidence for the ability of embryonic organs to exert neurotropic besides neuronotrophic control over their nerve supply.

ACKNOWLEDGEMENT

The work reported in this paper is supported by grants from the Swedish Natural Science Research Council.

REFERENCES

1. Adler, R., Landa, K. B., Manthorpe, M., and Varon, S. (1979): Science, 204:1434-1436.
2. Ebendal, T. (1979): Dev. Biol., 72:276-290.
3. Ebendal, T., and Hedlund, K.-O. (1975): Zoon, 3: 33-47.
4. Ebendal, T., and Jacobson, C.-O. (1977): Exp. Cell Res., 105:379-387.
5. Ebendal, T., and Jacobson, C.-O. (1977): Brain Res., 131:373-378.
6. Ebendal, T., Jordell-Kylberg, A., and Söderström, S. (1978): In: Formshaping Movements in Neurogenesis, edited by C.-O. Jacobson, and T. Ebendal, pp. 235-243. Almqvist & Wiksell International, Stockholm.
7. Ebendal, T., Makonnen, B., Jacobson, C.-O., and Porath, J. (1979): Neurosci. Lett., 14:91-95.
8. Edgar, D., Barde, Y.-A., and Thoenen, H. (1979): Exp. Cell Res., 12:353-361.
9. Helfand, S. L., Riopelle, R. J., and Wessells, N.K. (1978): Exp. Cell Res., 113:39-45.
10. Higgins, D., and Pappano, A. J. (1979): J. Mol. Cell. Cardiol., 11:661-668.
11. Levi-Montalcini, R. (1966): Harvey Lect., 60:217-259.
12. Levi-Montalcini, R., Menesini Chen, M. G., and Chen, J. S. (1978): In: Formshaping Movements in Neurogenesis, edited by C.-O. Jacobson, and T. Ebendal,

pp. 201-212. Almqvist & Wiksell International, Stockholm.
13. Lindsay, R. M., and Tarbit, J. (1979): Neurosci. Lett., 12:195-200.
14. McLennan, I. S., and Hendry, I. A. (1978): Neurosci. Lett., 10:269-273.
15. Noden, D. (1978): Dev. Biol., 67:313-329.
16. Rickenbacher, J., and Müller, E. (1979): Anat. Embryol., 155:253-258.

Control Mechanisms in Animal Cells,
edited by L. Jimenez de Asua et al.
Raven Press, New York © 1980.

Hormonal Controls of Collagen Substratum Formation by Cultured Mammary Cells: Implications for Growth and Differentiation

W. R. Kidwell, M. S. Wicha, D. Salomon, and L. A. Liotta

Laboratory of Pathophysiology, National Cancer Institute, National Institutes of Health, Bethesda, Maryland 20205, U.S.A.

Although it is clear that cloned cells and pure growth factors represent the most unambiguous system for studying growth control, more complicated interactions are probably involved in proliferative responses in vivo, such as in tissues in which a variety of cell types respond co-ordinately. An example of this is exemplified by the interactions of myoepithelial and epithelial cells of the mammary gland which we have studied in cultures of ducts and alveoli.

By means of limited collagenase digestion and Ficoll gradient sedimentation ducts and alveoli have been isolated (8). Electronmicrographic analysis shows that the isolated structures maintain the same anatomical relationship as is observed in vivo. These organoids consist of epithelial and myoepithelial cells attached to a partially intact basement membrane (8). In culture, as is true of the intact ducts and alveoli, the epithelial cells are largely attached to the myoepithelial cells. This is reminiscent of the behavior of myoepithelial and epithelial cells reported for mammary tumors (1).

Autoradiographic analysis of cells pulse labeled with ^{3}H-thymidine presented the first clue that some interaction cell types might be occurring. A non-random distribution of labeling was seen with foci of labeled epithelial cells clustered at the colony periphery (Fig. 1). At higher magnification (Fig. 2), the dividing epithelial cells were seen to be in the neighborhood of large, flat, fibroblast-like cells. These latter cells were especially visible when the cultures were allowed to cool to room temperature upon which the cells were seen to be localized at the rim of the cultures, or sometimes, the entire epithelial colony rested on top of the fibroblast-like cells.

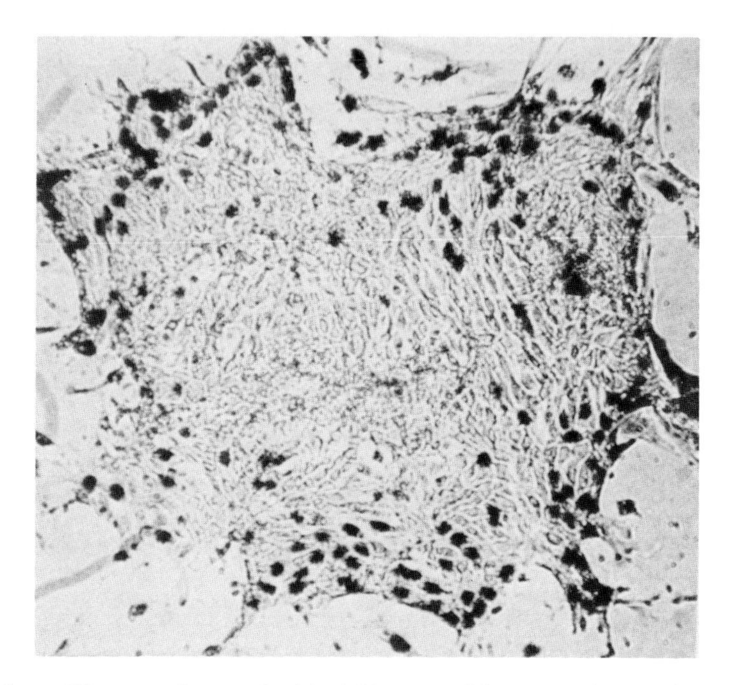

FIG. 1. Clustering of dividing cells at the colony periphery. Cells were labeled with ^{3}H-thymidine (5 μCi/ml) for 24 hr, then processed for autoradiography.

In addition to the clustering of the labeled cells near the fibroblast-like cells, a second zone of division was seen in association with the remnants of basement membrane to which the cells of the ducts and alveoli were attached at the same time of plating (Fig. 3). Division at the two loci accounted for more than 80% of the total dividing population.

When the ductal and alveolar cells were analyzed for the presence of type IV collagen by indirect immunofluorescence, it became apparent that there was a common entity in the basement membrane remnants and in the fibroblast-like cells around which cell division was localized. Type IV collagen, the collagen which is characteristic of basement membranes, stained both the basement membrane remnant and the fibroblast-like cells. The presence of type IV collagen in these cells was conclusive proof that they were not fibroblasts because we have shown that the cells of the stroma (the fibroblasts) synthesize types I and III collagen and not type IV collagen (4). This is the

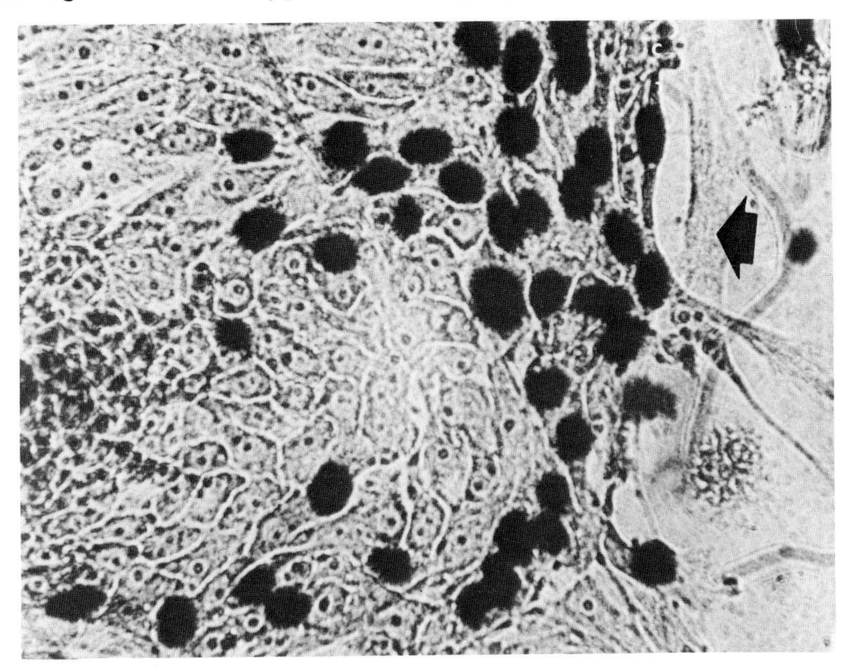

FIG. 2. Fibroblast-like cell in the vicinity of the cluster of dividing epithelial cells. Labeling conditions were as in Fig. 1. The fibroblast-like cell is indicated by the arrow.

case in vivo and in vitro. In proliferating mammary glands the myoepithelial cells have been shown to stain very intensely for type IV collagen and the epithelial cells only weakly (4). Thus we conclude that the fibroblast-like cells are the myoepithelial cells. Not only the normal cell cultures but also the cells derived from 7,12-dimethylbenz(α)anthracene-induced mammary tumors contained myoepithelial cells at the colony periphery. An immunofluorescent stain for type IV collagen in the tumor cell cultures is shown in Fig. 4. The staining of the myoepithelial cells seen here is very similar to the normal myoepithelial cells except that the latter appear to be more irregularly shaped. As is seen in Fig. 4, there is an extracellular deposit of type IV collagen. This matrix material appears to extend back and under the rim of the epithelial cell

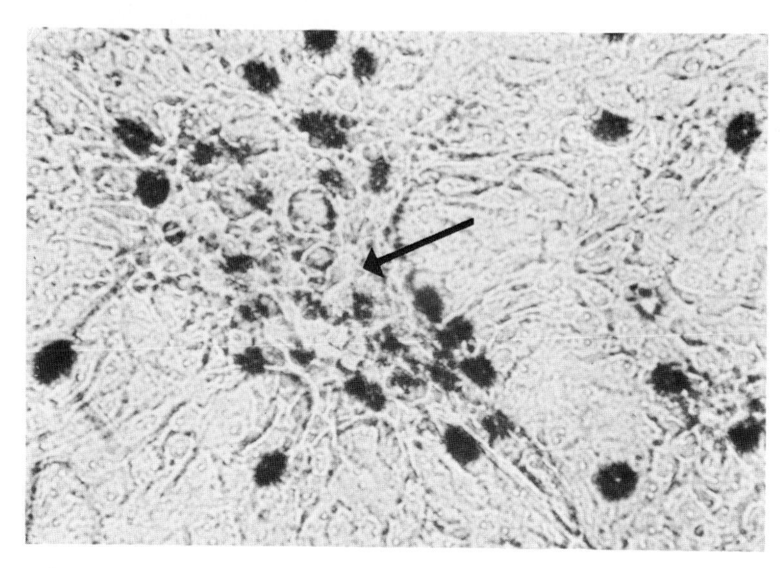

FIG. 3. Autoradiographic analysis of cultured ductal and alveolar cells labeled with ^{3}H-thymidine. Note the clustering of labeled cells around the remnants of basement membrane (arrow). When stained with periodic acid-Schiff, this area was found to be intensely positive, indicating a glycoprotein composition. It was also positive for type IV collagen.

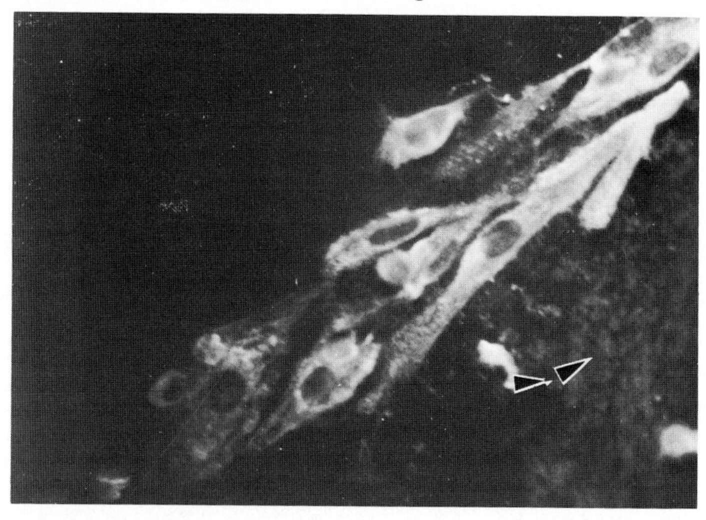

FIG. 4. Myoepithelial cells at the colony periphery. The cultures are stained for type IV collagen by indirect immunofluorescence (4). Collagen is localized with these cells and extracellularly (arrow).

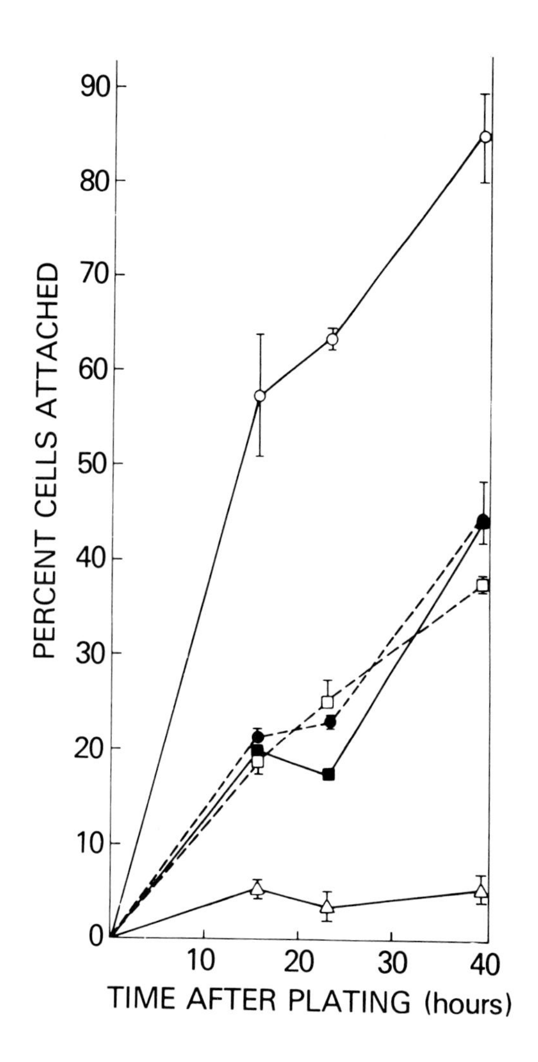

FIG. 5. Attachment of cells of ducts and alveoli on various substrata. 0———0, type IV collagen; ●———●, type I collagen; □———□, type III collagen, ■-----■, type II collagen; Δ----Δ, uncoated bacterial plastic dishes. Cells were plated in medium 199 containing 5% fetal calf serum, prolactin (1 ng/ml), hydrocortisone (300 ng/ml), estradiol and progesterone (1 ng/ml).

colony. Presumably this is the substratum collagen which the myoepithelial cells have laid down as they migrated out from the colony.

Because of the localization of type IV collagen in the two zones of the culture where cell division is particularly active, the effect of collagen substrata on the attachment and growth of the ductal and alveolar cells was assessed. As a control, comparison was made of the interaction of the cells on the type of collagen present in the stroma (types I and III) or that of cartilage (type II collagen). Freshly isolated ducts and alveoli were plated onto uncoated plastic dishes (bacteriological) or onto dishes coated with 10 $\mu g/cm^2$ of the collagen types (8). To assure that attachment per se was measured, 1 mM hydroxyurea was included in the growth medium to block cell division. The results of the cell attachment analyses are depicted in Fig. 5. In the first 15 hr about 60% of the cells have attached on the dishes coated with type IV collagen and only 1/3 as many have attached on the other collagen substrata. Few cells attach on the uncoated dishes.

When growth of the cells was measured on the collagen-coated dishes following attachment, the doubling time was found to be virtually identical (about 48 hr). However, when a specific inhibitor of collagen synthesis was added to the cultures, cell attachment on the type I collagen-coated dishes was nearly eliminated. If allowed to pre-attach before adding the inhibitor (cis-hydroxyproline, 25 $\mu g/ml$), cell division was completely blocked and cell death ensued. Entirely different results were seen on the type IV collagen-coated dishes. The cell growth rate was reduced by about 50% but cell attachment was not affected at all. These results have been interpreted to mean that before the mammary cells can proliferate on the type I collagen-coated dishes, the cells must lay down a type IV collagen substratum over the type I collagen (8). This is precisely what happens in the proliferating mammary glandular epithelium in vivo. Presumably then, the situation which is created in the cultures represents a good system for studying mammary cell proliferation.

In addition to the ability of the cells to specifically recognize the collagen of basement membranes, there are other observations which demonstrate the importance of collagen in cell proliferation. For example, we have recently found that when the hormonal support is removed from the cultures, there is a drop in the cell proliferation rate and newly synthesized collagen is very rapidly degraded by a collagenase activity which is latent in the presence of the hormonal support but activated by hormone removal (4).

The behavior of the cells in culture is thus analogous to that of the involuting mammary gland where there is an early degradation of the basement membrane and a degeneration of the epithelial cells (5). In fact, the mammary gland can be converted to a situation which is histologically and ultrastructurally indistinguishable from that of involution by administering cis-hydroxy-proline to block basement membrane collagen synthesis during glandular proliferation. In the case of mammary tumors which are synthesizing type IV collagen, there is a growth arrest or regression when cis-hydroxypro-line is administered.

From the studies which we have presented it is apparent that we view the myoepithelial cell as the key cell in the proliferative response of the mammary gland. The following steps in the overall process are viewed as follows. A stimulus to proliferate activates the myoepithelial cell to elaborate new basement membrane (or at least the collagen component therein) as it migrates outward over the stromal collagen. Thus a new substratum is provided on which the epithelial cells can proliferate. This latter process may be active (stimulated by growth factors or hormones) or passive (proliferation occurring simply in response to the new substratum on which cell flattening can occur). The withdrawal of the stimulus for basal lamina production would then limit the proliferative response. The epithelial cells remain viable so long as the basement membrane is intact but upon degradation of this element (catalyzed at least in part by a collagenase which becomes active on withdrawal of hormonal support) viability is impaired and the cells die.

Because the epithelial cells, not the cells of the stroma or even the myoepithelial cells, degenerate following dissolution of the basement membrane, there is obviously something very special about the interrelationship between the epithelial cells and the basement membrane. Such an interaction cannot yet be defined in molecular terms. The key to understanding this phenomenon lies, we think, in defining the components by which the selective attachment of the epithelial cell to type IV collagen is made possible. In our quest of this goal we have recently been able to simplify the in vitro culture system by plating the isolated ducts and alveoli directly into medium in the absence of serum. A cell doubling time of about 40 hr has been obtained in IMEM-Zinc Option (6) supplemented with epidermal growth factor, insulin, dexamethasone, transferrin and fetuin. In this system epidermal growth factor has been shown to facilitate attachment and growth of ductal and alveolar cells on type I collagen but only slightly when the cells are placed on type IV collagen substrata. It thus appears that this growth factor may function by stimulating the cells' production of extracellular matrix. Observations consistent with this possibility are that the action of this factor is enhanced by ascorbic acid, a known cofactor in the formation of the hydroxyproline by cells (2). Also, epidermal growth factor has been found to stimulate collagen production (3). In the case of the mammary ductal and alveolar cells, the effects of the growth factor are reduced by exogenous addition of type IV collagen.

REFERENCES

1. Bennett, D. C., Peachey, L. A., Durbin, H., and Rudland, P. S. (1978): Cell, 15:283-298.
2. Carpenter, G., and Cohen, S. (1976): J. Cell. Physiol., 88:227-238.
3. Lembach, K. J. (1976): Proc. Natl. Acad. Sci., 73: 183-187.
4. Liotta, L. A., Wicha, M. S., Foidart, J. M., Rennard, S. I., Garbisa, S., and Kidwell, W. R. (1979): Lab Invest. (in press).
5. Martinez-Hernandez, A., Fink, L. M., and Pierce, G. B. (1976): Lab. Invest., 34:455-462.
6. Richter, A., Sanford, K., and Evans, V. (1972): J. Natl. Cancer Inst., 49:1705.
7. Wicha, M. S., Liotta, L. A., and Kidwell, W. R. (1979): Cancer Res., 39:426-435.
8. Wicha, M. S., Liotta, L. A., Garbisa, S., and Kidwell, W. R. (1979): Exp. Cell Res. (in press).

Control Mechanisms in Animal Cells,
edited by L. Jimenez de Asua et al.
Raven Press, New York © 1980.

Differentiation of a Rat Mammary Stem Cell Line in Culture

P. S. Rudland, *D. C. Bennett, **M. A. Ritter, **R. A. Newman, and M. J. Warburton

*Ludwig Institute for Cancer Research, Haddow Laboratories, Sutton, Surrey SM2 5PX, England; *Salk Institute, San Diego, California 92138, U.S.A.; and **Imperial Cancer Research Fund, London WC2A 3PX, England*

INTRODUCTION

The mammary gland consists of two basic cellular structures, epithelium and mesenchyme. The epithelial components are embedded in a fatty stroma or mesenchyme and, when fully developed, consist of a branching system of ducts terminating in clusters of alveoli which secrete lipid and milk-specific proteins, notably caseins, during lactation. Three main types of mammary epithelial cells are usually distinguished: those lining the alveoli, those lining the ducts, and the myoepithelial cells which form a layer around ducts and alveoli (19). The development of these epithelial structures occurs both by processes of cell multiplication and differentiation (13). Certain mammotrophic hormones affecting these processes in normal glands and in carcinogen-induced tumours have been identified in rodents by means of a series of endocrine gland-ablation and hormone-replacement experiments; these include prolactin, growth hormone, estrogens, glucocorticoids and progesterone (17,21,27,30,40). However, both the relationships between the different cell types, including any programme of cellular interconversions, and the primary targets (whether epithelial or mesenchymal) for the mammotrophic hormones in mammary development are largely unknown.

The problem of mammary development has been tackled by maintaining, in organ culture, large fragments of mammary gland from immature rats and mice (9). These fragments can be induced to grow and develop alveolar-like structures upon addition of the hormones prolactin, growth hormone, insulin, estrogen, progesterone, and glucocorticoids (2). In this system however, the interpretation of the effects of hormones on proliferation and differentiation of epithelial cells is complicated by the presence of different cell types which may interconvert, affect each other, and/or act in different ways. In view of this difficulty we initially developed a system for obtaining short term cell cultures of relatively pure epithelial cells from both normal rat mammary glands and dimethylbenz{a}anthracene(DMBA)-induced mammary adenocarcinomas (15,35) and then separated epithelial cells by developing clonal cell lines (4). In this chapter we describe the properties of an established lining epithelial cell line, Rama 25, from a rat mammary tumour which can undergo differentiation in culture forming either a myoepithelial-like or an alveolar-like (secretory) cell (4,33). Morphologically similar established cell lines can also be obtained from normal glands (P.S.R.unpubl.). We have proposed that this type of cell line represents a mammary stem cell, and as such is a suitable *in vitro* model to study mammary epithelial cell differentiation (4,33,34).

MORPHOLOGICAL CLASSIFICATION AND RELATEDNESS OF DIFFERENT CELL TYPES

Both the isolation and initial characterisation of the cell line Rama (rat mammary) 25 have been extensively described elsewhere (4,33,34). Rama 25 cells when grown on plastic petri dishes resemble low, cuboidal epithelium, and cells of this appearance were termed "cuboidal".(Fig.1a). "Cuboidal" cells were cloned by selecting single cells and, after subculturing, these repeatedly gave rise not only to "cuboidal" cells, but also to a second cell type, somewhat like fibroblasts, which we have termed "elongated" cells. One such "elongated" cell line was designated Rama 29 (4) (Fig.1b). The frequency of conversion of "cuboidal" cells to "elongated" cells could be increased by allowing Rama 25 cells to attain high cell densities, notably at confluency. On the other hand Rama 25 cells could be maintained as an essentially homogenous line of "cuboidal" cells by frequent passage at low cell densities. A third distinct type of cell was cloned from sparse, or more frequently from dense cultures of Rama 25; this had a morphology between that of the "cuboidal" and "elongated" cell types (34). These were termed "intermediate" cells (P.S.R.unpubl.) (Fig.1d). Dense cultures of "cuboidal" cells formed not only "intermediate" and "elongated" cells, but groups of cells with a fourth morphology: small, dark and polygonal with many small vacuoles or "droplets" at their peripheries. These were termed "droplet" cells (Fig.1c); unlike the other three cell types they could not be cloned (Table 1). The main morphological cell types could also be divided into subclasses. Thus "elongated" cell lines were "bipolar", "multipolar" or a mixture of the two, as with Rama 29 (Fig.1b), and the "intermediate" cells ranged in morphology from nearly "cuboidal" (Rama 25-I2) to nearly "elongated" (Rama 25-I1) or a mixture of the two (Rama 25-I4) (Fig.1d). These subclasses were relatively stable and could be cloned within certain limits. However, subclasses of the "cuboidal" and "droplet" cells were unstable. This behaviour was typified by Rama 25 cells on approaching

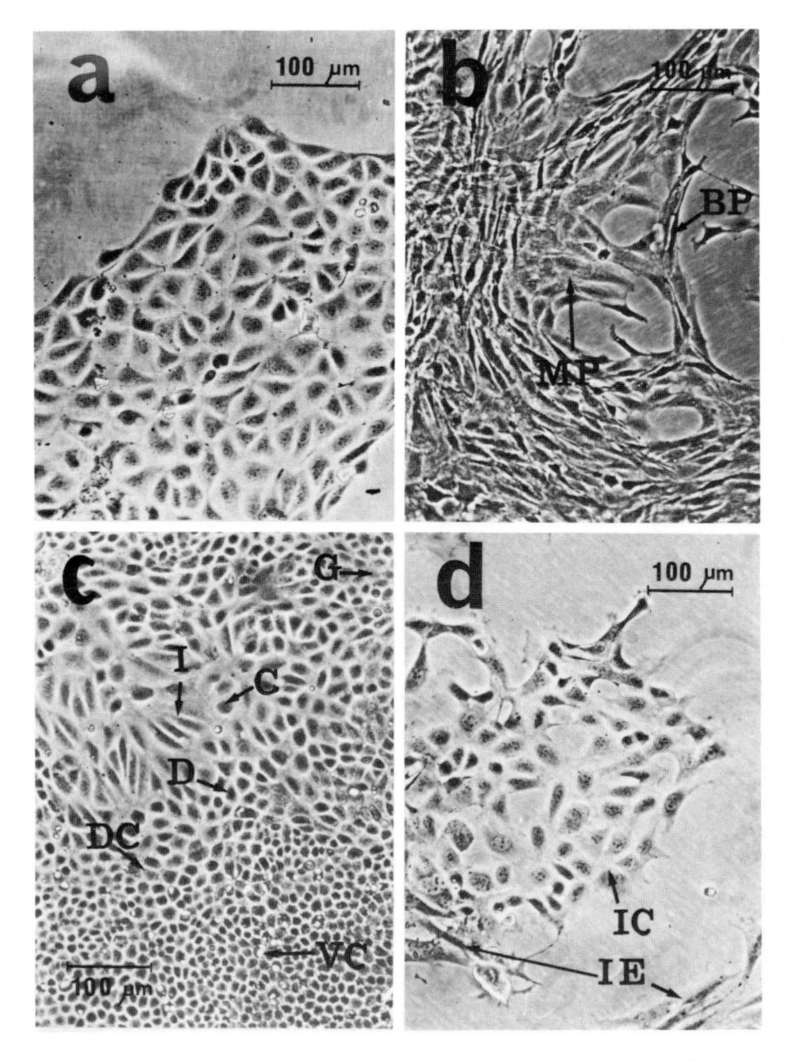

FIG. 1. Major morphological cell types generated in culture Living cells were photographed with phase contrast optics a. colony of growing "cuboidal" cells, Rama 25. b. "elongated" cells, Rama 29, cloned from Rama 25 showing both "multipolar"(MP) and "bipolar"(BP) forms. c. Culture of Rama 25 cells at saturation density showing confluent "cuboidal" cells (C) "grey" cells (G) "dark" cells (D), "droplet" cells (DC), and "droplet" cells with large vacuoles (VC). Cells labelled "I" may be "intermediate" cells. d. "Intermediate" cells cloned from Rama 25 (Rama 25-I4) showing a spectrum of morphologies from near cuboidal (IC) to near elongated (IE).

TABLE 1. Different cellular morphologies

Cell type[a]	Clonable	Subclasses
Cuboidal	yes	standard grey dark
Droplet	no[b]	standard vacuolated
Elongated	yes	multipolar bipolar
Intermediate	yes	spectrum of intermediates

[a]Different cell types can be obtained in reasonably pure form under the appropriate conditions although both "droplet" and "intermediate" forms may not be considered stable. [b]So far.

and being maintained at their saturation density[1]. Standard "cuboidal" cells gave rise to small "grey" cells, small "dark" polygonal cells, "dark" cells with small "droplets" ("droplet" cells), and "dark" cells with large vacuoles which occupied nearly the entire volume of the cell ("vacuolated" cells) (Fig.1c).

In addition to the various cell types formed in culture, multicellular structures were repeatedly observed. These were most apparent at high local cell densities. In some regions of "droplet" cells hemispherical blisters ("domes") were formed by local detachment of the cell monolayer (25) (Fig.2a). Confluent cultures of standard "cuboidal", "intermediate" and "elongated" cells failed to form such structures. In other regions "elongated" cells were formed; these frequently aggregated in a series of interconnecting ridges (Fig.2b,c), often connecting circular amorphous areas ("lumps") which were shedding cells (Fig.2b). Large "lumps" seemed to contain mainly elongated cells on replating, but smaller, looser aggregates often contained "cuboidal" cells. In a few of these, branching "tubules" could be discerned (3), perhaps an attempt to make primitive duct-like structures, as found when Rama 25 cells were grown on floating collagen gels (33). Finally, "cuboidal" cells could grow on top of "intermediate" (Fig.2d) or "elongated" cells (4), but not on mammary fibroblast lines. This ability of "cuboidal" cells to grow on top of "elongated" cells led to a pattern of growth similar to the multi-cell-layered colonies found in the original primary cultures of both normal and tumourous glands (35). No cell line tested could grow on top of the "cuboidal" cell lines at saturation density.

[1]Saturation density: highest cell density attainable for a given set of conditions.

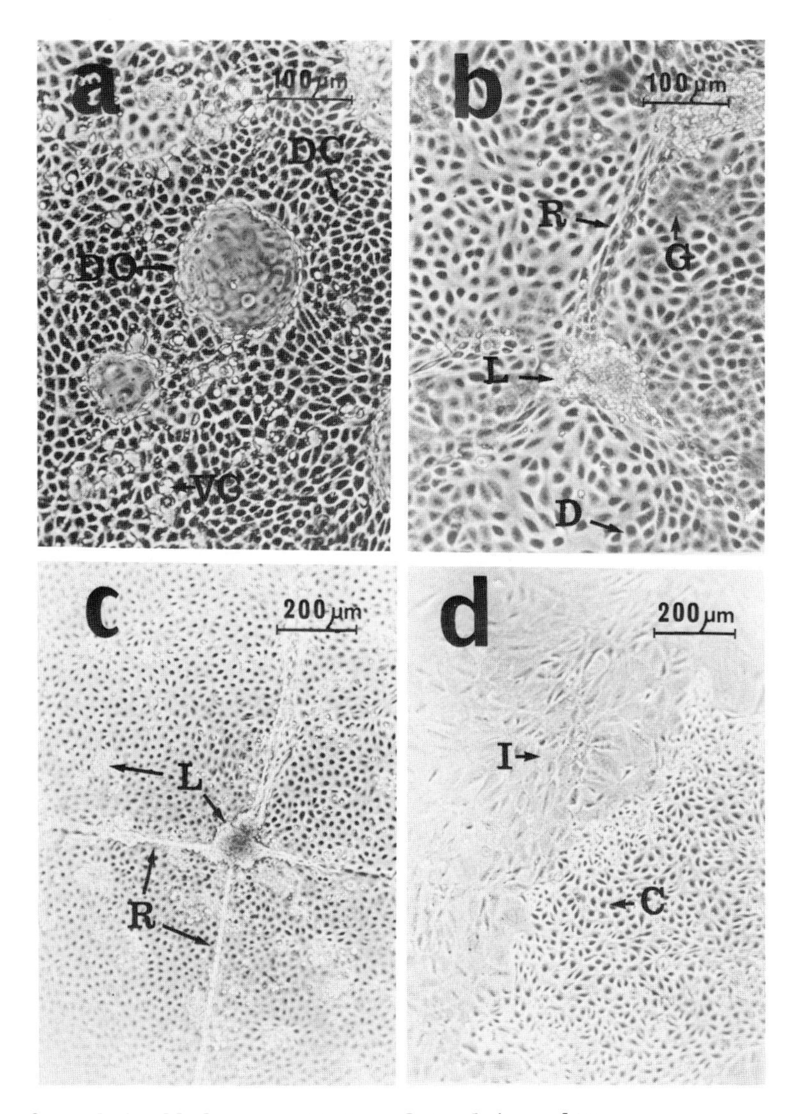

FIG. 2. Multicellular structures formed in culture
Living cells growing on plastic petri dishes were photographed with phase contrast optics. a. Cultures of Rama 25 cells at saturation density showing "domes"(DO) in an area of "droplet" cells (DC) and "vacuolated" cells (VC). b. Confluent cultures of Rama 25 cells showing "lumps"(L) and "ridges"(R); "dark" cells (D) and "grey" cells (G) are also visible. c. Lower magnification showing extended network of ridges (R) emanating from a central lump (L). d. "Intermediate" cells cloned from Rama 25 (Rama 25-I1) which have yielded "cuboidal" cells. The "cuboidal" cells (C) are growing on top of the "intermediate" cells (I).

The two main pathways of cellular interconversions from "cuboidal" cells to "elongated" or "droplet" cells are shown in Fig. 3.

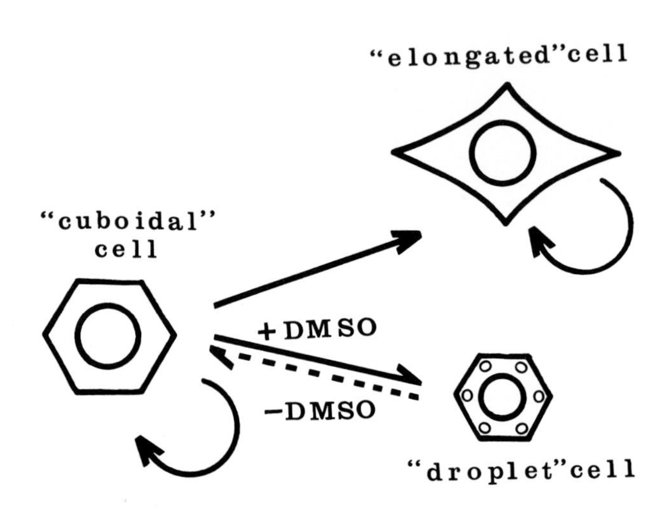

FIG. 3. Summary of cellular interconversions
The intercellular conversions of "cuboidal", "droplet" and "elongated" cells are shown by arrows. The broken line signifies that reversion of "droplet" cells to "cuboidal" cells can be seen under certain conditions. The presence of dimethyl sulphoxide (DMSO) accelerates the rate of conversion to "droplet" cells. From ref.33 with permission.

Conversion to "elongated" cells is probably irreversible, although a "cuboidal" variant of Rama 29 was obtained which could give rise to both "elongated" and "cuboidal" cells (H. Durbin and D.C.B. unpubl.). Whether this was a contaminant cell or a very low frequency event is uncertain. The "intermediate" cell types not shown in Fig.3 may represent true intermediates in the conversion pathway. Not only do they have a range of morphologies between that of their parent "cuboidal" cells and the "elongated" cells, but they also convert to "elongated" cells at higher rates than their parent cells. In addition, at least one clone of "intermediate" cells (Rama 25-I1) can form "cuboidal" cells as well as "elongated" cells (P.S.R.unpubl.). Thus it may be the final step from "intermediate" to "elongated" cell which is irreversible. However, an alternative explanation that the "intermediate" cell is merely a cell variant cannot be discounted at present. The second pathway to "droplet" cells shown in Fig.3 seems to be completely reversible, and the subclasses of cells observed during the change of standard "cuboidal" to "droplet" and "vacuolated" cells probably represent a real sequence of cellular events, since time-lapse cinematography has established that some single, standard "cuboidal" cells undergo the sequence of changes: "grey", "dark", "droplet", "vacuolated" (P.S.R. and P. Riddle unpubl.). We have not as yet, however, shown this to be an obligatory pathway. Finally, pure "cuboidal" Rama 25 cells when injected into immunodeficient mice, or the intrascapular fat pad of rats, formed tumours whereas the "elongated" cell converts, Rama 29, did not. These tumours derived from Rama 25 contained three major cell types, "cuboidal", "elongated" (4) and "secretory"[2] cells (P.S.R., D.C.B. and R.A.N.unpubl.), which correspond in morphology to the cells observed *in vitro*. Thus the conversion of the "cuboidal" cell to an "elongated" and to a "droplet" ("secretory") cell occurred *in vivo* as well as *in vitro*.

With regard to the possible existence of alternative conversion pathways, this is complicated by the fact that subclones of Rama 25, and even Rama 25 at different passage numbers have different frequencies for the formation of "elongated" cells on the one hand and "droplet" cells/"domes" on the other, when measured under approximately the same conditions (P.S.R.unpubl.). Thus, although we have isolated single cell "cuboidal" clones from Rama 25 in which conversion to "droplet" cells/"domes" or in which conversion to both "droplet" cells/"domes" and "elongated" cells has not been observed (P.S.R.unpubl.) these may not represent new cuboidal epithelial cell types or new pathways, but merely the extreme ends of a spectrum of "cuboidal" cells with differing conversion frequencies.

STRUCTURAL REORGANISATION AND BIOCHEMICAL DIFFERENTIATION TO MYOEPITHELIAL CELLS

The "cuboidal" Rama 25 cells and their "elongated" cell converts, Rama 29, were suggested to be mammary lining epithelial cells and myoepithelial-like cells, respectively, based on ultrastructural and histological studies of the cultured cells and mammary tumours formed from Rama 25 in rodents (4). In culture, Rama 29 cells differed from Rama 25 cells and more closely resembled a rat mammary fibroblast line (Rama 27) using the above techniques, together with histochemical staining for the Na^+/K^+-ATPase; fluorescently labelled serological stains for cytoskeletal proteins; actin (20) (Fig.4a,b), myosin (43) (Fig.4c,d); and an extracellular matrix glycoprotein, LETSP[3] (12) (Fig.4e,f) (33). In the latter three cases the peripheral organisation of the fluorescently labelled fibres round the edges of Rama 25 cells,

[2]Secretory cells: cells capable of binding anti-casein serum and peanut lectin (see later).
[3]LETSP. Large External Transformation Sensitive Protein.

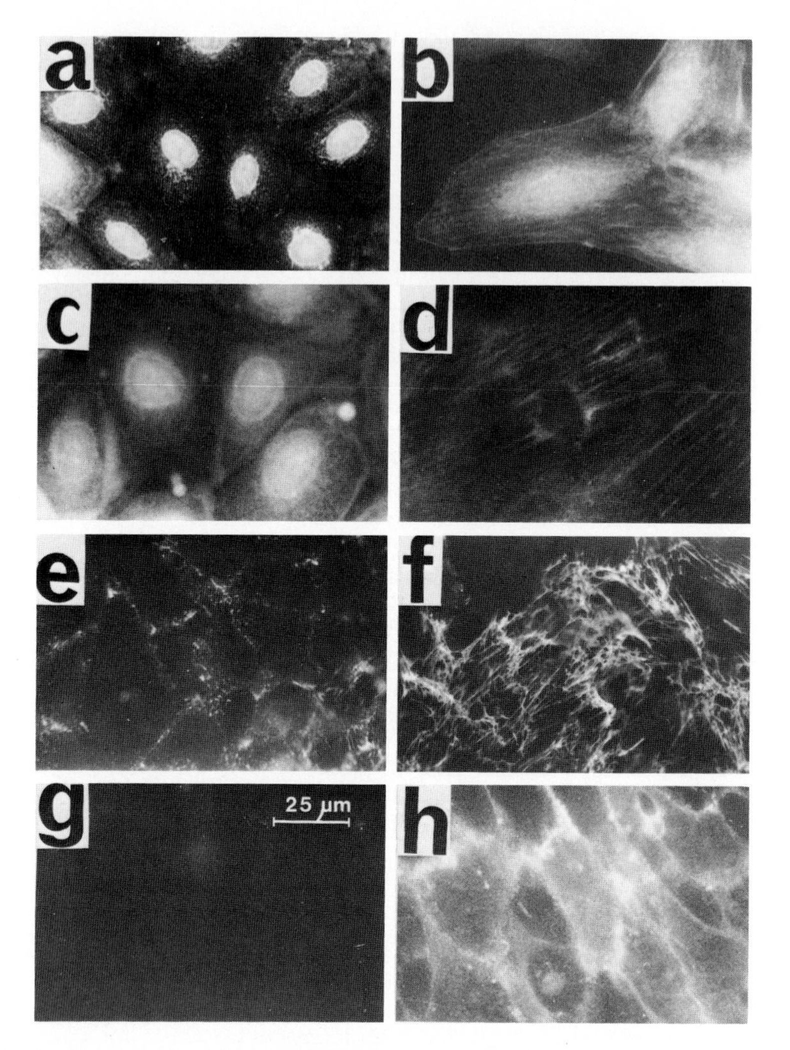

FIG. 4. Antigenic differences between the "cuboidal" and "elongated" cell lines Cultures of "cuboidal" (Rama 25) cells (left hand panels: a,c,e,g) and cultures of "elongated" (Rama 29) cells (right hand panels: b,d,f,h) were fixed, stained with the given antisera and then developed with a fluorescein conjugated second antibody. The fluorescent light emitted from the treated cultures after exposure to ultravoilet light was photographed. Antisera raised against the following antigens were tested: a,b: pig smooth muscle actin; c,d: pig smooth muscle myosin; e,f: hamster fibroblasts LETSP; and g,h: pure Thy-1.1 from rat's brain. Magnification is the same in all photographs.

probably part of structures that resemble the brush borders of intestinal epithelial cells (5,26), were replaced in Rama 29 cells by either unidirectionally organised cytoplasmic fibres of actin and myosin or the more randomly organised extracellular filaments of LETSP. These were similar to the structures found in Rama 27 and mouse-fibroblasts. In addition another marker, the thymocyte differentiation antigen Thy-1 was found on the surface of both Rama 29 (8) (Fig.4g,h) and other "elongated" cells as well as cultured mammary fibroblasts but not on the parent, Rama 25, cells (33). As in all other rat systems studied only the Thy-1.1 alloantigen was expressed (7). Actin, myosin, LETSP (M.J.W. and P.S.R.unpubl.) and Thy 1.1 (M.A.R. and P.S.R.unpubl.) could also be identified on myoepithelial and fibroblast cells in wax embedded histological sections of the rat mammary gland. Actin, myosin and Thy 1.1 were associated exclusively with the cells, while LETSP seemed to be additionally associated with components of the basement membrane.

The biochemical differences between Rama 25 and Rama 29 cells were investigated further. Radioactive labelling of the surfaces of Rama 25 and Rama 29 cells by lactoperoxidase catalysed iodination indicated that the major exposed components of these two cell lines were different (Fig.5).

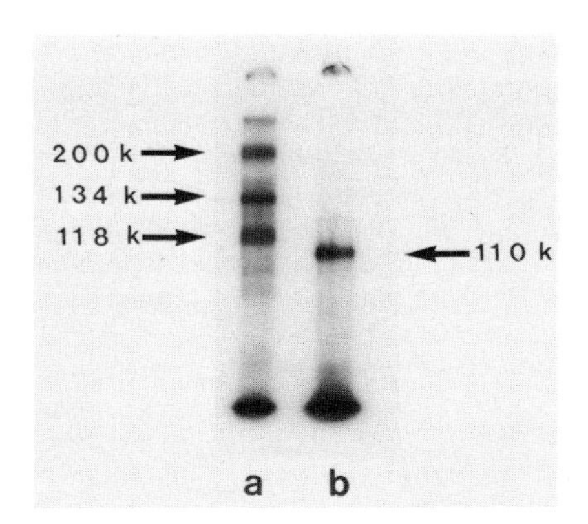

FIG. 5. Major surface components labelled by lactoperoxidase catalysed iodination Lactoperoxidase catalysed iodination of confluent cell monolayers of either (a) Rama 29 or (b) Rama 25 was performed in 5cm dishes (17). Iodinated cell products were separated by electrophoresis through sodium dodecyl sulphate 7% (w/v) polyacrylamide slab gels, visualised by radioautography and the apparent molecular weights of the major components were estimated from known standards. Results are expressed in kilodaltons (k).

The pattern of iodination of Rama 29 closely resembled Rama 27 (33); the major identifiable surface component being LETSP at 200,000 daltons. However, although LETSP was secreted into the medium by Rama 25 cells in nearly the same amounts as by Rama 29 cells, it failed to adhere as long filaments on the upper surface of or between Rama 25 cells (M.J.W.unpubl.). Moreover, in transition from the "cuboidal" to the "elongated" cell actin, myosin, Thy-1.1 and LETSP changes seemed to occur in a co-ordinated fashion. Thus in the more cuboidal "intermediate" cells (Rama 25-I2), these proteins were organised as in the "cuboidal" cells, whereas in the more elongated "intermediate" cells (Rama 25-I1) the reorganisation of the distributions of these proteins occurred together. Preliminary studies suggested that all these changes also occur in the conversion of "cuboidal" (Rama 401, Rama 402) to "elongated" cells which were isolated from normal mammary glands (P.S.R., M.A.R. and M.J.W.unpubl.), and thus are possibly not just the consequence of a change from a potentially neoplastic to a non-neoplastic cell.

All the above changes from the "cuboidal" to the "elongated" cell could be ascribed to molecular rearrangements rather than a true differentiation process. Even the appearance of Thy-1.1 on the surface of Rama 29 could be explained by reorganisation of the molecules in the cell (32,42) so making them accessible and/or easier to bind anti-Thy-1.1 antibodies. In addition all the above markers have failed to differentiate between "elongated" (Rama 29) and fibroblast (Rama 27) lines. However ultrastructural studies have shown that Rama 29 cultures make small strips of electron dense material between cells, reminiscent of pieces of basement membrane (4). This was again suggested by biochemical analysis: Rama 25 synthesised very small amounts of a collagen resembling basement membrane-specific type IV collagen while Rama 29 (M.J.W.unpubl.) and fresh "elongated" cell converts from Rama 25 (Fig.6) synthesised appreciable quantities. In contrast, Rama 27 cultures only synthesised type I collagen. (33,41). This confirmed the epithelial origins of the "elongated" cells and the fact that a true differentiation process had taken place. Probably, however, premyoepithelial (14,39), and not myoepithelial cells, were formed in culture, since typical myofilaments were not observed in Rama 29 or other lines of "elongated" cells grown on plastic dishes, although tumours produced by inoculation of Rama 25 cells into rodents contained "elongated" cells with identifiable myofilaments (4). The reason for the failure of Rama 25 cells to differentiate completely *in vitro* is unknown.

Finally no simple hormone combination has been found to affect the change from "cuboidal" to "elongated" cells in culture. Whether any or all of the changes observed during this conversion, increased synthesis of type IV collagen, formation of long, adherent filaments of LETSP, appearance of surface Thy-1.1 molecules and/or the rearrangement of the cytoskeleton (Table 2) play a causative role in this step remains to be established.

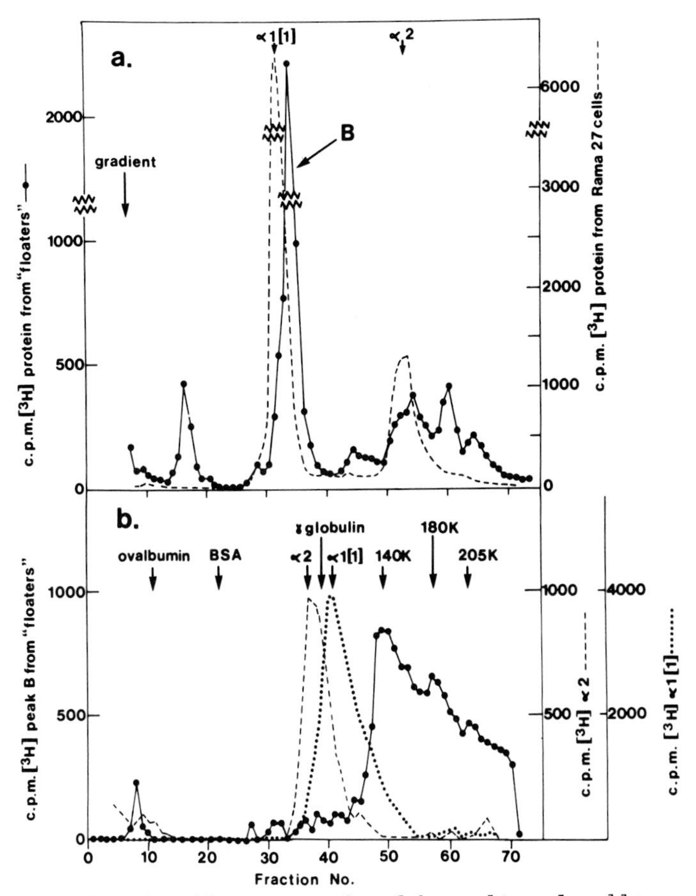

FIG. 6. Analysis of collagens produced by cultured cells
Fresh "elongated" cells from Rama 25 were obtained from the culture medium of confluent Rama 25 cultures and were designated as "floaters". Confluent monolayers of either "floaters" or a rat mammary fibroblast line, Rama 27, were radioactively labelled with $\{^3H\}$ proline (10μCi/ml) for 24 hours in the presence of β-aminoproprionitrile fumarate and ascorbic acid. The medium was removed, digested with pepsin and chromatographed with carrier rat tail collagen on columns of carboxymethylcellulose at 42°C (24,41).
a. Typical column chromatograms of 400ml, 0 to 0.14M NaCl gradient starting at fraction 7, 5ml fractions being collected. The cpm of $\{^3H\}$ protein per 200μl of each fraction were recorded separately for the processed medium from "floaters" (—●—) and from Rama 27 cells (----). b. Molecular weight analysis of the peak column fractions on sodium dodecyl sulphate 7% (w/v) polyacrylamide gels. The cpm of $\{^3H\}$ proteins per gel slice were recorded separately for gel fractions corresponding to α1{1}(....), α2(---) from Rama 27 cells and peak B (—●—) from the "floaters". Apparent molecular weights calculated from known standards were recorded, that of α1{1} and α2 were 105,000 and 95,000 respectively.

TABLE 2. Serological and biochemical properties of "cuboidal" and "elongated" cells

Reagent	"Cuboidal" (Rama 25)	"Elongated" (Rama 29)
Actin antiserum) Myosin antiserum)	±[a] (cell periphery only)	++ (intracellular fibres)
LETSP antiserum	± (very faint and patchy)	++ (patches of filaments)
Anti-Thy-1.1 serum	-	++ (cell periphery)
Peanut Lectin	± (cell periphery only)	-
+ DMSO + Hormones	++	-
Rat casein antiserum	± (cytoplasm)	-
+ DMSO + Hormones	++	-
Lactoperoxidase catalysed Iodination of cell surface — major M.W. bands	110,000	200,000 118,000
Lactoperoxidase catalysed Iodination of cell surface — minor M.W. bands		134,000 55-90,000
Analysis of collagen type	probably type IV ?	probably type IV

[a] symbols: ± small increase; + larger increase etc.; - no change;

From ref.33 with permission.

DMSO AND HORMONE-INDUCED DIFFERENTIATION TO ALVEOLAR-LIKE CELLS

Unlike the conversion of "cuboidal" cells to "elongated" cells, that to "droplet"/"vacuolated" cells and "doming" systems could be greatly accelerated with a combination of the Friend erythroleukemic cell differentiating agent, dimethyl sulphoxide (11), in the presence of the mammotrophic hormones prolactin, hydrocortisone, and insulin (4,33). Since antiserum against the major alveolar cell products, the caseins, showed specific fluorescence using an immunofluorescence assay (Fig.7) and appreciably increased levels of immunoreactive material in a radioimmune assay, some of which had the same molecular weight as the major casein, we have suggested that these "droplet" and "doming" cultures formed after treatment with DMSO and hormones contained alveolar-like cells (4,33). Peanut lectin also stained mainly secretory alveolar cells *in vivo*; this lectin also specifically stained the "doming" cultures, but only after they had been treated with neuraminidase (28). Anti-casein serum and peanut lectin reacted only weakly with the parent "cuboidal" cells and showed initially no reaction with the myoepithelial-like cells {Rama 29 (Fig.7)} or the mammary fibroblasts (Rama 27) with or without neuraminidase and/or DMSO treatment. Finally, increased numbers of very small fat droplets which could be stained by oil red O were seen in the "doming" cultures (3), although whether these contained the specific mammary fat produced by alveolar cell *in vivo* (6) remains to be established.

The fact that the amount of casein produced was about a hundred times less than that reported in mouse mammary primary cultures (10); that neuraminidase treatment was required to expose peanut lectin binding sites; and that the ultrastructure of these cultures showed little evidence of secretory activity (4) suggested that either only a small subpopulation of cells was fully differentiated alveolar cells, or more likely, that most of the cells were in a pre-secretory state, and that further steps were required before these cultures could differentiate into secretory alveolar cells. On the other hand mammary tumours formed by inoculation of Rama 25 cells into rats contained cells in comparatively rare, well-differentiated areas which showed specific fluorescence with anti-casein serum or peanut lectin alone. The intensity of this fluorescence was as high as in normal glands or in tumours induced by DMBA, providing the rats prolactin levels were elevated for several days before removing the Rama 25-induced tumours for analysis. Failure to maintain the elevated levels of prolactin resulted in tumours with lower specific fluorescence for casein and virtually no binding ability for peanut lectin unless the tissue sections were first treated with neuraminidase (R.A.N. and P.S.R.unpubl.). Thus a few of the Rama 25 cells inoculated into rats were capable of giving rise to functional alveolar cells although, if the animals were not primed with hormones, they were closer to the "pre-secretory" alveolar cell of the "droplet"/ "doming" cultures, at least as judged by the two criteria above.

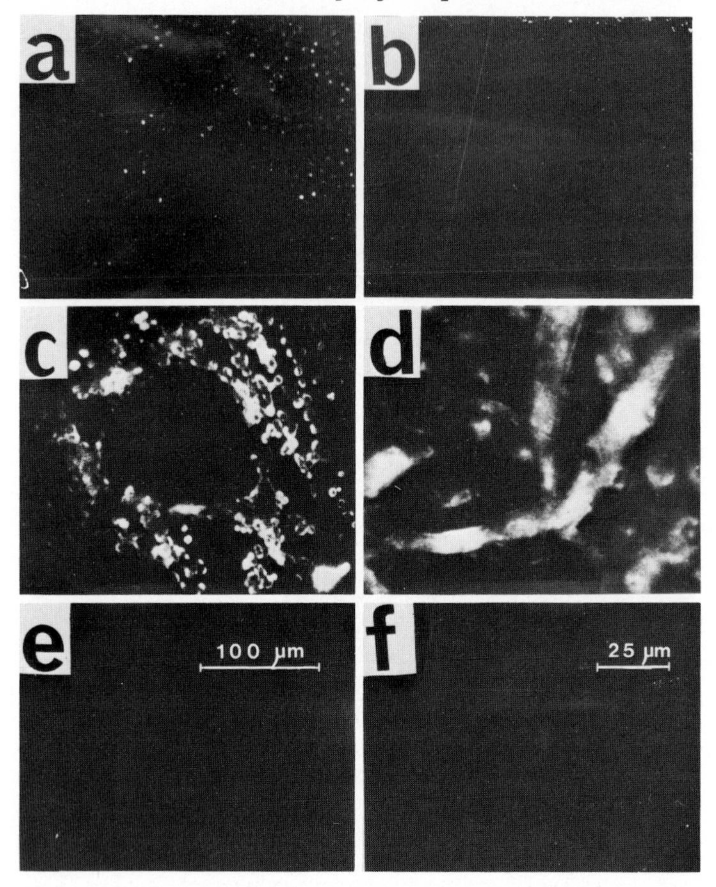

FIG. 7. Antigenic changes associated with differentiation to "droplet" cell and "doming" cultures. a,b: Cultures of "cuboidal" Rama 25 cells at saturation density, c,d: Similar cultures treated for two days with DMSO, hydrocortisone, insulin, and prolactin (Fig.8), e,f: "Elongated" Rama 29 cultures treated as above. They were fixed and then either stained directly with fluorescein conjugated peanut lectin (a,c,e) or indirectly with anti-rat casein serum and fluorescein conjugated goat anti-rabbit IgG serum (b,d,f). The fluorescent light emitted from the treated cultures after exposure to ultravoilet light was photographed. Magnification is the same in photographs a,c,e and in photographs b,d,f.

The reasons for the failure of Rama 25 cells to differentiate completely into active secretory alveolar cells in culture is unknown.

The effect of agents, notably DMSO, which were known to promote Friend erythroleukemic cells to differentiate, was investigated in detail by adding them in fresh medium containing 10% fetal calf serum and the mammotrophic hormones (hydrocortisone, insulin and prolactin) to Rama 25 cells at their saturation density. At this stage the cultures were mainly composed of "dark" cells with a few "droplet" cells and "grey" cells, rather than the original standard "cuboidal" cells. The kinetics of the increase in the number of "domes" and the amount of casein-like material produced for 3 different concentrations of DMSO are shown in Fig.8. Without DMSO there was a very slow rate of "dome" formation and of production of casein-like material. For increasing concentrations of DMSO there was a constant lag of about 8 hours before elevated rates of formation of "domes" and of production of casein-like material were observed; higher rates corresponded to increasing concentrations of DMSO. Since more than 90% of the casein-like material was released into the culture medium, a small period (perhaps less than 1 hour) of the lag might be due to the time required for the export of the increased amounts of casein synthesised in the cell (33). The reason for the majority of the lag is unknown. The number of "domes" and the amount of casein-like material for a given concentration of DMSO rose with time thereafter, reaching the top shoulder of the curve for 250mM DMSO after 24 hours.

The effect of other Friend cell differentiation-inducing-agents (31) is shown in detail in Table 3. All the "Friend-cell-inducers" tested were capable of increasing the number of "droplet" cells (not shown), "domes" and the total amount of immunoreactive casein produced after 24 hours. The concentrations which produced the maximum effects are shown; higher concentrations caused a reduction when measured after 24 hours, probably owing to their toxic effects. The most active of the "Friend-cell-inducers" (Table 3) tested was hexamethylenebisacetamide, although the most active inducer in our system was the vitamin A metabolite, retinoic acid. This was active at near physiological concentrations (37). The fractional increase in the number of "domes" and the amount of casein-like material was constant for different concentrations of the same agent, and for different agents, suggesting that all these diverse agents were interacting with the cell in the same way, possibly at the level of plasma membrane (22).

The effects of the three mammotrophic hormones with the optimal concentration of DMSO, on the number of "domes" and on the amount of casein-like material were different (Table 4). Most of the effect on "doming" was due to DMSO and the glucocorticoid; this combination produced only a moderate increase in the amount of casein. On the other hand DMSO and prolactin could produce near maximal amounts of casein-like material without increasing the number of "domes" at all. There was little effect of any of the hormones or of DMSO alone on either parameter of alveolar

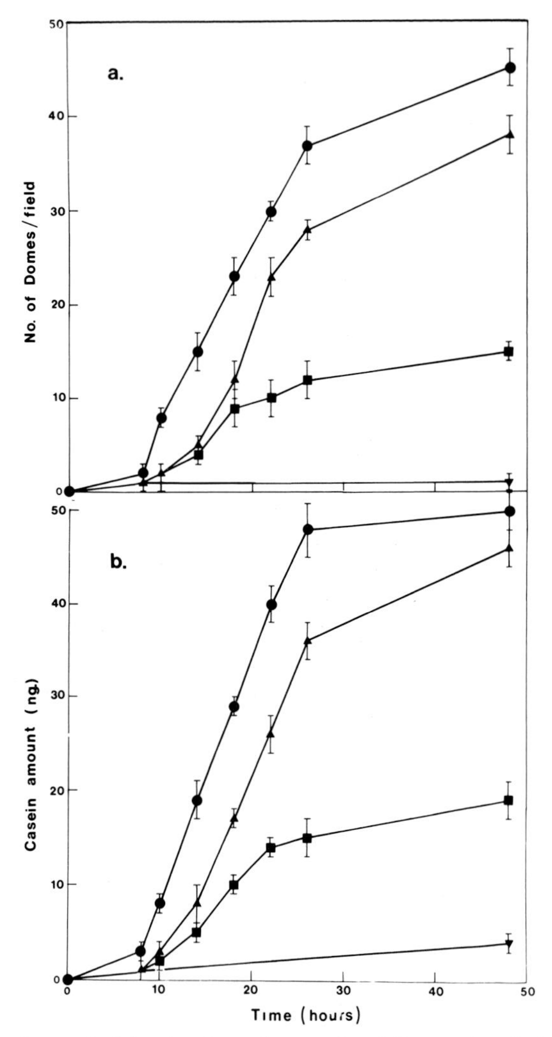

FIG. 8. Kinetics of "doming" and casein-like material production after addition of DMSO. Rama 25 cultures were grown to confluency (5 days) in Dulbecco's modified Eagle's medium, 10% fetal calf serum, 50ng/ml insulin, 50ng/ml hydrocortisone. The medium was changed on day 5 to one with 500ng/ml prolactin. On day 7 the medium was removed and replaced with fresh medium containing hydrocortisone, insulin, prolactin and in addition 250mM (—●—), 150mM (—▲—), 75mM (—■—), dimethyl sulphoxide (DMSO). Fresh medium without DMSO but with the hormones was also added (—▼—). In (a) the number of "domes" per $28mm^2$ field or (b) the ng of immunoreactive casein produced were followed with time (hrs.). Results are the average of 4 dishes ± SEM. For "doming" 3 microscopic fields were counted per culture dish, while for casein assay duplicate radioimmune assays were performed on the medium from each culture dish. Cellular protein content was approximately constant at 0.5mg in all cultures.

TABLE 3. Effect of different inducers on "doming" and casein-like material production

Compound	Concentration	No."domes"/field ± SEM	Total casein (ng) ± SEM
None	-	0 ± 0	2 ± 1
Dimethylsulphoxide	250 mM	50 ± 2	41 ± 2
Dimethylformamide	150 mM	52 ± 3	32 ± 2
N-methylacetamide	50 mM	55 ± 3	36 ± 1
N,N-dimethylacetamide	30 mM	51 ± 2	38 ± 1
1-methyl-2-piperidone	10 mM	44 ± 2	27 ± 2
Tetramethyl urea	10 mM	52 ± 2	31 ± 2
Hexamethylenebisacetamide	5 mM	46 ± 3	27 ± 1
Procaine	2.5 mM	32 ± 2	16 ± 1
6-thioguanine	10^{-4}M	24 ± 2	20 ± 1
Linoleic acid	3×10^{-6}M	16 ± 1	15 ± 2
Retinoic acid	10^{-6}M	59 ± 2	37 ± 2

Confluent culture of Rama 25 were set up as described in Fig.8. On day 7 the medium was then removed and replaced with fresh medium containing one of the compounds and hydrocortisone, insulin and prolactin. The number of "domes" per $28mm^2$ field and the total immunoreactive casein produced in 24 hours were recorded. Results are the means ± SEM from 4 cultures. Compounds tested which had little or no effect at the designated concentrations included: epidermal growth factor (10ng/ml), estradiol (5ng/ml), triiodothyronin (5ng/ml), prostaglandin E_1 (50ng/ml), prostaglandin $F_2\alpha$ (50ng/ml). Compounds which had a small effect included tetramethylenebisacetamide (5mM), tetramethylenebiformamide (5mM).

TABLE 4. Effect of hormones and DMSO on casein-like material secretion and "doming"

Additions	No. "domes" per field ± SEM	casein secreted (ng) ± SEM
None	0 ± 0	2 ± 1
P + HC + I	0 ± 0	4 ± 2
DMSO	1 ± 1	11 ± 4
DMSO + HC	25 ± 7	23 ± 3
DMSO + I	1 ± 1	12 ± 1
DMSO + P	1 ± 1	39 ± 7
DMSO + HC + I	21 ± 4	41 ± 6
DMSO + I + P	15 ± 4	38 ± 6
DMSO + HC + P	22 ± 5	39 ± 4
DMSO + HC + I + P	15 ± 5	36 ± 2
DMSO + HC + I + P + progesterone	37 ± 5	25 ± 2

Rama 25 cultures were grown as described in Fig.8 without hydrocortisone, insulin, prolactin in the medium. On day 7 fresh medium containing in addition 250mM dimethylsulphoxide (DMSO), 500 ng/ml prolactin (P), 50ng/ml hydrocortisone (HC), 50ng/ml insulin (I), or 1000ng/ml progesterone were added as indicated and the total amount of immunoreactive casein released into the medium and the number of "domes" per 28mm^2 were recorded after 48 hours. The casein amounts were the average of duplicate cultures ± SEM, and the number of "domes" were the average of a total of 10 fields from duplicate cultures ± SEM.

cell differentiation. Addition of progesterone to cultures growing in hydrocortisone, insulin, prolactin and DMSO increased the number of "domes" and reduced the amount of casein-like material. Thus the intracellular mechanisms which triggered formation of "domes" and production of casein-like material were at least partially different. The ability of progesterone to inhibit prolactin, hydrocortisone and insulin-stimulated production of casein-like material in our tissue culture cells was similar to its effect on production of casein messenger RNA in explant cultures (23) and on casein synthesis *in vivo* (16).

INTRACELLULAR REQUIREMENTS FOR DIFFERENTIATION TO ALVEOLAR-LIKE CELLS

The requirements for various intracellular processes for the appearance of two markers of alveolar cells, "doming" and casein-like material were tested. Each of the inhibitors cycloheximide, cordycepin, hydroxyurea and cytosine arabinoside when present for

24 hours with DMSO, hydrocortisone, insulin and prolactin, inhibited formation of "domes" and production of casein-like material almost completely (Table 5).

TABLE 5. Inhibitors of DMSO-induced "doming" and casein-like material production

Inhibitor	Concentration	No."domes"/field ± SEM	Total casein (ng) ± SEM
No DMSO	-	1 ± 1	3 ± 1
None	-	54 ± 2	30 ± 2
Cycloheximide	1μg/ml	0 ± 0	5 ± 2
Cordycepin	30μg/ml	0 ± 0	7 ± 2
Hydroxyurea	100μg/ml	1 ± 1	5 ± 1
Cytosine arabinoside	15μg/ml	5 ± 1	7 ± 1
Vinblastine	1μg/ml	0 ± 0	n.d.
Ouabain	1mM	9 ± 1	26 ± 2

Confluent cultures of Rama 25 were set up as described in Fig.8. On day 7 the medium was changed to one containing 250mM DMSO and different inhibitors at the given concentrations. The effect on the number of domes ± SEM/28mm field and the total amount of immunoreactive casein produced ± SEM were recorded after 24 hours as for Fig.8 and Table 3. n.d.; not determined.

Thus protein synthesis, poly(A)RNA synthesis, probably messenger RNA synthesis (38) and DNA synthesis were required during this 24 hour period for the appearance of both markers of differentiation. Protein synthesis was relatively unaffected (<15%) by the concentrations of hydroxyurea and cytosine arabinoside used while DNA synthesis was inhibited by >95%. Vinblastine inhibited the formation of "domes" which suggested that intact microtubules (44) might also be important in that process. Ouabain at 1mM, however, inhibited the formation of "domes" by more than 85% whilst barely affecting the production of casein-like material or protein synthesis, suggesting that even at this high concentration it was acting in a selective manner. Higher concentrations could completely inhibit the formation of "domes", but then production of the casein-like material and protein synthesis were also slightly reduced (M.J.W.unpubl.). Thus the activity of the membrane bound Na^+/K^+ ATPase was probably required for "dome" formation as it is in the MDCK line of kidney epithelial cells (1), but was not required for casein-like material production, once again confirming that the appearance of these two markers was caused by different intracellular pathways. The Na^+/K^+ ATPase possibly contributes to "dome" formation by pumping water under the cell monolayer, thereby forcing it up under hydrostatic pressure. The intracellular agent(s) which trigger the production of the casein-like material remain as yet unknown.

Since DNA synthesis was required for the appearance of both the markers for the alveolar cell, we investigated the relationship of cell DNA synthetic rates to the rate of appearance of one of the markers, "domes". Rama 25 cells at their saturation density had a very low DNA synthetic rate, (<3% of the cells were synthesising DNA in 4 hours). Adding fresh medium containing 10% fetal calf serum and the three hormones, hydrocortisone, insulin and prolactin caused an increase in the rate of incorporation of $\{^3H\}$-thymidine into DNA after a lag of about 8 hours, reaching a maximum value after about 16 hours (Fig.9a). This was reflected in the increase in cells synthesising DNA (14% in 4 hours). Cells then went into mitosis, divided, and often detached from the cell monolayer. Cells, therefore, were probably originally in G_1 and were stimulated to enter S phase and divide as when fresh medium is added to quiescent fibroblastic cell lines (36). Addition of fresh medium and the hormones, however, failed to increase the rate of "doming". Addition of 250mM DMSO together with the fresh medium and hormones caused a maximal increase in the rate of "doming" after an 8-10 hour lag period (Fig.9b) as described earlier, and a concomitant decrease in the rate of $\{^3H\}$thymidine incorporated into DNA. Hormones and DMSO when added without changing the medium caused just about 50% of the maximal increased rate of formation of "domes" with little or no change for the first 20 hours in the rate of DNA synthesis, and thereafter a sharp reduction (Fig.9a). Thus, although there is a requirement for DNA synthesis before increased rates of "doming" and of production of casein-like material occur, as observed in experiments with inhibitors of DNA synthesis, DMSO itself acts to reduce DNA synthetic rates at the same concentration and in about the same time that it stimulates the rate of differentiation. Presumably the minimal DNA synthetic rate observed in the presence of DMSO suffices to satisfy the requirement of DNA synthesis for differentiation, but whether this reduction in DNA synthetic rate plays

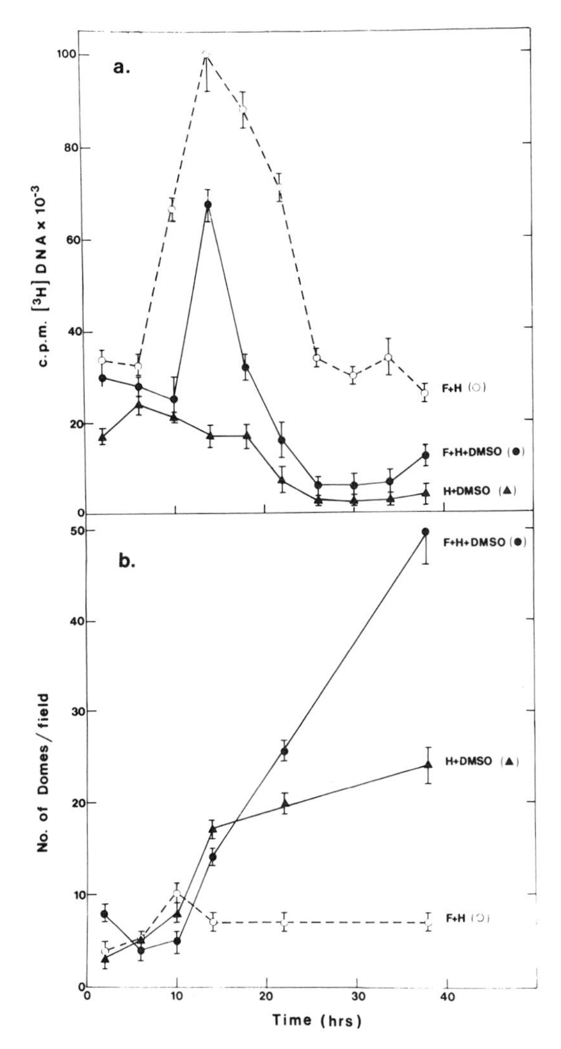

FIG. 9. Changes in rates of thymidine incorporation into DNA after addition of DMSO. Confluent monolayers of Rama 25 cells were set up as described in Fig.8. On day 7 the medium was removed and replaced with on of the following: fresh medium and hydrocortisone, insulin and prolactin (F + H) (--o--), fresh medium, hydrocortisone, insulin, prolactin and 250mM DMSO (F + H + DMSO) (—●—); or hydrocortisone, insulin, prolactin and 250mM DMSO were added to the original medium (H + DMSO) (—▲—) a. Cultures were exposed to {^{3}H}-thymidine (3μCi/ml, 3μM) for 4 hours, and the c.p.m. incorporated into DNA for ¼ culture were recorded (35). The time recorded represents the mid-point of the labelling period. b. "Domes" per 28mm^2 were recorded at the same time in parallel cultures. Results for DNA synthesis were the mean ± SEM of duplicate samples from duplicate cultures, results for "doming" were the mean ± SEM from counting 3 fields from each of 2 dishes.

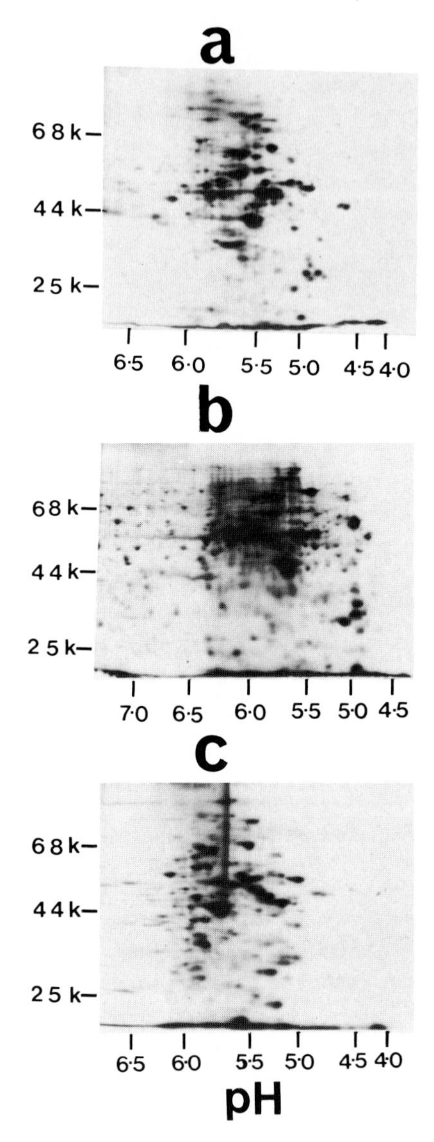

FIG. 10. Proteins synthesised by cultured cells. Confluent cultures of one of: a. "cuboidal" Rama 25 cells; b. Rama 25 cells at saturation density treated with hydrocortisone, insulin, prolactin, 250mM DMSO: c. "elongated" Rama 29 cells were exposed to {^{35}S}methionine (20μCi/ml) for 24 hours. The cell monolayers were then dissolved in 9M urea, 2% nonidet P40 and analysed on two-dimensional polyacrylamide gels, isoelectric focusing in the first dimension and electrophoresis through 10% polyacrylamide SDS gels in the second dimension (29). Radioactive proteins were visualised by radioauthography. The apparent pH values and molecular weights were estimated from known standards.

a role in altering the rate of differentiation remains to be determined.

Since protein synthesis was also required to increase the rate of appearance of both markers for the alveolar cells, we investigated the proteins synthesised during conversion of the "cuboidal" cells to alveolar-like cells. The radioactive proteins synthesised by the "cuboidal" cells and the "droplet" cell/"doming" cultures were quite similar when observed on two dimensional gel systems (29)(M.J.W.unpubl.) with only minor differences (Fig.10). A few radioactive protein spots were increased and others decreased in intensity. This suggested that no major reorganisation of the pattern of gene expression had taken place. Major changes, however, were observed when comparing the proteins synthesised by the "cuboidal" cells and the "elongated" myoepithelial-like cells (Fig.10). The protein changes involved in going from the "cuboidal" to either the "elongated" cells or the "droplet" cells/"doming" cultures, and their possible role in the differentiation processes, remain to be elucidated however.

The testing of the effects of hormones and other agents on the cellular interconversions of our stem cell line, coupled with the ability to identify different molecular changes concurrently occurring, should lay the foundations for the elucidation of the molecular mechanisms underlying the differentiation processes of the mammary gland.

ACKNOWLEDGEMENTS

We thank the following for excellent technical assistance in the order in which the work was completed; Helga Durbin, Linda Peachey, Linda Head, Ruth Thompson, Carol Pain and Mary Oliver. D.C.B. was in receipt of an Imperial Cancer Research Fund Scholarship during the course of this work.

REFERENCES

1. Abaza,N.A., Leighton,J., and Schultz,S.G. (1974): In vitro, 10: 172-183.
2. Banerjee,M.R. (1976): Int. Rev. Cytol., 47: 1-97.
3. Bennett,D.C. (1979): Ph.D. thesis. University of London, England.

4. Bennett,D.C., Peachey,L.A., Durbin,H., and Rudland,P.S. (1978): Cell, 15: 283-298.
5. Bretscher,A., and Weber,K. (1978): J. Cell Biol., 79: 839-845.
6. Dils,R., Clark,S., and Knudsen,J. (1977): In: Comparative Aspects of Lactation, edited by M. Parker, pp 43-55. Academic Press, New York.
7. Douglas,T.C. (1973): Transplant Proc., 5: 79-85.
8. Dulbecco,R., Bologna,M., and Unger,M. (1979): Proc. Natl. Acad. Sci. USA., 76: 1848-1852.
9. Elias,J.J.(1957): Science, 126: 842-844.
10. Emerman,J.T., Enami,J., Pitelka, D., and Nandi,S. (1977): Proc. Natl. Acad. Sci. USA., 74: 4466-4470.
11. Friend,C., Scher,W., Holland,J.G., and Sato,T. (1971): Proc. Natl. Acad. Sci. USA., 68: 378-382.
12. Graham,J.M., Hynes,R.O., Davidson,E.A., and Bainton,D.F. (1975): Cell, 4: 353-365.
13. Gros,R.J. (1967): In: Control of Cellular Growth in Adult Organisms, edited by H. Teir and T. Rytoma, pp 3-27. Academic Press, New York.
14. Hackett,A.J., Smith,H.S., Springer,E.L., Owens,R.B., Nelson-Rees,W.A., Riggs,J.L., Gardner,M.B. (1977): J. Natl. Cancer Inst., 58: 1795-1806.
15. Hallowes,R.C., Rudland,P.S., Hawkins,R.A., Lewis,D.J., Bennett,D.C., and Durbin,H. (1977): Cancer Res., 37: 2492-2504.
16. Houdebine,L-M. (1976): Europ. J. Biochem., 68: 219-225.
17. Huggins,C., Briziarelli,G., and Sutton,H. (1959): J. Expt. Med., 109: 25-42.
18. Hynes,R.O., and Humphryes,K.C. (1974): J. Cell Biol., 62: 438-448.
19. Kon,S.K., and Cowie,A.T. (1961): In: Milk, The Mammary Gland and its Secretions, ch.1. Academic Press, New York.
20. Lazarides,E., and Weber,K. (1974): Proc. Natl. Acad. Sci. USA 71: 2268-2272.
21. Lyons,R.W., Li,C.H., and Johnson,R.E. (1958): Rec. Prog. Horm. Res., 14: 219-254.
22. Mager,D., and Bernstein,A. (1978): J. Cell Physiol., 94: 275-286.
23. Matusik,R.J., and Rosen,J.M. (1978): J. Biol. Chem., 253: 2343-2347.
24. Mayne,R., Vail,M.S., and Miller,E.J. (1975) Proc. Natl. Acad. Sci., 72: 4511-4515.
25. McGrath,C.M. (1975): Amer. Zool., 15: 231-236
26. Moosekar,M.S., Pollard,F.D., and Fujiawa,K. (1978): J. Cell Biol., 79:444-453.
27. Nandi,S. (1958): J. Natl. Cancer Inst., 21: 1039-1055.
28. Newman,R., Klein,P.J., and Rudland,P.S. (1979): J. Natl. Cancer Inst. in press
29. O'Farrell,P.H. (1975): J. Biol. Chem., 250: 4007-4015.

30. Pearson,O.H., Molina,A., Butler,T.P., Llerena,L., and Nasr,H. (1972): In: Estrogen Target Tissues and Neoplasia. edited by T.L. Dao, p 287 Univ. of Chicago Press, Chicago.
31. Reuben,R.C., Wife,R.L., Breslow,R., Rifkind,R.A., and Marks,P.A. (1976): Proc. Natl. Acad. Sci. USA., 73: 862-866
32. Ritter,M.A., Morris,R.J., and Goldschneider,I. (1979): Immunology: in press.
33. Rudland,P.S., Bennett,D.C., and Warburton,M.J. (1979): Cold Spring Hb. Symp. Cell Prol. 6, Hormones and Cell Culture ch 47. in press.
34. Rudland,P.S., Bennett,D.C., and Warburton,M.J. (1980): First International Congress on Hormones and Cancer, in press, Raven Press, New York.
35. Rudland,P.S., Hallowes,R.C., Durbin,H., and Lewis,D. (1977): J. Cell Biol., 73: 561-577.
36. Rudland,P.S., and Jimenez de Asua,L. (1979): Biochem. Biophys. Acta., Cancer Res. Revs., 560: 91-133.
37. Sauberlich,H.E., Hodges,R.E., Wallace,D.L., Kolder,K., Canham,J.E., Hood,J., Raica,N., and Lowry,L.K. (1974): Vitamins and Hormones, 32: 251-275.
38. Sheldon,R., Jurale,C., and Kates,J. (1972): Proc. Natl. Acad. Sci. USA., 69: 417-421.
39. Tandler,B. (1965): Z. Zellforsch, 68: 852-863.
40. Talwalker,P.K., and Meites,J. (1961): Proc. Soc. Exp. Biol. Med., 107: 880-883.
41. Trelstad,R.L., and Slavkin,H.C. (1974): Biochem. Biophys. Res., Comm., 59: 443-449.
42. Trowbridge,I.S., Hyman,R., and Mazauskas,C. (1978): Cell, 14, 21-32.
43. Weber,K., and Groeschel-Stewart,U. (1974): Proc. Natl. Acad. Sci., 71: 4561-4564.
44. Wilson,L., Anderson,K., and Chin,D. (1976): Cold Spr. Hb. Symp. Cell Prolif., 3, Cell Motility, 1051-1064.

Control Mechanisms in Animal Cells,
edited by L. Jimenez de Asua et al.
Raven Press, New York © 1980.

Preparation and Initiation of DNA Synthesis in Eggs

J. Brachet

Department of Molecular Biology, University of Brussels, Genèse, Belgium; and Laboratory of Molecular Embryology, C.N.R., 80072 Naples, Italy

INTRODUCTION

DNA synthesis during the first stages of embryonic development presents several unusual features : during the very long period that precedes the formation of fertilizable eggs (oogenesis and maturation), there is no replication of chromosomal DNA; but the machinery required for DNA synthesis is built up during this early phase of development. Fertilization (or parthenogenesis) almost immediately induces very fast DNA replication and mitotic activity; cells become smaller and smaller during this period of cleavage. When the cells, at the blastula stage, have reached a normal nucleocytoplasmic volume ratio, the cell cycle becomes similar to that of somatic cells.

OOGENESIS

The last S-phase before the formation of the oocyte takes place in the oogonia; it is followed by entry into meiotic prophase. During its whole growth period, the oocyte has a tetraploid chromosome complement (4C). Its chromosomes remain in the diplotene stage of meiosis and stay in a highly decondensed state; these so-called "lampbrush" chromosomes are formed of chromomeres and loops; the latter are extremely active in transcription and are believed to synthesize between 10.000 and 20.000 different kinds of messenger RNA's (mRNA's). But neither the chromomeres nor the loops become labelled after repeated injections of radioactive thymidine : their DNA is thus very stable during a period of time which can last several years.

However, at the beginning of oogenesis (pachytene stage of meiosis), an important synthesis of extra-chromosomal DNA takes place in Amphibians : it is due to the extensive replication (amplification) of the 28S and 18S ribosomal genes, which later form the nucleolar organizers. In the toad Xenopus, somatic cells contain about 1.000 copies of the 28S and 18S rRNA genes; the oocytes build up as many as 2.10^6 copies of these genes (14).

The number of the nucleoli in the oocyte's large nucleus (germinal vesicle : GV) can vary from 500 to 2.400 for the same stage of oogenesis; but the total number of nucleolar organizers remains constant : fusion and fission of nucleoli thus occur during oogenesis (77). Amplification of the ribosomal genes takes place when replication of chromosomal DNA has stopped; it lasts for several days. Electron microscopy has disclosed (37,65) the presence of unusual DNA replication forms (circles, lariates) during the amplification process. The fact that ribosomal DNA (rDNA) is heavier than bulk chromosomal DNA and that it is so abundant, as a result of amplification, has allowed its isolation and a very detailed analysis of its molecular structure (especially in D. Brown's laboratory). It is outside the scope of the present review to discuss the results obtained by digestion with restriction enzymes of cloned fragments of rDNA (63;84,85). Electron microscopy (55;68) has also been a very powerful tool for the study of the transcription units present in the nucleolar organizers. The ribosomal genes are made of repeat units arranged in tandem and separated by spacers, whose length and base composition are highly variable even in the same nucleolus. It should be pointed out, that rDNA amplification is much less impressive in species where oocytes have a single nucleolus (worms, mollusks, echinoderms, mammals); for instance, there is only a 16-fold amplification in the oocytes of the echiuroid worm Urechis.

The number of mitochondria greatly increases during amphibian oogenesis, involving a considerable synthesis of mitochondrial DNA (19; 6). However, biogenesis of mitochondria progressively decreases and it almost stops during the second half of oogenesis, as recently shown by Callen and Mounolou (15) and by Webb and Camp (83). This decrease could be due to a negative control exerted by the G.V. on replication of the mitochondrial genome: Rinaldi et al. (64) have shown that mitochondrial DNA undergoes synthesis when anucleate fragments of sea urchin eggs are parthenogenetically activated while there is no multiplication of the mitochondria when whole eggs are treated in the same way. In Xenopus oocytes, mitochondrial DNA molecules are circular and have a M.W. of 4.7×10^6 daltons. Their main transcription products are the mitochondrial rRNA species (21S and 13S) and several t-RNA species which are not coded for by the nuclear genome; these RNA's associate to proteins in order to form mitochondrial "miniribosomes" (75). It is not yet known whether mitochondrial DNA, which represents about 10% of total DNA, codes for mRNA's; but there is little doubt that protein synthesis directed by endogenous mitochondrial DNA must be very small, if it exists at all. In fact, 80% of the mitochondrial DNA is made of spacers which have widely diverged during Evolution and which are devoid of coding significance. The remaining 20% of the mitochondrial genome are almost entirely occupied by the above-mentioned mitochondrial rRNA and tRNA genes. It is probable that, as in Mammals, mitochondrial DNA codes for only about 9 peptides of the respiratory chain enzymes, while for the synthesis of the others and for the formation and functioning of the mito-ribosomes, activity of more than 100 nuclear genes is required (Coote et al., (17).

Analyses of the DNA content of amphibian eggs by a variety of methods show that they contain a large excess of DNA over the expected tetraploid value. Mitochondria, as we have seen, represent only 10% of total egg DNA. Where is the major part of the extra DNA located in the oocyte ? Studies from this laboratory have shown that more than 65% of cytoplasmic DNA is present in the yolk platelets (35). Yolk DNA is made of linear double-stranded molecules somewhat smaller than those obtained by similar methods from Xenopus chromatin (respective M.W.'s 26.10^6 and 40.10^6 daltons). It is unlikely that yolk DNA is synthesized by the oocyte itself, since yolk platelets remain unlabelled in oocytes which were repeatedly injected with radioactive thymidine. It seems more probable that yolk DNA (if it is not just a preparation artefact) has an exogenous origin; like vitellogenin (the precursor of the yolk proteins), it might be present in the bloodstream and endocytosed by vitellogenic oocytes. In fact, at the time of vitellogenin synthesis by the liver, one observes extensive haemolysis and hepatocyte degeneration around the blood vessels. It is thus probable that yolk DNA arises from cytolysing somatic cells; a similar view has been expressed by Opresko et al. (59). It is unlikely that yolk DNA plays any genetic role; it is more probable that it is a reserve of deoxynucleotides to be used when the yolk platelets break down during embryonic development.

Xenopus oocytes contain three different types of DNA polymerases (Grippo et al.,29). Two of them (corresponding to the usual DNA polymerases - α and - β) are entirely located in the G.V. (47 ; 30); the third one is entirely cytoplasmic and probably mitochondrial (5). It is not known whether the nuclear DNA polymerases display any activity during oogenesis; if so, they should be more concerned with DNA repair than with semi-conservative DNA replication. When DNA's of various origins (ribosomal, 5S, viral, etc.) are injected directly into the V.G. of Xenopus oocytes, they are not replicated but they are accurately transcribed (54, 34) and they can be organized in nucleosomes (80). Factors which promote the unwrapping of the DNA which surrounds the nucleosomes (5) and which induce the formation of the latter (44) have been isolated from Xenopus oocytes.

In conclusion, the oocyte is highly specialized in RNA (and protein) synthesis; transcription is intense thanks to the presence of large amounts of the 3 RNA polymerases in the G.V. and of cytoplasmic factors which enhance the activity of these nuclear enzymes (18). On the other hand, there is no chromosomal DNA synthesis during months or years of growth. The oocyte thus behaves like a G_0-cell; it will not re-enter the cell cycle unless it is triggered to undergo maturation.

MATURATION

Amphibian oocytes are surrounded by a layer of follicle cells, which synthesize progesterone as a consequence of pituitary stimulation. Progesterone is the physiological inducer of maturation; in the laboratory, maturation can easily be obtained when full-grown oocytes dissected out of the ovary are treated in vitro with low concentrations of the hormone. The response of the oocyte to progesterone is not specific; besides other steroids, a number of agents which have in common to increase the intracellular Ca^{++} concentration (organomercurials, $LaCl_3$, etc.) also induce maturation, i.e. resumption of meiosis. The nuclear membrane surrounding the G.V. breaks down (GVBD), the lampbrush chromosomes condense and attach to a meiotic spindle, the first polar body is expulsed. Maturation ends when the chromosomes (whose number has decreased by half) are at metaphase II stage. The egg remains arrested at this stage until it is fertilized or parthenogenetically activated. From the biochemical viewpoint, the most important changes (besides the release of membrane-bound Ca^{++}) are an initial drop in the cAMP content and an increase in protein phosphorylation at the time of GVBD. Before GVBD, there is an increase in the rate of protein synthesis and the production of a factor (maturation promoting factor : MPF) which induces nuclear membrane breakdown and chromosome condensation (50). Interestingly, MPF has recently been detected in HeLa cells, where it is produced during the G_2 phase of the cell cycle and reaches a peak at mitosis (74): MPF might thus play a role in nuclear membrane breakdown and chromosome condensation in all cells. We still know little about the chemical nature of MPF : it is a protein which is inactivated by Ca^{++} and activated by Mg^{++}. Its intracellular activity could thus be modulated by changes in the balance of bivalent ions. Its synthesis takes place in oocytes enucleated before progesterone stimulation, but requires protein synthesis. When oocytes are injected with MPF, they synthesize more MPF by some kind of autocatalytic or amplification process.

There is no chromosomal or nucleolar DNA synthesis during maturation; all that can be detected is a low level of mitochondrial DNA synthesis, even if the oocyte has been enucleated before progesterone treatment (36). According to Barat et al. (6), the rate of mitochondrial DNA synthesis still decreases during progesterone-induced maturation in amphibian oocytes. The rDNA of the nucleolar organizers is eliminated in the cytoplasm where spherules of extrachromosomal Feulgen staining material can be seen under the microscope; this now cytoplasmic rDNA does not serve as a template for replication and it remains intact in unfertilized eggs and even during early cleavage (78).

Although there is no DNA synthesis during maturation (except for mitochondrial DNA) dramatic cytoplasmic changes, which have much to do with DNA synthesis take place. Gurdon (32) has injected, into Xenopus oocytes and unfertilized eggs, nuclei isolated from adult brain, where no DNA replication normally takes place. He found that, if the nuclei are injected into oocytes, they swell, form large nucleoli and synthesize RNA; there is no DNA synthesis. The swelling of the injected nuclei is largely due to the uptake of proteins located in the oocyte's cytoplasm; this uptake is selective, i.e. it does not depend on the size or charge of the cytoplasmic proteins; the reasons for this selectivity remain unknown. If the same adult brain nuclei are injected into oocytes undergoing maturation (at about the time of GVBD), their chromatin undergoes condensation, but there is no DNA or RNA synthesis. But, if the nuclei are injected into unfertilized eggs, they swell and synthesize DNA : there is no RNA synthesis and nucleoli do not make their appearance. These experiments suggest that the cytoplasm of the unfertilized egg contains a repressor of RNA synthesis and an inducer of DNA synthesis. As we shall see, there is experimental evidence for the view that factors which stimulate DNA synthesis become detectable during maturation. Regarding the shutdown of RNA synthesis, the only thing we know is that the factors which stimulate RNA polymerase activity in the oocytes are no longer detectable after maturation (18) : the decrease in RNA synthesis could thus be due to the degradation or the inactivation of stimulating factors rather than to the production of inhibitors; the two mechanisms could of course co-exist and cooperate in the shutdown of RNA synthesis. It is worth noting that many, if not all the changes which lead to the production of new proteins during maturation still take place when the oocyte has been enucleated before progesterone stimulation. Delicate controls of specific protein synthesis obviously exist, at the post-transcriptional level, in hormone-stimulated oocytes.

Other experiments by Gurdon (33) and Ford and Woodland (24) have shown that if single stranded DNA is injected into oocytes, a complementary strand can be built; but there is no net DNA synthesis. In sharp contrast, any kind of DNA (except circular Escherichia coli DNA) is replicated if it is injected into eggs.

It is thus clear that the machinery required for DNA synthesis is built up during maturation, but is not used for the replication of endogenous chromosomal and ribosomal DNA; they are protected against the DNA-synthesizing machinery which is present in the cytoplasm after GVBD. It could be that the 50 fold increase in histone H_1 synthesis which takes place during maturation is connected with the protection of endogenous DNA against this DNA synthesizing machinery at maturation. Interestingly, H_1 histone synthesis is selective again (i.e. the other histones are synthesized at a much lower rate)and is again controled at the post-transcriptional level (1 ; 66).

There is clear evidence (28; 29; 7) that DNA polymerase activity increases 4 fold during maturation and that a new form of DNA polymerase makes its appearance during this process. There is no further increase in DNA polymerase until the neurula

(10.000 cells) stage. However, DNA polymerase activity alone cannot explain the exceedingly fast DNA replication which occurs during cleavage. But, as shown by Benbow and Ford (7), amphibian eggs contain an initiation factor (FI) which stimulates DNA synthesis in nuclei isolated from adult Xenopus liver; this factor is present in minute amounts only in the oocytes; it greatly increases in activity during maturation, remains very active during cleavage and progressively decreases to low activity afterwards. It has been suggested that factor FI might be a specific endonuclease which recognizes long palindromic sequences. Similar factors, which stimulate DNA synthesis in homologous or heterologous isolated nuclei have been found in other fast replicating cells (21; 25) including fertilized sea urchin and clam eggs (38). These factors display no species-specificity, are soluble, thermolabile and non-dialyzable; more precise characterization is obviously needed before their mechanism of action and biological significance can be assessed.

The production of the necessary precursors for DNA synthesis also takes place during maturation : thymidine phosphorylation increases (86; 6) and ribonucleotide reductase (an enzyme required for the building up of a deoxynucleotide pool) is present in unfertilized Xenopus eggs, but not in oocytes (79). Activity of this enzyme is not required for maturation itself, since GVBD and polar body expulsion are unaffected by treatment of stimulated oocytes with hydroxyurea, an inhibitor of ribonucleotide reductase.

In summary, maturation is a very important and complex step in development. One of its major functions is the building up, in the cytoplasm, of the DNA-synthesizing machinery which will be used at later stages of development. The oocyte undergoing maturation lies in the G_1-phase of the cell cycle if one admits, with Rao et al. (62), that the role of the G_1-phase, when it exists, is the synthesis of factors which will allow DNA synthesis during the S-phase. In eggs, fertilization (or parthenogenetic activation) is required for the induction of the first S-phase in embryogenesis

FERTILIZATION

From now on, we shall shift from amphibian to sea urchin eggs since the latter have always been a favourite material for chemical embryologists interested in the analysis of fertilization and cleavage.

Artificial fertilization is quickly followed by dramatic morphological, physical and chemical changes (reviewed by Epel et al. 22). Among them are the cortical reaction (breakdown of the cortical granules, uplifting of the fertilization membrane) which is triggered by a release of Ca^{++}, an increase in oxygen consumption (due to the formation of H_2O_2 which is used by an ovoperoxidase for the hardening of the fertilization membrane), an increase in the rate of protein synthesis, the induction of DNA synthesis , an increase in the intracellular pH, etc... This increase in pH which results from an efflux of protons and an intake of Na^+ (22) is the major factor controlling protein (and probably DNA) synthesis (27).

As soon as the spermatozoon nucleus is injected into the egg, it loses its specific basic proteins : a recent study by Carroll and Ozaki (16) shows that 3 histone-like proteins typical of sperm are lost and replaced by those of the egg (maternal histones) in the zygote nucleus. This leads to decondensation of the sperm chromatin. DNA synthesis immediately follows; it takes place simultaneously in both male and female pronuclei and the fertilized egg reaches the 4C value within 10-20 min. It is unlikely that DNA synthesis is triggered by the histone changes in the male pronucleus since the female pronucleus of the egg, which contains maternal histones, does not replicate its DNA unless parthenogenesis is induced.

In both sea urchin (58 ;20) and Xenopus (79) eggs, fertilization is followed by a sharp increase in ribonucleotide reductase activity. This increase is inhibited by puromycin, but not by actinomycin D; this suggests that enzyme synthesis is directed by pre-existing maternal mRNA.

Interestingly, it is possible to dissociate the cortical reaction (uplifting of the fertilization membrane) from the stimulation of DNA and protein synthesis : in eggs treated with ammonia (52), procaine (81), urea or glycerol (53), there is no breakdown of the cortical granules and thus no fertilization membrane formation; nevertheless, the chromosomes condense and DNA is replicated. Ammonia induces the same increase in internal pH (0.3 - 0.5 units) as fertilization (39 ;71) as already mentioned the increase in pH is responsible for the increase in protein synthesis and probably of DNA synthesis as well. Other agents which activate sea urchin eggs, butyric acid for instance, induce DNA synthesis; however, in eggs treated with the classical method of J. Loeb for parthenogenesis, the rate of DNA synthesis decreases when, after an initial butyric acid treatment, the eggs are transfered to hypertonic sea water (57).

Unfertilized amphibian eggs contain a "cytostatic factor" which arrests cells in metaphase (and thus prevents later DNA synthesis) when it is injected into one of the blastomeres of fertilized eggs at the two-cell stage (49). It is believed that the role of this inhibitory factor is to arrest maturation at metaphase II. This cytostatic factor cannot be detected anymore in fertilized eggs; it is likely that it is inactivated by the release of Ca^{++} which takes place at fertilization (48). It is not known whether a similar cytostatic factor is present in sea urchin eggs.

CLEAVAGE

The large increase in number of the nuclei during cleavage has been known to embryologists since a long time; many of them believed that DNA (then called thymonucleic acid) is not synthesized during cleavage, but is translocated from a cytoplasmic reserve to the nuclei. Using the diphenylanine reaction specific for d-ribose, I could give the first demonstration that extensive DNA synthesis takes place during cleavage of sea urchin eggs. A recent paper by E. Parisi et al. (60) gives a curve for DNA synthesis which is not very different from the one I obtained more than 40 years ago; but their work brings to light interesting correlations between DNA synthesis and the mitotic index, which markedly drops at hatching : while 60% of the cells are in mitosis in the 6 hr blastula, there are only 11% at hatching and 4% in the mesenchyme blastula. Of great interest for embryologists is the fact that, from the 16-cell stage on, a gradient in mitotic activity, decreasing from the vegetal to the animal pole, can be detected (60). The 16-cell stage is the time of embryonic determination; it is interesting to note that, at this stage, quantitative changes in the pattern of histone synthesis can be detected between the micromeres and the other cells (70) and that two different histone genes become active at this time (70). In contrast to the extensive nuclear DNA replication which characterizes cleavage, there is apparently no multiplication of the mitochondria from maturation until the larval pluteus stage (64).

The cell cycle, in cleaving eggs (especially during the first mitotic cycles), is very different from that of ordinary somatic cells : there is no G_1 phase, DNA replication (S-phase) is exceedingly rapid (8-15 min. according to species) and the G_2-phase is confused with mitotic prophase. In large eggs such as those of the Amphibians, furrowing is a slow process, so that the nuclei already enter in prophase at a time when the first furrow has not yet reached the vegetal pole of the egg; there is thus, in such eggs, a marked asynchrony between karyo- and cytokinesis as compared with ordinary cells.

The absence of a G_1-phase can be explained by the fact that the whole machinery for DNA synthesis has been built up during maturation, a period of development which we have compared to a prolonged G_1-phase. However, there seems to be a need for increased production of deoxyribonucleotide precursors during cleavage : ribonucleotide reductase continues to greatly increase in activity during early cleavage in both sea urchin (58 ; 20) and amphibian (79) eggs. Inhibitors of this enzyme, such as hydroxyurea (10) and deoxyadenosine (11), stop cleavage of sea urchin eggs at the 8-16 cell stage; inhibition of cleavage by deoxyadenosine is completely reversed by addition of thymidine to the medium. This result is consistent with existing data about the size of the deoxyribonucleoside triphosphate pool in sea urchin eggs (51 ;26). Both the size of this pool and ribonucleotide reductase activity strongly decrease at the blastula stage when, as we have seen, mitotic activity falls to a low rate. Another enzyme, which has a more obscure function, might be involved in DNA replication during sea urchin egg cleavage: it is a DNA-topo-isomerase, which removes superhelical turns from supercoiled DNA; it is synthesized during cleavage, localized in the nuclei and has a maximal activity during the S-phase (61). On the other hand, the amount of DNA polymerase present in unfertilized sea urchin eggs is sufficient for the production of 1000 nuclei (56); there is no synthesis during cleavage of the enzyme, which is thought to move from the cytoplasm into the nuclei when the chromosomes swell at telophase (23).

The unusual speed of DNA replication during cleavage (less than 8 min.in Drosophila) is probably due to the presence of numerous initiation points; this is suggested by electron micrographs of DNA isolated from cleaving Drosophila eggs: they show that replication "eyes" are close to each other (87). The size of the replicons during the third cleavage of sea urchin eggs has been estimated as 4×10^4 daltons by Shimada and Teramaya (72), who showed that DNA pieces of this size are synthesized after a 15s pulse; neosynthesized DNA has become macromolecular after a 15 min. pulse. A recent electron microscope study by Baldari et al. (4) has demonstrated that DNA isolated during the S-phase (also at the third cleavage of sea urchin eggs) shows clusters of "microbubbles" made of single stranded DNA. These microbubbles - which presumably represent a stage of DNA replication - are still present in DNA isolated from blastulae, but are much less frequent.

A few other peculiarities of DNA replication during cleavage deserve a brief mention. For instance, Baker (3) has reported, on the basis of molecular hybridization data, that nuclei, during early cleavage of sea urchin eggs, contain a small (10S) DNA molecule; from blastulae, he could isolate a 60S DNA species containing 80 segments homologous to the 10S DNA species found in earlier stages; finally, in late blastulae, he found a 240S DNA species which contained an increased number of sequences which were present in the 10S and 60S DNA species. It thus looks as if low-molecular weight species were progressively integrated into macromolecular DNA during cleavage. Another peculiarity of DNA synthesis during cleavage is that labelled uridine is a good precursor (although not as good as thymidine); the label is recovered mainly in DNA cytidine and, to a lesser extent, in its thymidine (76). This efficient utilization of uridine for DNA synthesis is probably linked to the low level of RNA synthesis during early cleavage.

It is likely that the initiation factor FI discovered by Benbow and Ford (7) and similar factors (38) play a crucial role in the initiation, as many different points, of DNA replication during early cleavage. The slowdown of DNA replication and the appearance of a regular cell cycle in the blastula might be due to a decrease in the activity of FI. Many biochemical changes are known to take place in blastulae; among them is the appearance, in both sea urchins and amphibians, of basophilic, RNA-containing nucleoli. Their formation is probably not a mere consequence of the slowing down of mitotic activity since we found (12) that only "cleavage nucleoli", which are devoid of RNA, form in the nuclei of eggs arrested during cleavage by treatment with inhibitors of DNA synthesis. Another factor which might play a role in the slowing down of DNA replication and of mitotic activity in blastulae could be a change in histone synthesis. In contrast with ordinary somatic cells, there is no temporal coincidence between histone synthesis and DNA synthesis : anucleate fragments of sea urchin eggs synthesize histones after parthenogenetic activation; this synthesis is directed by mRNA's of maternal origin. After fertilization, there is an additional de novo synthesis of histone mRNA's (2). There is strong evidence that the regulation of histone synthesis changes, in sea urchins, at the blastula stage : according to Kunkel and Weinberg (41), new histone genes are switched on (activated) in the hatching blastula; for Grunstein (31), the genes coding for histone H_4 are different before and after hatching; Spinelli et al. (73) found similar electrophoretic patterns for histone mRNA's extracted from sea urchin oocytes, cleaving eggs and morulae; but blastulae gave a very different pattern. This shift from "cleavage" to "embryonic" histones could affect the condensation of chromatin and its accessibility to the DNA synthesizing machinery. However, it should be pointed out that, according to Keichline and Wasserman (40), the organization of chromatin in nucleosomes does not change during sea urchin egg development; this organization is very different in sea urchin sperm, which contains the specific histone species which disappear soon after fertilization.

Among the many factors which regulate DNA synthesis during cleavage might be the production of polyamines : the only enzyme, besides ribonucleotide reductase, which is known to be synthesized during cleavage, is ornithine decarboxylase (ODC), the key enzyme for polyamine biosynthesis. This is the case in amphibians (67), sea urchins (45 ; 42) and the Nudibranch Phestiella (42). According to Kusunoki and Yasumusu (43), there are cyclic changes related to the mitotic cycle, in ODC activity and polyamine content during sea urchin egg cleavage. The same authors have more recently (43) found that an inhibitor of ODC (α -hydrazinoornithine) slows down cleavage and arrests development at the morula stage in the sea urchins Hemicentrotus and Anthocidaris. These results are at variance with those we obtained (13) on two other sea urchin species, Paracentrotus and Arbacia : 3 inhibitors of ODC (α-methylornithine, α-difluoro methylornithine and α -hydrizino ornithine) have no effect on cleavage, but arrest development at the late blastula stage (thus when a regular cell cycle sets in). At that stage, DNA synthesis is more affected than RNA synthesis by the inhibitors. However, a third species (Sphaerechinus) behaves towards the three inhibitors like the japanese species studied by Kusunoki and Yasumusu (43). Experiments still in progress suggest that these species differences probably reflect differences in pool size and rate of synthesis of the polyamines: unfertilized eggs of Paracentrotus and Arbacia already have a high content of polyamines and there seems to be little, if any, synthesis during cleavage. Although the evidence is still indirect, these observations suggest that polyamines are involved in DNA synthesis during early development.

To summarize, there is no typical cell cycle during the very early stages which have been discussed in this review. The tetraploid oocyte is obviously in G_2; but it is out of the cell cycle, since all it can do is to grow or die : in this respect, it behaves like a G_o-cell. Maturation, which does not occur spontaneously (even in the very large oocytes obtained in vitro by Wallace and Misulovin (82), is a G_2-M transition. But it is an abnormal one in some respects : preparation to DNA synthesis, characteristic of G_1, takes place during maturation as well as the formation of MPF, a feature typical of G_2. Finally, during cleavage, there is no G_1-phase and the G_2-phase is confused with mitotic prophase. One has to wait until the blastula stage before a regular cell cycle sets in; the factors which control this shift remain poorly understand and deserve further study.

REFERENCES

1. Adamson, E.D. and Woodland, H.R. (1977): Dev. Biol.,57:136-149.
2. Arceci, R.J. and Gross, P.R. (1977): Proc. natl. Acad. Sci. US 74: 5016-5020.
3. Baker, R.F. (1972): J. Cell Sci. 11: 153-171.
4. Baldari, C.T., Amaldi, F. and Buongiorno-Nardelli, M. (1978): Cell, 15: 1095-1107.
5. Baldi, M.I., Mattoccia, E. and Tocchini-Valentini, G.P. (1978) Proc.natl.Acad.Sci. US, 75: 4873-4876.
6. Barat, M., Dupresne, C., Pinon, H., Tourte, M. and Mounolou, J.C. (1977): Dev. Biol.,55: 59-68.
7. Benbow, R.M. and Ford, C.C. (1975): Proc. Natl. Acad. Sci. US, 72: 3437-3441.
8. Benbow, R.M., Pestell, R.Q. and Ford, C.C. (1975): Dev. Biol., 43: 159-174.
9. Brachet, J. (1933): Arch. Biol., 44: 519.
10. Brachet, J. (1967): Nature, 214: 1132-1133.
11. Brachet, J. (1968): Curr. in modern Biol. 1: 314-319.
12. Brachet, J., O'Dell, D., Steinert, G. and Tencer, R. (1972): Exptl. Cell Res., 73: 463-468.
13. Brachet, J., Mamont, P., Boloukhère, M., Baltus, E. and Hanocq Quertier, J. (1978): C.R. Acad. Sci. Paris, 287: 1289-1292.
14. Brown, D.D. and Dawid, I. (1968): Science, 160: 272-280.
15. Callen, J.E. and Mounolou, J.C. (1978): Biol. Cell., 33: 5-14.
16. Carroll, A.G. and Ozaki, H. (1979): Exptl. Cell Res., 119: 307-315.
17. Coote, J.L., Szabados, G. and Work, T.S. (1979): Febs. Lett., 99: 255-260.
18. Crampton, J.M. and Woodland, H.R. (1979): Dev. Biol., 70: 453-466.

19. Dawid, I.B. (1966): Proc. Natl. Acad. Sci. US, 56: 269-276.
20. De Petrocellis, B. and Rossi, M. (1976): Dev. Biol., 48: 250-257.
21. Edelman, G.M. (1976): Science, 192: 218-226.
22. Epel, D., Nishioka, D. and Perry, G. (1978): Biol. Cell., 32: 135-140.
23. Fansler, B. and Loeb, L.A. (1969): Exptl. Cell Res., 57: 305-310.
24. Ford, C.C. and Woodland, H.C. (1975): Dev. Biol., 43: 189-199.
25. Fraser, J.M.K. and Huberman, J.A. (1978): Biochim. biophys. Acta, 520: 271-284.
26. Gourlie, B.B. and Infante, A.A. (1975): Biochem. Biophys. Res. Comm., 64: 1206-1214.
27. Grainger, J.L., Winkler, M.M., Shen, S.S. and Steinhardt, R.A. (1979): Dev. Biol. 68: 396-406.
28. Grippo, P. and Lo Scavo, A. (1973): Biochem. Biophys. Res. Comm., 64: 1206-1214.
29. Grippo, P., Caruso, A., Locorotondo, G. and La Bella, T.(1976): Cell Differentiation, 5: 129-137.
30. Grippo, P., Locorotondo, G. and Taddei, C. (1977): J. exptl. Zool., 200: 143-148.
31. Grunstein, M. (1978): Proc. Natl. Acad. Sci. US, 75: 4135-4139.
32. Gurdon, J.B. (1968): J. Embryol. exper. Morphol., 20: 401-414
33. Gurdon, J.B. (1969): Dev. Biol. Suppl.3: 59-82.
34. Gurdon, J.B. and Brown, D.D. (1976): "Symposium on Molecular Biology of the Gen. appar." Ed. North Holland, Amsterdam.
35. Hanocq, F., Kirsch-Volders, M., Hanocq-Quertier, J., Baltus, E. and Steinert, G. (1972): Proc. Natl. Acad. Sci. US,69: 1322-1326.
36. Hanocq, F., De Schutter, A., Hubert, E. and Brachet, J. (1974) Differentiation, 2: 75-90.
37. Hourcade, D., Dressler, D. and Wolfson, J. (1973). Proc. Natl. Acad. Sci. US, 70: 2926-2930.
38. Jimenez, R.N. and Grossman, A. (1979). Feder. Proc.,38: 666.
39. Johnson, J.D., Epel, D. and Paul, M. (1976): Nature,262: 661-664.
40. Keichline, L.D. and Wasserman, P.H. (1979): Biochemistry, 18: 214-219.
41. Kunkel, S. and Weinberg, E.S. (1978): Cell, 14: 313-326.
42. Kusunoki, S. and Yasumusu, I. (1976): Biochem. Biophys. Res. Comm., 68: 881-885.
43. Kusunoki, S. and Yasumusu, I. (1978): Dev. Biol., 67: 336-345.
44. Laskey, R.A., Honda, B.M., Mills, A.D. and Finch, J.T. (1978): Nature, 275: 416-420.
45. Manen, C.A. and Russell, D.H. (1973): J. Embryol. exper. Morphol., 29: 331-345.
46. Manen, C.A., Hadfield, M.G. and Russell, D.H. (1977): Dev. Biol.,57: 454-459.
47. Martini, G., Tato, F., Gandini Attardi, D. and Tocchini Valentini, G.P. (1976): Biochem. Biophys. Res. Comm., 72 : 875-879.

48. Masui, Y. (1974): J. exp. Zool., 187: 141-147.
49. Masui, Y. and Markert, C.L. (1971): J. exp. Zool., 177: 129-145.
50. Masui, Y. and Clarke, H.J. (1979): Intern. Rev. Cytol.,57: 185-282.
51. Mathews, C.K. (1975): Exptl. Cell Res., 92: 47-56.
52. Mazia, D. (1974): Proc. Natl. Acad. Sci. US, 71: 690-693.
53. Mazia, D., Schatten, G. and Steinhardt, R. (1975): Proc. Natl. Acad. Sci. US,72: 4469-4473.
54. Mertz, J.E. and Gurdon, J.B. (1977): Proc. Natl. Acad. Sci. US, 74: 1502-1506.
55. Miller, O. and Beatty, B.R. (1969): Genetics, Suppl. 1, 61: 133.
56. Morris, P.W. and Rutter, W.J. (1976): Biochemistry, 15: 3106-3113.
57. Nakashima, J. and Ishikawa, M. (1979): W. Roux's Arch. Dev. Biol., 185: 323-332.
58. Noronha, J.M., Sheys, G.H. and Buchanan, M. (1972): Proc. Natl. Acad. Sci. US, 69: 2006-2010.
59. Opresko, L., Wiley, H.S. and Wallace, R.A. (1979): J. exp. Zool. "(in press)".
60. Parisi, E., Filosa, S., De Petrocellis, B. and Monroy, A. (1978): Dev. Biol., 65: 38-49.
61. Poccia, D.L., Levine, D. and Wang, J.C. (1978): Dev. Biol., 64: 273-283.
62. Rao, P.N., Wilson, B.A. and Sunkara, P.S. (1978): Proc. Natl. Acad. Sci. US, 75: 5043-5047.
63. Reeder, R.H. (1974): in "Ribosomes" Cold Spring Harbor Symposium, 489-518.
64. Rinaldi, A.M., De Leo, G., Arzone, A., Salcher, I., Storace, A. and Mutolo, V. (1979): Proc. Natl. Acad. Sci. US,76: 1916-1920.
65. Rochaix, J.D., Bird, A. and Bakken, A. (1974): J. mol. Biol. 87: 473-487.
66. Ruderman, J.V., Woodland, H.R. and Sturgess, E.A. (1979): Dev. Biol.,71: 71-82.
67. Russell, D.H. (1971): Proc. Natl. Acad. Sci. US, 68: 523-527.
68. Scheer, U., Trendelenburg, M.F. and Franke, W.W. (1973): Exptl. Cell Res. 80: 175-190.
69. Senger, D.R., Arceci, R.J. and Gross, P.R. (1978): Dev. Biol., 65: 416-423.
70. Senger, D.R. and Gross, P.R. (1978): Dev. Biol. 65: 404-415.
71. Shen, S.S. and Steinhardt, R.A. (1978): Nature, 272: 253-254.
72. Shimada, H. and Teramaya, K. (1976): Dev. Biol., 54: 151-156.
73. Spinelli, G., Gianguzza, F., Casano, C., Acierno, P. and Burckhardt, J. (1979): Nucl. Acid Res. 6: 545-560.
74. Sunkara, P.S., Wright, D.A. and Rao, P.N. (1979): Proc. Natl. Acad. Sci. US, 76: 2799-2802.
75. Swanson, R.F. and Dawid, I.B. (1970): Proc. Natl. Acad. Sci. US, 66: 117-124.
76. Tencer, R. (1970): Biochim. Biophys. Acta, 204: 627-629.
77. Thiebaud, C.H. (1979): Chromosoma, 73: 37-44.

78. Thomas, C., Hanocq, F. and Heilporn, V. (1977): Dev. Biol., 56: 204-207.
79. Tondeur-Six, N., Tencer, R. and Brachet, J. (1975): Biochim. Biophys. Acta, 395: 41-47.
80. Trendelenburg, M.F. and Gurdon, J.B. (1978): Nature, 276: 292-294.
81. Vacquier, V. and Brandriff, B. (1975): Dev. Biol. 47: 12-31.
82. Wallace, R.A. and Misulovin, Z. (1978): Proc. Natl. Acad. Sci. 75: 5534-5538.
83. Webb, A.C. and Camp, C.J. (1979): Exptl. Cell Res. 119: 414-418.
84. Wellauer, P.K., Dawid, I.B., Brown, D.D. and Reeder, R.H. (1976): J. mol. Biol. 105: 461-486.
85. Wellauer, P.K., Dawid, I.B., Brown, D.D. and Reeder, R.H. (1976): J. mol. Biol. 105: 487-505.
86. Woodland, H.R. and Pestell, R.G.W. (1972): Biochem. J. 127: 597-605.
87. Wostenholme, R. (1973): Chromosoma 43 : 1-18.

Subject Index